Flow-Induced Structure in Polymers

ACS SYMPOSIUM SERIES **597**

Flow-Induced Structure in Polymers

Alan I. Nakatani, EDITOR
National Institute of Standards and Technology

Mark D. Dadmun, EDITOR
University of Tennessee

Developed from a symposium sponsored
by the Division of Polymeric Materials: Science
and Engineering, Inc., at the 208th National Meeting
of the American Chemical Society,
Washington, D.C.,
August 21–25, 1994

American Chemical Society, Washington, DC 1995

Library of Congress Cataloging-in-Publication Data

Flow-induced structure in polymers / Alan I. Nakatani, editor, Mark D. Dadmun, editor.

　　p.　cm.—(ACS symposium series; 597)

"Developed from a symposium sponsored by the Division of Polymeric Materials: Science and Engineering, Inc., at the 208th National Meeting of the American Chemical Society, Washington, D.C., August 21–25, 1994."

Includes bibliographical references and indexes.

ISBN 0–8412–3230–X

1. Polymers—Structure—Congresses. 2. Polymers—Mechanical properties. 3. Polymers—Rheology.   I. Nakatani, Alan I., 1957– . II. Dadmun, Mark D., 1964– .   III. American Chemical Society. Division of Polymeric Materials: Science and Engineering. IV. American Chemical Society. Meeting (208th: 1994: Washington, D.C.) V. Series.

QD381.9.S87F58 1995
620.1′92—dc20
　　　　　　　　　　　　　　　　　　　　　　95–16754
　　　　　　　　　　　　　　　　　　　　　　CIP

This book is printed on acid-free, recycled paper.

Copyright © 1995

American Chemical Society

All Rights Reserved. The appearance of the code at the bottom of the first page of each chapter in this volume indicates the copyright owner's consent that reprographic copies of the chapter may be made for personal or internal use or for the personal or internal use of specific clients. This consent is given on the condition, however, that the copier pay the stated per-copy fee through the Copyright Clearance Center, Inc., 222 Rosewood Drive, Danvers, MA 01923, for copying beyond that permitted by Sections 107 or 108 of the U.S. Copyright Law. This consent does not extend to copying or transmission by any means—graphic or electronic—for any other purpose, such as for general distribution, for advertising or promotional purposes, for creating a new collective work, for resale, or for information storage and retrieval systems. The copying fee for each chapter is indicated in the code at the bottom of the first page of the chapter.

The citation of trade names and/or names of manufacturers in this publication is not to be construed as an endorsement or as approval by ACS of the commercial products or services referenced herein; nor should the mere reference herein to any drawing, specification, chemical process, or other data be regarded as a license or as a conveyance of any right or permission to the holder, reader, or any other person or corporation, to manufacture, reproduce, use, or sell any patented invention or copyrighted work that may in any way be related thereto. Registered names, trademarks, etc., used in this publication, even without specific indication thereof, are not to be considered unprotected by law.

PRINTED IN THE UNITED STATES OF AMERICA

1995 Advisory Board

ACS Symposium Series
M. Joan Comstock, *Series Editor*

Robert J. Alaimo
Procter & Gamble Pharmaceuticals

Mark Arnold
University of Iowa

David Baker
University of Tennessee

Arindam Bose
Pfizer Central Research

Robert F. Brady, Jr.
Naval Research Laboratory

Mary E. Castellion
ChemEdit Company

Margaret A. Cavanaugh
National Science Foundation

Arthur B. Ellis
University of Wisconsin at Madison

Gunda I. Georg
University of Kansas

Madeleine M. Joullie
University of Pennsylvania

Lawrence P. Klemann
Nabisco Foods Group

Douglas R. Lloyd
The University of Texas at Austin

Cynthia A. Maryanoff
R. W. Johnson Pharmaceutical Research Institute

Roger A. Minear
University of Illinois at Urbana–Champaign

Omkaram Nalamasu
AT&T Bell Laboratories

Vincent Pecoraro
University of Michigan

George W. Roberts
North Carolina State University

John R. Shapley
University of Illinois at Urbana–Champaign

Douglas A. Smith
Concurrent Technologies Corporation

L. Somasundaram
DuPont

Michael D. Taylor
Parke-Davis Pharmaceutical Research

William C. Walker
DuPont

Peter Willett
University of Sheffield (England)

Foreword

THE ACS SYMPOSIUM SERIES was first published in 1974 to provide a mechanism for publishing symposia quickly in book form. The purpose of this series is to publish comprehensive books developed from symposia, which are usually "snapshots in time" of the current research being done on a topic, plus some review material on the topic. For this reason, it is necessary that the papers be published as quickly as possible.

Before a symposium-based book is put under contract, the proposed table of contents is reviewed for appropriateness to the topic and for comprehensiveness of the collection. Some papers are excluded at this point, and others are added to round out the scope of the volume. In addition, a draft of each paper is peer-reviewed prior to final acceptance or rejection. This anonymous review process is supervised by the organizer(s) of the symposium, who become the editor(s) of the book. The authors then revise their papers according to the recommendations of both the reviewers and the editors, prepare camera-ready copy, and submit the final papers to the editors, who check that all necessary revisions have been made.

As a rule, only original research papers and original review papers are included in the volumes. Verbatim reproductions of previously published papers are not accepted.

M. Joan Comstock
Series Editor

Contents

Preface xi

1. **An Introduction to Flow-Induced Structures in Polymers** 1
   Alan I. Nakatani

   ## POLYMER SOLUTIONS

2. **Enhancement of Concentration Fluctuations in Solutions Subject to External Fields** 22
   G. Fuller, J. van Egmond, D. Wirtz, E. Peuvrel-Disdier, E. Wheeler, and H. Takahashi

3. **String Phase in Semidilute Polystyrene Solutions Under Steady Shear Flow** 35
   Takuji Kume and Takeji Hashimoto

4. **Small Angle Neutron Scattering from Sheared Semidilute Solutions: Butterfly Effect** 48
   François Boué and Peter Lindner

5. **Structural Evolution and Viscous Dissipation During Spinodal Demixing of a Biopolymeric Solution** 61
   A. Emanuele and M. B. Palma-Vittorelli

6. **Flow-Induced Structuring and Conformational Rearrangements in Flexible and Semiflexible Polymer Solutions** 75
   A. J. McHugh, A. Immaneni, and B. J. Edwards

7. **Changes of Macromolecular Chain Conformations Induced by Shear Flow** 91
   M. Zisenis, B. Prötzl, and J. Springer

   ## POLYMER BLENDS AND HOMOPOLYMERS

8. **Effect of Flow on Polymer–Polymer Miscibility** 106
   M. L. Fernandez and J. S. Higgins

9. Effect of Shear Deformation on Spinodal Decomposition 122
   Tatsuo Izumitani and Takeji Hashimoto

10. Domain Structures and Viscoelastic Properties of Immiscible
    Polymer Blends Under Shear Flow 140
    Yoshiaki Takahashi and Ichiro Noda

11. Effects of Flow on the Structure and Phase Behavior
    of Polymer Blends 153
    Z. J. Chen, N. G. Remediakis, M. T. Shaw, and R. A. Weiss

12. Influence of Interface Modification on Coalescence in Polymer
    Blends 169
    K. Søndergaard and J. Lyngaae-Jørgensen

13. Orientation Behavior of Thermoplastic Elastomers Studied
    by ^{2}H-NMR Spectroscopy 190
    Alexander Dardin, Christine Boeffel, Hans-Wolfgang Spiess,
    Reimund Stadler, and Edward T. Samulski

14. Monte Carlo Simulations of End-Grafted Polymer Chains
    Under Shear Flow 204
    Pik-Yin Lai

BLOCK COPOLYMERS
AND MULTICOMPONENT POLYMER SYSTEMS

15. Effect of Shear on Self-Assembling Block Copolymers
    and Phase-Separating Polymer Blends 220
    M. Muthukumar

16. Shear-Induced Changes in the Order–Disorder Transition
    Temperature and the Morphology of a Triblock Copolymer 233
    C. L. Jackson, F. A. Morrison, A. I. Nakatani,
    J. W. Mays, M. Muthukumar, K. A. Barnes, and C. C. Han

17. Phase-Separation Kinetics of a Polymer Blend Solution
    Studied by a Two-Step Shear Quench 246
    A. I. Nakatani, D. S. Johnsonbaugh, and C. C. Han

18. Shear-Induced Structure and Dynamics of Tube-Shaped
    Micelles 263
    L. E. Dewalt, K. L. Farkas, C. L. Abel,
    M. W. Kim, D. G. Peiffer, and H. D. Ou-Yang

LIQUID CRYSTALLINE POLYMERS

19. **The Texture in Shear Flow of Nematic Solutions of a Rodlike Polymer** 274
    Beibei Diao, Sudha Vijaykumar, and Guy C. Berry

20. **Light Scattering from Lyotropic Textured Liquid-Crystalline Polymers Under Shear Flow** 298
    S. A. Patlazhan, J. B. Riti, and P. Navard

21. **Comparison of Molecular Orientation and Rheology in Model Lyotropic Liquid-Crystalline Polymers** 308
    Wesley Burghardt, Bruce Bedford, Kwan Hongladarom, and Melissa Mahoney

22. **Shear-Induced Orientation of Liquid-Crystalline Hydroxypropylcellulose in D_2O as Measured by Neutron Scattering** 320
    Mark D. Dadmun

23. **Shear-Induced Alignment of Liquid-Cystalline Suspensions of Cellulose Microfibrils** 335
    W. J. Orts, L. Godbout, R. H. Marchessault, and J. F. Revol

INDEXES

Author Index 351

Affiliation Index 351

Subject Index 352

Preface

FLOW-INDUCED STRUCTURES IN POLYMERIC SYSTEMS are an inevitable result of industrial processing and can have a profound effect on ultimate material properties. In polymer blend processing, shear-induced shifts in the phase boundary and shear-induced morphologies have been observed. Knowledge of such behavior would aid in the process design and control of these materials. The examination of structures produced in flow provides a unique probe into the physical chemistry of polymers. Investigations in this field are of great significance in understanding the influence of flow on phase separation in multicomponent systems, chain deformation in homopolymer melts and solutions, microphase separation behavior in block copolymers, and ordered phases in colloidal and liquid-crystalline materials.

Recent advances in experimental techniques that allow in situ measurements of materials under deformation have escalated research in this subject area. Two symposia have been sponsored in recent years by the ACS Division of Polymeric Materials: Science and Engineering, Inc. In 1991, the first symposium on flow-induced structures in polymers presented 11 contributed papers and attracted little representation from outside the United States. The second symposium, called "Flow-Induced Structures in Polymeric Systems," was held in 1994. To demonstrate the growth in the field in three years, the 1994 symposium consisted of five plenary lectures and 30 contributed talks with representatives from the United Kingdom, France, Germany, Denmark, Italy, Japan, Taiwan, Hong Kong, Russia, Kuwait, and the United States. The goal of these symposia was to bring together experts from around the world in the areas of homopolymer melts and blends, polymer solutions, block copolymers, liquid-crystalline polymers, colloidal suspensions, and multiphase systems to discuss recent advances in flow-induced structures for a broad range of materials. This volume was developed around the presentations made at the 1994 symposium.

The extensive international representation provided an opportunity for attendees to meet and discuss current progress and issues in this area with internationally renowned researchers. The plenary lectures were given by Takeji Hashimoto of Kyoto University; Julia S. Higgins of the Imperial College of Science, Technology and Medicine in London; Gerald G. Fuller from Stanford University; M. Muthukumar from the University of Massachusetts, Amherst; and Guy C. Berry from Carnegie Mellon

University. In view of the activity and accomplishments in these fields by the European and Asian scientific communities, inclusion of foreign scientists was essential to the success of this symposium and to help make this book valuable. All of the plenary speakers and approximately half of the other speakers who contributed to the symposium have submitted manuscripts for this volume. In addition, several authors who were unable to attend the symposium but are active in the field also contributed to the book. The participation of the international scientists will encourage future interactions and exchange of information between U.S. and international laboratories.

A number of other recent reviews and monographs concerned with the topic are available. Thus, this book concentrates on recent advances in the field rather than reviewing prior work. The initial overview chapter serves as a guide and reference source for readers new to the field. Subsequent chapters represent recent results and summaries from researchers active in the field. Contributions from academic, governmental, and industrial research are represented in the book so that readers may benefit from the efforts and experience of a wide variety of contributors.

Acknowledgments

We thank all plenary speakers and contributors for their time and effort in making the symposium a success and bringing this volume to fruition. We also thank the donors of the Petroleum Research Fund, administered by the American Chemical Society, and the ACS Division of Polymeric Materials: Science and Engineering, Inc., for partial support of this activity. We are pleased to acknowledge Karen Nakatani for assistance on the cover design.

ALAN I. NAKATANI
Polymers Division
National Institute of Standards and Technology
Building 224, Room B210
Gaithersburg, MD 20899

MARK D. DADMUN
Chemistry Department
University of Tennessee
Knoxville, TN 37996–1600

January 25, 1995

Chapter 1

An Introduction to Flow-Induced Structures in Polymers

Alan I. Nakatani

Polymers Division, National Institute of Standards and Technology, Building 224, Room B210, Gaithersburg, MD 20899

An overview of flow-induced structures in polymers is presented and recent references to research in the areas of polymer solutions, homopolymer melts, polymer blends, block copolymers and multi-component systems, and liquid crystalline polymers are given. The research area of flow-induced structures in polymers has expanded with the advent of in-situ measuring techniques. Studies of this type are useful for probing fundamental issues such as chain deformation, shifts in phase behavior, and the formation of novel shear-induced structures. The applications addressed include the kinetics of morphology evolution and the relationship between ultimate properties and morphology.

Flow-induced changes in polymeric materials are the focus of many recent investigations. Such studies are highly relevant because polymer processing involves complex deformation histories which can influence ultimate material properties. Fundamental studies have been designed to address such basic questions as the effect of shear on phase behavior, microstructure formation, chain deformation, and macroscopic domain structure. These studies concentrate on in-situ examination of materials subject to an imposed deformation. In the subsequent chapters, results from a number of research groups are reported to familiarize readers with the current state-of-the-art in flow-induced structures in polymers based on the 1994 symposium on "Flow-Induced Structures in Polymers" (1). The materials to be considered include colloidal suspensions, micellar systems, liquid crystalline materials, block copolymers, polymer solutions, homopolymer melts, and networks. The effect of shear on these materials has been reported separately in a wide range of journals of various disciplines.

The interdisciplinary nature of the research reviewed in this chapter suggests that some important analogues between materials may exist. Some of these analogies are illustrated in Figures 1-7. Figure 1 demonstrates the shear rate dependence of light scattering patterns from a suspension of hard spheres (2). Figure 2 shows the

This chapter not subject to U.S. copyright
Published 1995 American Chemical Society

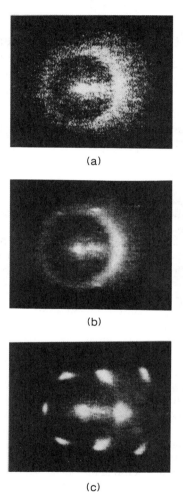

Figure 1. - Shear rate dependence of light scattering patterns from a suspension of hard spheres. a) Zero shear scattering pattern showing equilibrium liquid-like ordering. b) Steady shear scattering pattern (Peclet number (Pe) = reduced shear stress = $\dot{\gamma}\eta_E R^3/k_B T = 30$, where $\dot{\gamma}$ is the shear rate, η_E is the suspension viscosity at $\dot{\gamma}$, R is the radius of the suspended particles, k_B is Boltzmann's constant and T is the temperature) showing weak intensity maxima identified with possible string like ordering. c) Oscillatory strain-induced crystalline ordering with strain amplitude of approximately four. Reproduced with permission from ref. 2. Copyright 1990, American Institute of Physics.

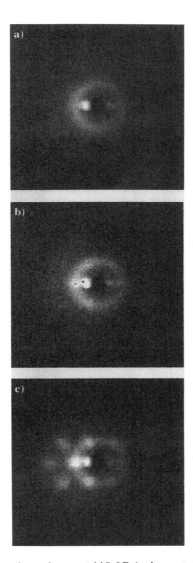

Figure 2. - Shear rate dependence at 118 °C (quiescent disordered state) of end-view SANS patterns from a polystyrene-d_8/polybutadiene/polystyrene-d_8 triblock copolymer (23 % PSD, cylindrical ordered phase, T_{ODT} = 110 °C). a) 0.26 s^{-1} showing almost no ordered structure. b) 0.65 s^{-1} showing slight orientation. c) 1.30 s^{-1} showing hexagonally packed cylindrical structure induced by shear. Adapted from ref. 3.

shear rate dependence of small angle neutron scattering patterns from a triblock copolymer melt (*3*). Figure 3 shows the light scattering pattern from a semidilute, polystyrene solution under shear (*4*) while Figure 4 shows the SANS pattern from a swollen, stretched polymer gel (*5*). Figure 5a shows the predicted scattered intensity for an electrorheological fluid (ER fluid) (*6*) and Figure 5b shows the small angle x-ray scattering pattern from an aligned polymer nematic liquid crystal (*7*). Finally, Figure 6 shows the temporal evolution of light scattering from an ER fluid after the inception of an electric field (*8*), whereas Figure 7 shows the temporal evolution of light scattering after cessation of a shear field (*9*). From Figures 1-7, it is apparent that for vastly different materials with different characteristic length scales, rescaling the data to the appropriate size scale of the incident radiation yields striking similarities in the scattering patterns. It is hoped that these similarities will be appreciated by researchers active in this field and provoke novel theoretical and experimental approaches.

Flow can affect changes in polymers on three basic length scales: a) molecular level (~1-10 nm); b) mesoscopic level (~10-100 nm); c) macroscopic level (>0.5 μm). At each length scale, different concentration regimes may exhibit different behavior. These concentration regimes fall into the dilute solution regime, semidilute regime, and concentrated or bulk regimes. Within each of these concentration regimes, one might also consider the nature of the polymer interactions, i.e. whether the chains are entangled or unentangled and whether the entanglement interactions are temporary or permanent (networks and gels). The analogy between some universal aspects of rubber network behavior and polymer melt behavior have been noted by Wagner and Schaeffer (*10*). If one considers deformation to be an external field, then experiments involving the effects of other external fields on polymers such as pressure, electric fields, and magnetic fields must also be considered. Similarly, any unified theories dealing with deformation induced structures in polymers must consider different forms of the deformation tensor such as simple shear, elongational flow (uniaxial extension), or biaxial flow. Finally, extension of these theories and experiments toward understanding the behavior of these materials in complex flow geometries which are encountered in extrusion or molding processes must be considered.

Current theories and significant predictions will be referred to in the overview, however detailed discussions are beyond the scope of this chapter. The thermodynamic treatment of flowing systems has been discussed extensively in a treatise by Beris and Edwards with particular emphasis on systems with internal microstructure (*11*). Other thermodynamic treatments of flow-induced phase behavior are given by Vrahopoulou-Gilbert and McHugh (*12*) and by Kammer et al. (*13*). A monograph edited by Søndergaard and Lyngaae-Jørgensen contains general information and reviews on the rheo-physics of multiphase systems (*14*). A review by Larson covers flow-induced mixing, flow-induced demixing, polymer migration in shearing flows, stress-induced diffusion, and flow effects on crystalline, liquid crystalline, and block copolymer ordering transitions (*15*). Other reviews are available by Rangel-Nafaile et al. on the stress-induced phase behavior of polymer solutions (*16*) and Tirrell on the phase behavior of flowing polymer mixtures (*17*). The interested reader is referred to these reviews and the references

1. NAKATANI   *An Introduction to Flow-Induced Structures in Polymers*

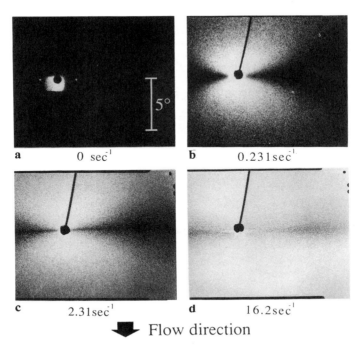

Figure 3. - a) Small angle light scattering pattern of a semidilute polymer solution (6% polystyrene, $M_w = 5.48 \times 10^6$, in dioctyl phthalate) under shear. a) $0\ s^{-1}$. b) $0.231\ s^{-1}$. c) $2.31\ s^{-1}$. d) $16.2\ s^{-1}$. Flow direction is parallel to the vertical axis. Reproduced with permission from ref. 4. Copyright 1994, The American Physical Society.

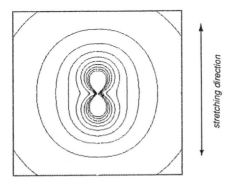

Figure 4. - Small angle neutron scattering contour of a stretched, swollen gel ($\lambda = 1.2$, λ is the deformation ratio $= L/L_0$). Deformation direction is parallel to the vertical axis. Reproduced with permission from ref. 5. Copyright 1994, American Chemical Society.

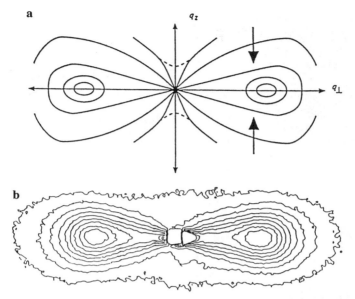

Figure 5. - a) Predicted scattering intensity contour of an aligned, electrorheological fluid. Reproduced with permission from ref. 6. Copyright 1992, American Physical Society. b) Small angle x-ray scattering contour of an aligned, nematic polymer liquid crystal. Reproduced with permission from ref. 7. Copyright 1991, Elsevier Science Publishers, B. V. North-Holland.

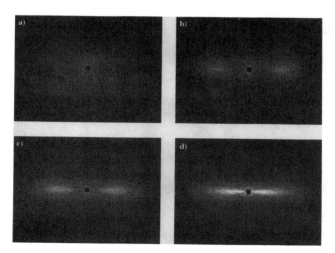

Figure 6. - Temporal evolution of light scattering patterns from an electrorheological fluid after inception of an electric field. a) Before inception of electric field. b); c); and d) show development of scattering after field is turned on. Field direction is parallel to the vertical axis. Reproduced with permission from ref. 8. Copyright 1992, Science.

cited therein for extensive background material on these topics. Here we endeavor to highlight recent developments since the publication of these reviews.

Experimental and Instrumental Techniques

A wide variety of experimental methods are available for studying flow-induced structures in polymers. The following provides a brief synopsis of the available methodology and references for each. Specific references to in-situ flow techniques have also been included. Optical techniques have been the method of choice for studying many different aspects of polymer orientation. A review of optical rheometry techniques has been written by Fuller (18). A definitive review of flow birefringence techniques has been written by Janeschitz-Kriegl (19). The photophysics of polymers such as fluorescence, phosphorescence, and luminescence have been reviewed in a book edited by Phillips (20). Light scattering has long been utilized in the characterization of polymers and there is an enormous amount of literature available on the subject. A recent book by Chu is a good source for other reference material and contains a detailed discussion of light scattering techniques (21). A new monograph on applications of small angle neutron scattering to polymers has been published by Higgins and Benoit (22).

Numerous instrumental devices have become available to examine flow-induced behavior in polymers. The most commonly available instruments are light scattering shear cells. Transparent cone and plate geometry light scattering shear cells have been constructed by Hashimoto and coworkers (23) and by Nakatani et al. (24). Couette geometry light scattering cells have been constructed by Wolfle and Springer to examine polymer chain dimensions under shear (25). Wu et al. have also built a couette geometry shear cell to examine shear enhanced concentration fluctuations in semidilute polymer solutions (26). Menasveta and Hoagland have used an extensional flow geometry to examine semidilute polymer solutions (27). A number of other research groups are utilizing similar light scattering instrumentation to investigate various aspects of flow-induced behavior in polymers, which will be discussed below.

Because of the limited number of neutron scattering facilities worldwide, neutron scattering shear cells are much less common. Kalus and coworkers have reported on a shear apparatus for small angle neutron scattering experiments (28). Another shear cell for SANS has been constructed by Cummins et al. (29). Baker and coworkers have created a shear cell for examining liquid-solid interfaces by small angle neutron scattering (30). Lindner and Oberthur (31,32), Hayter and Penfold (33), and Nakatani and coworkers (34) have also reported on shear cells for SANS experiments. With the availability of x-ray synchrotron sources, SAXS studies of in-situ fiber spinning to examine the structure development during the spinning process are also in progress (Hsiao, B. S., E. I. Dupont, Co., personal communication, 1994). Safinya and co-workers (35,36) have developed a couette geometry shear cell for utilization in with the x-ray synchrotron source and Bubeck and co-workers have examined fiber spinning processes with synchrotron radiation (37).

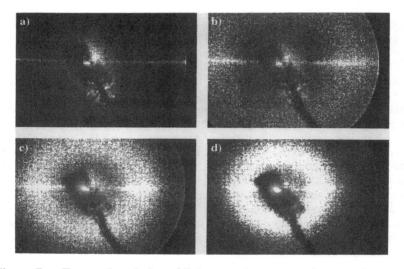

Figure 7. - Temporal evolution of light scattering patterns from a polymer blend solution (8% 50:50 polystyrene/ polybutadiene in dioctyl phthalate) after cessation of shear. Times after cessation: a) 2.3 s. b) 8.7 s. c) 13.5 s. d) 19.9 s. Original flow direction is parallel to the vertical axis. Reproduced with permission from ref. 9. Copyright 1992, Butterworth-Heinemann Ltd.

Nuclear magnetic resonance techniques have proven to be useful for examining polymer melts under flow. Nakatani et al. (*38*) have utilized "rheo-NMR" in an attempt to examine the change in dipolar interactions of the polymer chains due to changes in the polymer chain conformations during flow. Magnetic resonance imaging techniques have also been developed to examine particle and velocity distributions in flowing suspensions, as will be discussed below.

Polymer Solutions

The materials that have attracted the most interest in this field are semidilute solutions of polymers. VerStrate and Philippoff (*39*) showed that clear, single-phase, semidilute solutions of polystyrene in dioctyl phthalate become turbid above a certain shear rate. This observation has initiated many theoretical and experimental studies which are reviewed by Larson and Rangel-Nafaile et al. (*15,16*). One common scattering observation from sheared semidilute solutions is the so-called "butterfly" pattern (*4,40,41,42,43*). The contribution by Boué will discuss some recent results of SANS butterfly patterns from semidilute polymer solutions. SANS experiments on polystyrene/dioctyl phthalate solutions have also been reported by Nakatani et al. (*44*). Shear-induced turbidity has also been observed in extensional flow by van Egmond and Fuller (*45*). As noted by Beris and Mavrantzas (*46*), a wide number of theories exist (*16,39,47,48*) which describe the same phenomenon. However, the theories are widely disparate in the nature of the governing equations. The reason that these different approaches are all able to predict the observed phenomena remains to be determined.

For semidilute solutions of polymer blends, the opposite behavior is observed under shear, in most cases. A sample which is phase separated at rest, becomes optically transparent (single phase) with shear. One intuitively expects this type of behavior, instead of the type of behavior described in the previous paragraph. This type of system has been studied extensively by Hashimoto and coworkers (*49*) as well as by Nakatani and coworkers (*50*). Both research groups will discuss recent results on these systems in this volume.

Molecular level understanding in polymer solutions is relevant to the problem of drag reduction. Ng and Leal (*51*) have used flow birefringence techniques on semidilute solutions of polystyrene undergoing extensional flow. Callaghan and coworkers have utilized NMR to examine the velocity distribution in shear thinning polymer solutions (*52,53*). Measures of the single chain polymer dimensions in dilute solution as a function of shear rate have been reported by Zisenis and associates (*54*).

Polymer Blends and Homopolymers

The behavior of polymer blends under the influence of flow is extremely important. Flow can influence blends in two way: a) The phase boundary may shift with shear, possibly affecting the optimum processing conditions. b) Anisotropic domains which are produced with shear may create anisotropy in the final bulk material properties. A discussion of the shift in the phase behavior of polymer blends with shear rate has been published by Douglas (*55*). The scaling behavior for polymer blends which undergo shear-induced mixing is discussed in a review by Hobbie et al. (*56*). Recently, a series of papers by Horst and Wolf have proposed a theory for the phase behavior of polymer blends under shear by introducing an additional term to the equation for the Gibbs energy of mixing (*57,58,59,60*). A comparison of the predictions of Horst and Wolf with experimental results will be made in the chapter by Higgins in this book. Kammer has also compiled some recent results on polymer blends (*61,62*) and presented a theory for shear-induced shifts in the phase boundary based on a decoupled mode theory (*63*). Flow-induced mixing behavior in a blend of polystyrene and polyisobutylene has been observed by Wu et al. (*64*) using a light scattering apparatus. Earlier results by Nakatani et al. (*65*) on blends of polystyrene/poly(vinyl methyl ether) (PS/PVME) and polystyrene/polybutadiene using small angle neutron scattering (SANS) showed a suppression of concentration fluctuations along the flow direction for both systems but no effect perpendicular to flow for the high molecular weight PS/PVME system, indicating no true shift in the phase boundary with shear for a mean-field system in accord with Douglas (*55*). Mani et al. have utilized fluorescence techniques to examine the behavior of a PS/PVME blend under shear (*66*). Mani et al. have also examined the viscoelastic behavior of the same blend around the critical point and observed an increase in the fluorescence intensity at the phase transition associated with a sharp change in the low frequency viscoelastic behavior (*67*).

Studies on the addition of compatibilizing agents to polymer blends have been conducted for many years. With the recent capabilities to examine polymeric materials under flow, compatibilized blends are materials that will be of increasing interest in these types of experiments. Germain et al. (*68*) have examined the rheological and morphological behavior of a compatibilized polypropylene/polyamide blend by capillary extrusion.

Examination of the flow behavior of homopolymers is of particular interest in testing current theories of molecular dynamics. Predictions concerning the chain configuration in homopolymer melts under flow are now being tested with some techniques that have been modified to address this problem. The difficulty in studying homopolymers is the high sample viscosities which necessitate the use of high torque drive mechanisms. The validity of results from deformed, and subsequently quenched samples, which are then tested in separate equipment has been questioned. The newer in-situ shear capabilities of many instruments have removed these doubts. Predictions for the behavior in homopolymer melts under shear using non-equilibrium molecular dynamics simulations have been proposed by Kröger and coworkers (*69*). Included in these predictions are forms for the scattering structure factor for polymer chains under shear. Spenley and Cates have modified a tube model of high molecular weight polymer fluids to predict the behavior of entangled fluids under strong shear (*70*).

Archer, Huang and Fuller have examined the orientational dynamics of a polymer melt by polarization-modulated laser Raman spectroscopy (*71*). Their results were consistent with constitutive theories such as the "temporary network" theory or the "partially extending strand convection" theory but inconsistent with predictions of the Doi-Edwards tube model theories. The behavior of bidisperse temporary networks has been studied by Seidel and coworkers (72) using simultaneous infrared dichroism and birefringence techniques. The temporary networks were composed of polybutadiene chains modified with urazole groups ("stickers"). The temperature and composition dependence of the relaxation behavior of the temporary networks is not accounted for in current theories (73). Deformation of permanent networks, as well as gels also results in the appearance of "butterfly" type scattering patterns (42,74,75). A recent discussion of these results is presented by Ramzi et al. (76).

The behavior of polymers under shear near surfaces has also been important in understanding lubrication effects of polymers. The adsorption of diblock copolymer chains to surfaces leads to the formation of polymer brushes (77). The shear behavior of these brushes has been examined by Klein and coworkers (78,79) and Granick and coworkers (80,81,82). Using a surface force apparatus, Pelletier et al. have examined the rheological properties of polymers adsorbed to surfaces (83). Monte Carlo simulations of polymer brushes adsorbed to a surface in the presence of a flow field are discussed in this book by Lai.

Texture evolution in polymer melts as a result of shear effects have been examined by Bartczak et al. (84). These textures also may affect the ultimate material properties. Flow-induced crystallization has been studied extensively by the research group of McHugh (85) and a report on the extensional flow-induced crystallization of molten polyethylene has recently appeared (86).

Block Copolymers and Multicomponent Systems

Extensive research is in progress in the area of flow-induced changes in block copolymers (87). Of particular interest are novel microstructural morphologies induced by shear and shear-induced shifts in the order-disorder transition (ODT) temperature. Many theories predicting the shear behavior of block copolymers around the ODT temperature have been proposed. Theories by Cates and Milner (88) and subsequently Marques and Cates (89) have predicted shear-induced effects in block copolymers. Fredrickson has recently discussed the alignment of block copolymers in steady shear near the ODT (90). The current state of theories on shear induced effects in block copolymers will be discussed in the chapter by Muthukumar in this volume.

There are also a number of experimental efforts aimed at uncovering the shear behavior of block copolymers. Bates and coworkers have numerous reports on the behavior of deformed poly(ethylene-propylene)/poly(ethylethylene) (PEP/PEE) block copolymers (91,92,93,94) by small angle neutron scattering and rheological techniques. Kannan and Kornfield have elucidated the conditions for alignment of lamellar domains of PEP/PEE parallel and perpendicular to the flow direction by rheo-optical techniques (95,96). Balsara et al. (97) have observed shear-induced structures in a block copolymer solution under shear by SANS. Morrison et al. (98) have reported on a shear-induced order-order transition in triblock copolymers. Details of this work will be presented in the chapter by Jackson et al. The ordering by flow near the disorder-order transition of a triblock copolymer has also been studied by small angle x-ray scattering (SAXS) by Winter et al. (99). Dynamic SAXS experiments on sphere-forming block copolymer under oscillatory shear have been performed by Okamoto et al. (100) to examine the real-time response of the material to oscillatory deformation. Lee and coworkers have examined the orientation of soft segments and hard segments of poly(butylene terephthalate-co-tetramethylene oxide) in uniaxial extension by infrared dichroism and polarizing microscopy (101). The oscillatory shear flow behavior of star diblock copolymers has also been examined by Archer and Fuller (102) using birefringence and rheological techniques. The same anomalous low frequency behavior which is observed in linear diblock and triblock copolymers was also observed for the star diblock materials. This indicates the low frequency behavior is independent of the molecular architecture of the block copolymer sample.

The scattering from the ordered microstructures observed in block copolymers is reminiscent of the scattering behavior observed from colloidal suspensions. Therefore, familiarity between the results on block copolymers with suspensions and micelles may provide a deeper understanding for the behavior of all three types of materials. A broad review for the structure of dispersions under shear has been given by Ackerson (2,103). Chen et al. (104,105,106) have examined the effect of microstructural transitions on the rheological behavior of colloidal suspensions by small angle neutron scattering. The rheological and small angle neutron scattering behavior of concentrated dispersions of electrostatically stabilized styrene-ethyl acrylate copolymer spheres in glycol or water have been examined by Laun and coworkers (107). Similar SANS experiments have been performed on adhesive hard sphere dispersions by Wouterson and coworkers (108). SANS has also been used by deKruif and coworkers to examine the behavior of concentrated colloidal dispersions under shear (109,110). Chow et al. have used nuclear magnetic resonance imaging techniques (NMRI) to examine the microstructure of hard sphere suspensions during a falling-sphere experiment (111,112). Other research groups have performed similar NMRI experiments on flowing suspensions to measure the velocity and concentration distributions of the particles (113,114). The shear-induced migration of particles discussed theoretically in the review by Larson (15), has been examined for suspensions of rods by Mondy and coworkers (115,116). No difference between the measured concentration profiles for the rods and spheres was observed. The motion of rod-like particles in pressure driven flow between two plates has been studied by Stover and Cohen (117) where the particles were observed to undergo a "pole-vaulting" trajectory.

The behavior of micelles under flow is of increasing importance with relation to recent applications. Particularly, the effects of flow on micelles has important ramifications with respect to the use of these materials in drug delivery systems. Hu et al. (118) have examined the rheological and birefringence behavior of a micelle forming surfactant solution and found that shear-induced coagulation occurs with the production of large scale micelles which are stable over long periods of time. Kalus and coworkers have reported extensively on the shear behavior of micelles studied by SANS (119,120,121,122,123). The flow behavior and shear-induced, isotropic-nematic transition of a worm-like micelle by SANS has been reported by Schmitt and coworkers (124). Schmitt et al. claim the shear-induced alignment is different from the shear-induced isotropic-nematic transition observed in other systems. The shear-induced crystallization and melting of a micellar solution observed by SANS has been reported by Phoon et al. (*125*). The behavior of rodlike micelles in shear flow by SANS has been reported by Penfold et al. (*126*). Hamilton et al. have reported on the formation of a shear-induced hexagonal phase near a surface from a polyionic fluid which forms threadlike micelles (*127*).

Liquid Crystalline Polymers

Liquid crystalline polymers are anisotropic and easily oriented in a flow field, which is desirable in certain applications. Recent examples of work in the area of flow effects on liquid crystalline polymers have been published as part of the Proceedings of the Boston Syposium on Liquid Crystals at the Society of Rheology in the Journal of Rheology (*128*). A basic review of liquid crystalline polymers has been given by Jackson and Shaw (*129*). A general overview of theories for rodlike polymers in the nematic phase under flow has been given by Marrucci (*130*). A theory for the formation of domain structures in liquid crystalline polymers undergoing oscillatory shear flow has recently been proposed by Semenov (*131*). Larson and Doi have also presented a mesoscopic domain theory for liquid crystalline polymers (*132*).

Berry and coworkers have performed extensive rheological and rheo-optical studies on nematic solutions of rodlike polymers (*133*). Recent work from this group will be presented in this book. Burghardt and coworkers have examined various aspects of flowing liquid crystals by optical techniques (*134,135*). A summary of these results will be presented in the chapter by Burghardt. De'Nève and coworkers have examined the rheology and shear-induced textures in a thermotropic liquid crystalline polymer (*136*). They found three different textures (thread, worm and ordered), which were a function of strain and shear rate, corresponding to samples with a low defect density, high defect density created by tumbling, and oriented molecules due to cessation of tumbling, respectively. Yan et al. (*137*) have also examined shear-induced textures in a lyotropic poly(γ-benzyl-L-glutamate) solution by optical microscopy. A rheo-optic examination of hydroxypropylcellulose solutions in Poiseuille flow has been reported by Guido and coworkers (*138*). They observed an elongation of domains at the onset of Region II in the viscosity curve (Newtonian regime). From transient rheological measurements, Gu et al. found that a flow-aligning nematic solvent undergoes a transition to director-tumbling behavior upon addition of a side-chain liquid crystalline polymer (*139*). Conversely, a tumbling small molecule nematic becomes aligned in flow with addition of a main chain liquid crystalline polymer (*140*). Rheo-NMR has also been utilized to examine the director orientation in side-chain liquid crystalline polymers by Grabowski and Schmidt (*141*).

The behavior of defect structures in liquid crystalline materials has also been examined. A theory for the shape and motion of defect loops in nematic liquid crystals has been proposed by Rey (*142*). Experimental reports on the defect behavior under flow are given in the chapter by Patlazhan and coworkers. The deformation and relaxation behavior of a nematic comb-like liquid crystalline polymer by SANS has been reported by Brulet and coworkers (*143*).

Related Research

The gelation behavior of a photoinitiated crosslinkable material has been examined by Khan (*144*). Shear was found to enhance the gelation process, presumably by enhancing the diffusion of trapped free radicals as well as promoting the number of encounters the trapped radicals make with other reactive sites. The structure of stretched poly(vinyl alcohol) gels has been measured with small angle x-ray scattering (SAXS) by Hirai et al. (*145*). The physical properties of some materials may become anisotropic due to the underlying anisotropy of the constituent molecules or the result of a shear field. Van den Brule and Slikkerveer have reported on the anisotropic thermal conductivity caused by molecular orientation in a flowing polymeric liquid (*146*).

The analogy between mechanical deformation and other external fields has been mentioned previously. One of the most striking analogies is observed in the field of electrorheological (ER) fluids. Typical studies on ER fluids involve the superposition of an external electric field and a mechanical deformation. The similarity between the experiments performed on these types of fluids and the predictions of Ji and Helfand (*147*) for an oscillatory shear flow superimposed upon a steady shear flow may prove interesting. Ceccio and Wineman have developed a theoretical model for the influence of electric field orientation on the shear behavior of electrorheological fluids (*148*).

Some less common techniques have been applied to the problem of flow-induced structures in polymers. Fodor and Hill have used microdielectrometry to probe the structure of flowing liquid crystals near a wall (*149*) and the flow induced fractionation of homopolymer melts (*150*). Adachi and coworkers have constructed an instrument for rheodielectric measurements of polymeric emulsions (*151*). Adachi et al. were able to determine from a decrease in the dielectric constant, ε', with increasing shear rate, that the droplets were elongating and orienting in the flow direction.

Outlook

Advances in experimental methods to the field of flow-induced structures in polymers have produced a wealth of new data. Typically, researchers have focussed on results in well-defined, simple flow fields. The logical progression is the extension of many of these techniques to on-line, commercial applications. Further studies are also required to better understand the similarities and possible universality across various disciplines of material behavior under flow.

Despite the effort expended on the examination of flow-induced behavior in polymers, but many questions still remain. The theoretical treatment of flow-induced behavior is an exceedingly complex and difficult problem and further improvement of current theories is warranted. Experimentally, further studies which examine the effects of shear on phase transitions such as the spinodal temperature in polymer blends, and the order-disorder temperature in block copolymers are necessary before true understanding of this phenomenon can be achieved. The extension of these results to complex flow patterns is essential for commercial application of this technology. Many of the current techniques are best suited for solution studies, however, instrumentation for examining polymer melt behavior is limited and further development of equipment with the goal of examining melt behavior is required. The prospect of successfully resolving these unanswered questions is bright due to the number of high quality researchers contributing to this promising area of research.

Literature Cited

1. *ACS Polym. Matls.: Sci. and Eng. Preprints*, **1994**, *71*, 39-52, 111-124, 185-195, 241-252, 323-335.
2. Ackerson, B. J. *J. Rheol.* **1990**, *34*, 553.
3. Nakatani, A. I.; Morrison, F. A.; Muthukumar, M.; Mays, J. W.; Douglas, J. F.; Jackson, C. L.; Han, C. C. *Macromolecules* to be submitted.
4. Moses, E.; Kume, T.; Hashimoto, T. *Phys. Rev. Lett.* **1994**, *72*, 2037.
5. Bastide, J.; Leibler, L.; Prost, J. *Macromolecules* **1990**, *23*, 1821.
6. Kamien, R. D.; Le Doussal, P.; Nelson, D. R. *Phys. Rev. A* **1992**, *45*, 8727.
7. Ao, X.; Wen, X.; Meyer, R. B. *Physica A* **1991**, *176*, 63.
8. Halsey, T. C. *Science* **1992**, *258*, 761.
9. Waldow, D. A.; Nakatani, A. I.; Han, C. C. *Polymer* **1992**, *33*, 4635.
10. Wagner, M. H.; Schaeffer, J. *J. Rheol.* **1993**, *37*, 643.
11. Beris, A. N.; Edwards, B. J. *Thermodynamics of Flowing Systems*; Oxford University Press: New York, NY, 1994.
12. Vrahopoulou-Gilbert, E.; McHugh, A. J. *Macromolecules*, **1984**, *17*, 2657.
13. Kammer, H. W.; Kummerlowe, C.; Kressler, J.; Melior, J. P. *Polymer* **1991**, *32*, 1488.
14. *Rheo-Physics of Multiphase Polymeric Systems. Application of Rheo-optical Techniques in Characterization*; Søndergaard, K.; Lyngaae-Jørgensen, J., Eds.; Technomic Publishing: Lancaster, PA, 1993.
15. Larson, R. G. *Rheol. Acta* **1992**, *31*, 497.
16. Rangel-Nafaile, C.; Metzner, A. B.; Wissbrun, K. F. *Macromolecules* **1984**, *17*, 1187.
17. Tirrell, M. *Fluid Phase Equilibria* **1986**, *30*, 367.
18. Fuller, G. G. *Ann. Rev. Fluid Mech.* **1990**, *22*, 387.
19. Janeschitz-Kriegl, H. *Polymer Melt Rheology and Flow Birefringence*; Polymers/Properties and Applications, Vol. 6; Springer-Verlag, Berlin, 1983.
20. *Polymer Photophysics;* Phillips, D., Ed.; Chapman and Hall: New York, NY, 1985.
21. Chu, B. *Laser Light Scattering: Basic Principles and Practice*; Academic Press, Inc.: San Diego, CA, 1991; Second edition.
22. Higgins, J. S.; Benoit, H. C. *Polymers and Neutron Scattering;* Oxford Series on Neutron Scattering in Condensed Matter; Oxford University Press: New York, NY, 1994.
23. Hashimoto, T.; Takebe, T.; Suehiro, S. *Polymer J.* **1986**, *18*, 123.

24. Nakatani, A. I.; Waldow, D. A.; Han, C. C. *Rev. Sci. Instr.* **1992**, *63*, 3590.
25. Wolfle, A.; Springer, J. *Colloid Polym. Sci.* **1984**, *262*, 876.
26. Wu, X.-L.; Pine, D. J.; Dixon, P. K. *Phys. Rev. Lett.* **1991**, *66*, 2408.
27. Menasveta, M. J.; Hoagland, D. A. *Macromolecules* **1991**, *24*, 3427.
28. Kalus, J.; Neubauer, G.; Schmelzer, U. *Rev. Sci. Instr.* **1990**, *61*, 3384.
29. Cummins, P. G.; Staples, E.; Millen, B.; Penfold, J. *Measurement Sci. Technol.* **1990**, *1*, 179.
30. Baker, S. M.; Smith, G.; Pynn, R.; Butler, P.; Hayter, J.; Hamilton, W.; Magid, L. *Rev. Sci. Instr.* **1994**, *65*, 412.
31. Lindner, P.; Oberthur, R. C. *Colloid Polym. Sci.* **1985**, *263*, 443.
32. Lindner, P.; Oberthur, R. C. *Revue Phys. Appl.* **1984**, *19*, 759.
33. Hayter, J. B.; Penfold, J. *J. Phys. Chem.* **1984**, *88*, 4589.
34. Nakatani, A. I.; Kim, H.; Han, C. C. *J. Res. Natl. Inst. Stand. Technol.* **1990**, *95*, 7.
35. Plano, R. J.; Safinya, C. R.; Sirota, E. B.; Wenzel, L. *J. Rev. Sci. Instr.* **1993**, *64*, 1309.
36. Safinya, C. R.; Sirota, E. B.; Bruinsma, R. F.; Jeppesen, C.; Plano, R. J.; Wenzel, L. J. *Science*, **1993**, *261*, 588.
37. Radler, M. J.; Landes, B. G.; Nolan, S. J.; Broomall, C. F.; Chritz, T. C.; Rudolf, P. R.; Mills, M. E.; Bubeck, R. A. *ACS PMSE Preprints*, **1994**, *71*, 328.
38. Nakatani, A. I.; Poliks, M. D.; Samulski, E. T. *Macromolecules* **1990**, *23*, 2686.
39. VerStrate, G.; Philippoff, W. *J. Polym. Sci.: Polym. Lett. Ed.* **1974**, *12*, 267.
40. Boué, F.; Lindner, P. *Europhys. Lett.* **1994**, *25*, 421.
41. Moldenaers, P.; Yanase, H.; Mewis, J. Fuller, G. G.; Lee, C. S.; Magda, J. J. *Rheol. Acta* **1993**, *32*, 1.
42. Boué, F.; Bastide, J.; Buzier, M.; Lapp, A.; Herz, J.; Vilgis, T. A. *Colloid Polym. Sci.* **1991**, *269*, 195.
43. Yanase, H.; Moldenaers, P.; Mewis, J.; Abetz, V.; van Egmond, J.; Fuller, G. G. *Rheol. Acta* **1991**, *30*, 89.
44. Nakatani, A. I.; Douglas, J. F.; Ban, Y.-B.; Han, C. C. *J. Chem. Phys.* **1994**, *100*, 3224.
45. van Egmond, J. W.; Fuller, G. G. *Macromolecules* **1993**, *26*, 7182.
46. Beris, A. N.; Mavrantzas, V. G. *J. Rheol.* **1994**, *38*, 1235.
47. Helfand, E.; Fredrickson, G. H. *Phys. Rev. Lett.* **1989**, *62*, 2468.
48. Milner, S. T.; *Phys. Rev. Lett.* **1991**, *66*, 1477.
49. Takebe, T.; Hashimoto, T. *Polymer* **1988**, *29*, 261.
50. Nakatani, A. I.; Waldow, D. A.; Han, C. C. *Polymer Preprints* **1991**, *65*, 266.
51. Ng, R. C.-Y.; Leal, L. G. *J. Rheol.* **1993**, *37*, 443.
52. Rofe, C. J.; Lambert, R. K.; Callaghan, P. T. *J. Rheol.* **1994**, *38*, 875.
53. Xia, Y.; Callaghan, P. T. *Macromolecules* **1991**, *24*, 4777.
54. Zisenis, M.; Springer, J. *Polymer* **1994**, *35*, 3156.
55. Douglas, J. F. *Macromolecules* **1992**, *25*, 1468.
56. Hobbie, E. K.; Nakatani, A. I.; Han, C. C. *Mod. Phys. Lett. B* **1994**, *8*, in press.
57. Horst, R.; Wolf, B. A. *Rheol. Acta* **1994**, *33*, 99.

58. Horst, R.; Wolf, B. A. *Macromolecules* **1993**, *26*, 5676.
59. Horst, R.; Wolf, B. A. *Macromolecules* **1992**, *25*, 5291.
60. Horst, R.; Wolf, B. A. *Macromolecules* **1991**, *24*, 2236.
61. Kammer, H. W.; Kressler, J.; Kummerlowe, C. *Adv. Polym. Sci.* **1993**, *106*, 31.
62. Kammer, H. W. *Acta Polymerica* **1991**, *42*, 571.
63. Kammer, H. W. *Macromol. Symp.* **1994**, *78*, 41.
64. Wu, R.; Shaw, M. T.; Weiss, R. A. *J. Rheol.* **1992**, *36*, 1605.
65. Nakatani, A. I.; Kim, H.; Takahashi, Y.; Matsushita, Y.; Takano, A.; Bauer, B. J.; Han, C. C. *J. Chem. Phys.* **1990**, *93*, 795.
66. Mani, S.; Malone, M. F.; Winter, H. H.; Halary, J. L.; Monnerie, L. *Macromolecules*, **1991**, *24*, 5451.
67. Mani, S; Malone, M. F.; Winter, H. H. *J. Rheol.* **1992**, *36*, 1625.
68. Germain, Y.; Ernst, B.; Genelot, O.; Dhamani, L. *J. Rheol.* **1994**, *38*, 681.
69. Kröger, M.; Loose, W.; Hess, S. *J. Rheol.* **1993**, *37*, 1057.
70. Spenley, N. A.; Cates, M. E. *Macromolecules* **1994**, *27*, 3850.
71. Archer, L. A.; Huang, K.; Fuller, G. G. *J. Rheol.* **1994**, *38*, 1101.
72. Seidel, U.; Stadler, R.; Fuller, G. G. *Macromolecules* **1994**, *27*, 2066.
73. Leibler, L.; Rubinstein, M.; Colby, R. H. *Macromolecules* **1991**, *24*, 4701.
74. Rabin, Y.; Bruinsma; R. *Europhys. Lett.* **1992**, *20*, 79.
75. Mendes, E.; Lindner, P.; Buzier, M.; Boué, F.; Bastide, J. *Phys. Rev. Lett.* **1991**, *66*, 1595.
76. Ramzi, A.; Mendes, E.; Zielinski, F.; Rouf, C.; Hakiki, A.; Herz, J.; Oeser, R.; Boué, F.; Bastide, J. *J. de Physique IV* **1993**, *3*, 91.
77. Milner, S. T. *Science* **1991**, *251*, 905.
78. Klein, J.; Kamimiya, Y.; Yoshizawa, H.; Israelachvili, J. N.; Fredrickson, G. H; Pincus, P.; Fetters, L. J. *Macromolecules* **1993**, *26*, 5552.
79. Klein, J.; Perahia, D.; Warburg, S. *Nature* **1991**, *352*, 143.
80. Hu, H. W.; Granick, S. *Science*, **1992**, *258*, 1339.
81. Granick, S. *Science* **1991**, *253*, 1374.
82. Peachey, J.; van Alsten, J.; Granick, S. *Rev. Sci. Instr.* **1991**, *62*, 463.
83. Pelletier, E.; Montfort, J. P.; Lapique, F. *J. Rheol.* **1994**, *38*, 1151.
84. Bartczak, Z.; Argon, A. S.; Cohen, R. E. *Polymer* **1994**, *35*, 3427.
85. McHugh, A. J.; Spevacek, J. A.; *J. Polym. Sci: Polym. Phys. Ed.* **1991**, *29*, 969.
86. McHugh, A. J.; Guy, R. K.; Tree, D. A. *Colloid Polym. Sci.* **1993**, *271*, 629.
87. *ACS Polym. Preprints* **1994**, *35*, 620-629.
88. Cates, M. E.; Milner, S. T. *Phys. Rev. Lett.* **1989**, *62*, 1856.
89. Marques, C. M.; Cates, M. E. *J. Phys. Paris* **1990**, *51*, 1733.
90. Fredrickson, G. H. *J. Rheol.* **1994**, *38*, 1045.
91. Koppi, K. A.; Tirrell, M.; Bates, F. S.; Almdal, K.; Mortensen, K. *J. Rheol.* **1994**, *38*, 999.
92. Almdal, K.; Koppi, K. A.; Bates, F. S. *Macromolecules* **1993**, *26*, 4058.
93. Koppi, K. A.; Tirrell, M.; Bates, F. S. *Phys. Rev. Lett.* **1993**, *70*, 1449.
94. Koppi, K. A.; Tirrell, M.; Bates, F. S.; Almdal, K.; Colby, R. H. *J. Phys. II France* **1992**, *2*, 1941.

95. Kannan, R. M.; Kornfield, J. A. *Macromolecules* **1994**, *27*, 1177.
96. Kannan, R. M.; Kornfield, J. A. *J. Rheol.* **1994**, *38*, 1127.
97. Balsara, N. P.; Hammouda, B.; Kesani, P. K.; Jonnalagadda, S. V.; Straty, G. C. *Macromolecules* **1994**, *27*, 2566.
98. Morrison, F. A.; Mays, J. W.; Muthukumar, M.; Nakatani, A. I.; Han, C. C. *Macromolecules* **1993**, *26*, 5271.
99. Winter, H. H.; Scott, D. B.; Gronski, W.; Okamoto, S.; Hashimoto, T. *Macromolecules* **1993**, *26*, 7236.
100. Okamoto, S.; Saijo, K.; Hashimoto, T. *Macromolecules* **1994**, *27*, 3753.
101. Lee, H. S.; Lee, N. W.; Paik, K. H.; Ihm, D. W. *Macromolecules* **1994**, *27*, 4364.
102. Archer, L. A.; Fuller, G. G. *Macromolecules* **1994**, *27*, 4804.
103. Ackerson, B. J.; Pusey, P. N. *Phys. Rev. Lett.* **1988**, *61*, 1033.
104. Chen, L. B.; Ackerson, B. J.; Zukoski, C. F. *J. Rheol.* **1994**, *38*, 193.
105. Chen, L. B.; Chow, M. K.; Ackerson, B. J.; Zukoski, C. F. *Langmuir* **1994**, *10*, 2817.
106. Chen, L. B.; Zukoski, C. F. *J. Chem. Soc. Faraday Trans.* **1990**, *86*, 2629.
107. Laun, H. M.; Bung, R.; Hess, S.; Loose, W.; Hess, O.; Hahn, K.; Hädicke, E.; Hingmann, R.; Schmidt, F.; Lindner. P. *J. Rheol.* **1992**, *36*, 743.
108. Wouterson, A. T. J. M.; May, R. P.; deKruif, C. G. *J. Rheol.* **1993**, *37*, 71.
109. deKruif, C. G.; Vanderwerff, J. C.; Johnson, S. J.; May, R. P. *Phys. of Fluids A: Fluid Dyn.* **1990**, *2*, 1545.
110. Vanderwerff, J. C.; Ackerson, B. J.; May, R. P.; deKruif, C. G. *Physica A*, **1990**, *165*, 375.
111. Sinton, S. W.; Chow, A. W. *J. Rheol.* **1991**, *35*, 735.
112. Chow, A. W.; Sinton, S. W.; Iwamiya, J. H. *J. Rheol.* **1993**, *37*, 1.
113. Altobelli, S. A.; Givler, R. C.; Fukushima, E. *J. Rheol.* **1991**, *35*, 721.
114. Abbott, J. R.; Tetlow, N.; Graham, A. L.; Altobelli, S. A.; Fukushima, E.; Mondy, L. A.; Stephens, T. S. *J. Rheol.* **1991**, *35*, 773.
115. Mondy, L. A.; Brenner, H.; Altobelli, S. A.; Abbott, J. R.; Graham, A. L. *J. Rheol.* **1994**, *38*, 444.
116. Husband, D. M.; Mondy, L. A.; Ganani, E.; Graham, A. L. *Rheol. Acta* **1994**, *33*, 185.
117. Stover, C. A.; Cohen, C. *Rheol. Acta* **1990**, *29*, 192.
118. Hu. Y.; Wang, S.Q.; Jamieson, A. M. *J. Rheol.* **1993**, *37*, 531.
119. Munch, C; Hoffman, H.; Ibel, K.; Kalus, J.; Neubauer, G.; Schmelzer, U.; Selbach, J. *J. Phys. Chem.* **1993**, *97*, 4514.
120. Kalus, J.; Lindner, P.; Hoffmann, H.; Ibel, K.; Munch, C; Sander, J.; Schmelzer, U.; Selbach, J. *Physica B*, **1991**, *174*, 164.
121. Munch, C.; Hoffman, H.; Kalus, J.; Ibel, K.; Neubauer, G.; Schmelzer, U. *J. Appl. Crystallography* **1991**, *24*, 740.
122. Kalus, J. *Mol. Cryst. Liq. Cryst.* **1990**, *193*, 77.
123. Jindal, V. J.; Kalus, J.; Pilsl, H.; Hoffmann, H.; Lindner, P. *J. Phys. Chem.* **1990**, *94*, 3129.
124. Schmitt, V.; Lequeux, F.; Pousse, A.; Roux, D. *Langmuir* **1994**, *10*, 955.

125. Phoon, C. L.; Higgins, J. S.; Allegra, G.; Vanleeuwen, P.; Staples, E. *Proc. Royal Soc. London Ser. A: Mathematical and Physical Sci.* **1993**, *442*, 221.
126. Penfold, J.; Staples, E.; Cummins, P. G. *Adv. Colloid Interface Sci.* **1991**, *34*, 451.
127. Hamilton, W. A.; Butler, P. D.; Baker, S. M.; Smith, G. S.; Hayter, J. B.; Magid, L. J.; Pynn, R. *Phys. Rev. Lett.* **1994**, *72*, 2219.
128. *J. Rheol.* **1994**, *38*, 1471-1638.
129. Jackson, C. L.; Shaw, M. T. *Intl. Matls. Revs.* **1991**, *36*, 165.
130. Marrucci, G. *Rheol. Acta* **1990**, *29*, 523.
131. Semenov, A. N.; *J. Rheol.* **1993**, *37*, 911.
132. Larson, R. G.; Doi, M. *J. Rheol.* **1991**, *35*, 539.
133. Berry, G. C. *J. Rheol.* **1991**, *35*, 943.
134. Burghardt, W. R.; Hongladarom, K. *Macromolecules* **1994**, *27*, 2327.
135. Hongladarom, K.; Burghardt, W. R. *Macromolecules* **1994**, *27*, 483.
136. De'Nève, T.; Navard, P.; Kléman, M. *J. Rheol.* **1993**, *37*, 515.
137. Yan, N. X.; Labes, M. M.; Baek, S. G.; Magda, J. J. *Macromolecules* **1994**, *27*, 2784.
138. Guido, S.; Frallicciardi, P.; Grizzuti, N.; Marrucci, G. *Rheol. Acta* **1994**, *33*, 22.
139. Gu, D.-F.; Jamieson, A. M.; Wang, S. Q. *J. Rheol.* **1993**, *37*, 985.
140. Gu, D.-F.; Jamieson, A. M. *Macromolecules* **1994**, *27*, 337.
141. Grabowski, D. A.; Schmidt, C. *Macromolecules* **1994**, *27*, 2632.
142. Rey, A. D. *Mol. Cryst. Liq. Cryst.* **1993**, *225*, 313.
143. Brulet, A.; Boué, F.; Keller, P.; Davidson, P.; Strazielle, C.; Cotton, J. P. *J. de Physique II* **1994**, *4*, 1033.
144. Khan, S. A. *J. Rheol.* **1992**, *36*, 573.
145. Hirai, M.; Hirai, T.; Ueki, T. *Macromolecules* **1994**, *27*, 1003.
146. van den Brule, B. H. A. A.; Slikkerveer, P. J. *Rheol. Acta* **1990**, *29*, 175.
147. Ji, H.; Helfand, E. *Macromolecules* **1995**, *xx*, yyyy.
148. Ceccio, S. L.; Wineman A. S. *J. Rheol.* **1994**, *38*, 453.
149. Fodor, J. S.; Hill, D. A. *J. Rheol.* **1994**, *38*, 1071.
150. Fodor, J. S.; Hill, D. A. *Macromolecules* **1992**, *25*, 3511.
151. Adachi, K.; Fukui, F.; Kotaka, T. *Langmuir* **1994**, *10*, 126.

RECEIVED February 6, 1995

POLYMER SOLUTIONS

Chapter 2

Enhancement of Concentration Fluctuations in Solutions Subject to External Fields

G. Fuller[1], J. van Egmond[2], D. Wirtz[3], E. Peuvrel-Disdier[4], E. Wheeler[1], and H. Takahashi[5]

[1]Chemical Engineering Department, Stanford University, Stanford, CA 94305-5025
[2]Chemical Engineering Department, University of Massachusetts, Amherst, MA 21218
[3]Chemical Engineering Department, Johns Hopkins University, Baltimore, MD 21218
[4]Centre de Mise en Forme des Materiaux, Ecole des Mines, Sophia Antipolis, B.P. 207, F-06904 Valbonne, France
[5]Mechanical Engineering Department, Nagaoka University of Technology, Nagaoka, Niigata, Japan

> This paper discusses the phenomena of the coupling of concentration fluctuations in complex fluids to external hydrodynamic and electric fields. The use of modified Cahn-Hilliard models to describe this phenomena is presented, along with experimental methods to measure flow-induced structure. It is demonstrated that the application of flow to polymer solutions can enhance concentration fluctuations in the direction perpendicular to the principal axis of strain, and produces butterfly shaped structure factors. Electric fields, on the other hand, distort fluctuations parallel to the applied field, and can induce a real shift of the coexistence curve. In viscoelastic surfactant solutions subject to flow, concentration fluctuations under go a transition from perpendicular flow alignment to parallel flow alignment on time scales associated with the breakage time of wormlike micelles.

This paper considers the coupling of external fields (flow and electrical) to concentration fluctuations that are present in polymeric and surfactant liquids. Such couplings are known to produce dramatic effects and alterations of the microstructure in these systems. In some instances these lead to irreversible phenomena, such as flow induced crystallization and gelation. In this paper, however, we are concerned with reversible phenomena where the sample returns to its initial state once the field is removed. A common observable is the simple turbidity of a liquid, and how this can be affected by the presence of a field. Polymer solutions subject to flow, for example, can be transformed from a transparent to an opaque appearance (1). Flow induced structural changes can also occur in some micellar systems that are formed from concentrated surfactant molecules (2). The first interpretations of these phenomena was that they signaled the occurrence of an actual field-induced phase transition, and theories based on modifications of the free energy of polymer liquids through the addition of flow-induced decreases in chain entropy were advanced. However, more recent understanding of these observations suggests that in the case of flow, concentration fluctuations couple to stress and become enhanced along particular directions (*3-5*). This alternative description has been supported by several experimental observations using light scattering (*6*, *7*) and scattering dichroism (*8*). However, when electric fields are applied, true shifts in the phase boundary can be induced (*9*, *10*). In the following sections, theories describing these phenomena are reviewed and optical measurements of flow-induced structure are presented for both flow and electric fields.

Theoretical Background

General Developments. We seek here a description of the dynamics of concentration fluctuations in multicomponent liquids. Such fluctuations arise spontaneously due to a variety of stochastic phenomena present in liquids. The result is that bulk properties, such as temperature and concentration, will fluctuate about a mean value as a function of both time and space. For example, the concentration of the monomer species of a polymer, $\phi(\mathbf{r}, t)$, will generally be given by:

$$\phi(\mathbf{r}, t) = \bar{\phi} + \delta\phi(\mathbf{r}, t), \tag{1}$$

where $\bar{\phi}$ is the average, bulk concentration and $\delta\phi$ is the fluctuation. These fluctuations will, in turn, induce fluctuations in the optical properties of the liquid (e.g. its refractive index), and cause light to be scattered.

The equation of motion for the concentration fluctuations is the following statement of continuity:

$$\frac{\partial}{\partial t}\delta\phi + \nabla \cdot \mathbf{j} = \vartheta(\mathbf{r}, t), \tag{2}$$

where the function $\vartheta(\mathbf{r}, t)$ is a noise term accounting for random, stochastic interactions between molecules in the system that induce the fluctuations. $\mathbf{j}$ is a flux that arises from the combined influence of convection and diffusion down a chemical potential gradient, $\nabla\mu$. The flux is then,

$$\mathbf{j} = (\mathbf{G} \cdot \mathbf{r})\delta\phi - \Xi a^3 \nabla\mu, \tag{3}$$

where $\mathbf{G}$ is the velocity gradient tensor of the flow, Ξ is the friction factor, and a is the monomer size. If f is the free energy density of the system, the chemical potential is $\mu = \dfrac{\partial f}{\partial(\delta\phi)}$. What is needed is a relationship between the free energy density and the concentration fluctuations present in the system. The approach used here is to use the Cahn-Hilliard energy functional, (*11*)

$$F\{\delta\phi\} = \int d\mathbf{r}\left[\frac{1}{2}K(\nabla\delta\phi(\mathbf{r}))^2 + f(\delta\phi(\mathbf{r}))\right], \tag{4}$$

and the chemical potential is calculated as the functional derivative, $\mu \approx \dfrac{\delta F\{\delta\phi\}}{\delta(\delta\phi)}$.
The parameter K is a phenomenological constant multiplying a term that measures the strength of gradients in the fluctuations.

The Cahn-Hilliard approach has been used extensively in studies of spinodal decomposition of binary mixtures. In general, the function $f(\delta\phi(\mathbf{r}))$ is nonlinear and numerical procedures are needed to solve for the resulting phase diagram and dynamical response. In this description, however, a solution is sought in the vicinity of small fluctuations. In this limit,

$$\frac{\partial}{\partial t}\delta\phi + \nabla \cdot (\mathbf{G} \cdot \mathbf{r})\delta\phi = \Xi a^3 \nabla^2 (f'' - K\nabla^2)\delta\phi + \vartheta(\mathbf{r}, t). \tag{5}$$

The evolution of concentration fluctuations is normally measured using light scattering. In such an experiment, the observable is proportional to the structure factor, $S(\mathbf{q})$, where $\mathbf{q}$ is the scattering vector. The structure factor is defined as,

$$S(\mathbf{q}) = \int_{-\infty}^{\infty} d\mathbf{r}\, e^{i\mathbf{q}\cdot\mathbf{r}} \langle \delta\phi(\mathbf{r})\delta\phi(0) \rangle. \tag{6}$$

In other words, the structure factor is the Fourier transform of the spatial autocorrelation function of concentration fluctuations.

Equation 5 can be converted to an equation of motion for the structure factor by first Fourier transforming this equation and then multiplying it by its complex conjugate. The result is then averaged over the function describing the distribution of concentration fluctuations. To complete the analysis, it is assumed that the noise term is a Gaussian random function with the following moments:

$$\langle \vartheta(\mathbf{r}, t) \rangle = 0;\ \langle \vartheta(\mathbf{r}, t)\vartheta(\mathbf{r}', t) \rangle = k_B T \Xi \delta(\mathbf{r} - \mathbf{r}')\delta(t). \tag{7}$$

The result is,

$$\frac{\partial S}{\partial t} - \mathbf{q} \cdot \mathbf{G} \cdot \nabla_\mathbf{q} S = -2\Xi a^3 q^2 (f'' + Kq^2) S + 2D_t q^2, \tag{8}$$

where $D_t = k_B T \Xi$ is the translational diffusion coefficient. At steady state, and in the absence of flow, the structure factor is predicted to have the "Ornstein-Zernike" form,

$$S(\mathbf{q}) = \left(\frac{k_B T}{f'' a^3}\right)\frac{1}{1 + \xi^2 q^2}, \tag{9}$$

where the correlation length has been defined as $\xi = \sqrt{K/f''}$.

Coupling of Concentration Fluctuations to Stress. When the transport coefficients of a mixture are a strong function of composition, the application of mechanical stress from hydrodynamic forces can induce an additional flux of concentration, $\mathbf{j}_M$ (3-5,

12). This flux is driven by the force density, $\nabla \cdot \underline{\tau}$, where $\underline{\tau}$ is the stress tensor, and is taken to be $j_M = -\Xi \nabla \cdot \underline{\tau}$. To insert this term within the equation of motion for the structure factor, the stress tensor is first expanded as a function of concentration, in the limit of small fluctuations. This leads to,

$$\frac{\partial S}{\partial t} - \mathbf{q} \cdot \mathbf{G} \cdot \nabla_\mathbf{q} S = -2\Xi a^3 q^2 \left(f'' + K q^2 - \hat{\mathbf{q}}\hat{\mathbf{q}} : \frac{\partial}{\partial \phi} \underline{\tau} \right) S + 2D_t q^2, \qquad (10)$$

where $\hat{\mathbf{q}} = \mathbf{q}/|\mathbf{q}|$.

Equation 10 is a first order partial differential equation that can be solved using the method of characteristics. At steady state and in the limit where the convection term can be neglected, a qualitative prediction for structure factor is obtained as,

$$S \approx \left(\frac{k_B T}{a^3 f''} \right) \frac{1}{1 + \xi^2 q^2 - \frac{\xi^2}{f''} \hat{\mathbf{q}}\hat{\mathbf{q}} : \frac{\partial}{\partial \phi} \underline{\tau}}. \qquad (11)$$

An explicit form requires a specification of the rheological constitutive representing the sample. For example, if the material is Newtonian and subject to a simple shear flow in the (x,y) plane, the term involving the stress tensor is $\hat{\mathbf{q}}\hat{\mathbf{q}} : \frac{\partial}{\partial \phi} \underline{\tau} = 2\hat{q}_x \hat{q}_y G \eta'$, where G is the shear rate, and $\eta' = \frac{\partial \eta}{\partial \phi}$ is the differential viscosity relative to concentration. In this case, the structure factor is predicted to have a "butterfly" shaped pattern, with the wings of the butterfly oriented parallel to the rate of strain tensor of the simple shear flow. This axis is at an angle of 45° relative to the flow direction. Because the structure factor is the Fourier transformation of the actual pattern of concentration fluctuations, this simple model predicts that the fluctuations grow perpendicular to the rate of strain tensor, at an angle of 135° to the flow.

For viscoelastic polymeric liquids, complex, nonlinear rheological constitutive equations must be used. A simple model that has been employed for the purpose of examining the role of normal stresses is the second order fluid model. Several papers have discussed the use of this model in predicting the structure factor for both shear (*3, 4*) and extensional flows (*13*). In the case of shear flow, the presence of normal stresses causes the butterfly pattern to rotate with the vorticity of the flow. At sufficiently high flow rates, the wings of the pattern rotate from the 45° orientation towards an orientation of −90°. In extensional flow, the presence of normal stresses superimposes a four fold symmetry on top of the butterfly shaped pattern. The second order fluid model, however, assumes the material has an instantaneous response, and more realistic models incorporating relaxation time phenomena have also been used.

Coupling of Concentration Fluctuations to Electric Fields. The application of electric fields onto a system of concentration fluctuations will cause them to distort, since such fluctuations will also cause the dielectric constant to fluctuate in space. This effect can also be incorporated within the Cahn-Hilliard model presented earlier. It is assumed that the electric field will augment the free energy expression given in equation 4 to include terms that explicitly couple the electric field to concentration fluctuations. In the present theory, these terms are *ad hoc* and are chosen to posses the correct symmetry and scalar properties. We write the free energy functional as (*9*, *10*)

$$F\{\delta\phi\} = \int d\mathbf{r} \left[\frac{1}{2} K (\nabla \delta\phi(\mathbf{r}))^2 + f(\delta\phi(\mathbf{r})) + K_2 \varepsilon_0 |\mathbf{E}\delta\phi|^2 + K_3 \varepsilon_0 (\mathbf{E} \cdot \nabla \delta\phi)^2 \right],$$

(12)

The constants K_2 and K_3 are two-body interaction parameters that gauge the strengths of the couplings between the electric field and the fluctuations and gradients of the fluctuations, respectively.

This functional is easily incorporated within the model, and taking the electric field to be parallel to the x axis, we have:

$$\frac{\partial S}{\partial t} = -2\Xi a^3 q^2 (f'' + Kq^2 + K_2 \varepsilon_0 E^2 + K_3 \varepsilon_0 q_x^2 E^2) S + 2D_t q^2,$$

(13)

in the absence of flow. At steady state, the structure factor is

$$S(\mathbf{q}, \mathbf{E}) = \frac{1}{\kappa(E) + K_1 q^2 + K_3 \varepsilon_0 E^2 q_x^2},$$

(14)

where $\kappa(E) = f'' + K_2 \varepsilon_0 E^2$. Defining a field dependent correlation length, $\xi_E = \sqrt{K_1/\kappa(E)}$, we have,

$$S(\mathbf{q}, \mathbf{E}) = \frac{1/\kappa(E)}{1 + \xi_E^2 q_y^2 + \xi_E^2 q_x^2 \left(1 + \frac{K_3}{K_1} E^2\right)}.$$

(15)

This model predicts that the application of an electric field will distort the structure factor into an elliptical shape, with the long axis of the ellipse perpendicular to the direction of the field. This is consistent with a deformation of concentration fluctuations parallel to the field. This result should be contrasted with the effects of flow, where butterfly shaped patterns oriented parallel to the rate of strain are produced.

The presence of an electric field not only distorts the concentration fluctuations, but also produces a real shift of the phase boundary. This can be seen by examining the temperature dependence of the correlation length. We start by specifying a form for the free energy density function, f''. A form that has been used for semidilute polymer solutions is $f'' \approx 1 - 2\chi$, where χ is the Flory-Huggins interaction parameter. The χ parameter depends on temperature, and a mean field treatment leads to the following result for the correlation length:

$$\xi = \xi_0 \left(\frac{T_c}{T-T_c}\right)^{\frac{1}{2}}. \tag{16}$$

Comparing this result for the correlation length to the case when an electric field is applied, the critical temperature, T_C, is predicted to have the following dependence on the field strength,

$$\frac{T_C - T_C(E)}{T_C} \approx AE^2, \tag{17}$$

where A is a constant that can be shown to be $\frac{1}{2k_BT}\left(\frac{\partial n}{\partial \phi}\right)^2$. Here n is the refractive index.

This result indicates that the imposition of an electric field will cause the critical temperature to decrease. This depression will occur for both upper critical and lower critical temperature solutions, since the parameter $A > 0$.

Calculation of Scattering Dichroism. In addition to measurements of the structure factor using small angle light scattering, it is often convenient to measurement the moments of this function with respect to the scattering vector, $\mathbf{q}$. This can be accomplished by measuring the scattering dichroism of the sample. Formally, this optical anisotropy is defined as the tendency of a material to attenuate light in a manner that is specific to its polarization and is linked to the imaginary part of its refractive index. For example, if an anisotropic material has its principal directions parallel to the x and y axes, the dichroism in the (x,y) plane is $\Delta n'' = n''_x - n''_y$. In a system of anisotropic, oriented scatterers, the scattering process contributes to the imaginary part of the refractive index and causes attenuation of light transmitted by the sample. If the wavelength of light is removed from any absorption region, the dichroism is referred to as conservative, or scattering dichroism. Onuki and Doi have developed the relationship between the scattering process and the dielectric tensor, from which the scattering dichroism can be calculated. They determined that *(14)*

$$\Delta n'' = \frac{k^3}{32\pi^2 \varepsilon_0^{3/2}} \left(\frac{\partial \varepsilon_0}{\partial \phi}\right)^2 Im\int d\mathbf{q}\, \frac{q_y^2 - q_x^2}{q^2 - k^2 - i0} S(\mathbf{q}), \tag{18}$$

where k is the wavevector of the incident light. From this expression, it is evident that the dichroism is a measure of anisotropy in the second moments of the structure factor.

Materials

Flow-Induced Structure in Polymer Solutions. The polymer solution used in the study of concentration fluctuation enhancement in flow consisted of polystyrene (PS) dissolved in dioctyl phthalate (DOP). The molecular weight was 1.86×10^6 with a polydispersity index of $M_W/M_N = 1.06$ and was present at a volume concentration of 6%. The PS/DOP system has a theta temperature of 22 °C and the cloud point for this concentration and molecular weight was 10.5 °C.

Electric Field-Induced Structure in Polymer Solutions. The experiments exploring the coupling of concentration fluctuations to electric fields used solutions of polystyrene dissolved in cyclohexane. The molecular weight of the PS was 400,000 with $M_W/M_N = 1.01$. The concentration was 5.89%, which is the critical concentration for this molecular weight. The cloud point for this solution is 19.06 °C.

Flow Induced Structure in Surfactant Solutions. The surfactant system studied was a mixed system of 40 mmol/L cetylpyridinium chloride/40 mmol/L sodium salicylate in water. This composition forms a system of entangled, wormlike micelles that behave much like a concentrated solution of flexible polymer chains. Furthermore, the rheology can be adequately described using a single relaxation time, Maxwell fluid over a considerable range of frequencies.

Instrumentation

Small Angle Light Scattering and Scattering Dichroism. The instrument used to measure small angle light scattering (SALS) from flowing systems was described in detail in reference *13*. The experiment was designed to simultaneously measure scattering dichroism, and an account of the measurement of this property can also be found in this reference. Light from a HeNe laser is first polarized and then sent through a collection of optics (a photo-elastic modulator and a quarter wave plate) that modulate its polarization. This modulation is used in the measurement of scattering dichroism, and is not used for the SALS measurements. The light is then sent through the sample, which either resides in a flow cell or a electric field cell. Beneath the sample, a flat, white screen is used to collect the scattered light. An aperture in the screen passes the unscattered, transmitted light to a photodetector placed below the screen. This detector is used to record light for the purposes of measuring the scattering dichroism. The pattern of light intensity on the screen represents the SALS and is measured using a black and white video camera. The images from the camera were digitized using a frame grabber that passed the data to the memory of a computer.

Flow and Electric Field Cells. Both simple shear flow and extensional flow were examined. To generate extensional flow, a four roll mill was used, and is shown in Figure 1. This device consists of four cylinders arranged on the corners of a square. The cylinders were 0.484 inches in diameter and were separated by a center-to-center distance of 0.625 inches on the corners of the square. The cylinders were 3 inches in length. Rotation of the cylinders produces a two-dimensional extensional flow in the center region. At the geometric center, a stagnation point exists where there is no motion of the fluid. The light from the laser is sent through the stagnation point.

The flow cell used to generate simple shear flow consisted of a pair of parallel quartz disks with the upper disk rotating relative to the lower plate. This cell was part of a Rheometrics Dynamic Stress Rheometer and optical experiments were performed simultaneously with the acquisition of mechanical rheometric measurements. The figure below shows the geometry of the disks and the direction of the light beam when this flow cell was used.

As seen in Figure 2, the light beam is directed along the velocity gradient axis in the parallel disk cell. The SALS and scattering dichroism measurements will provide measures of both the degree of orientation and distortion, as well as the average angle, θ, along which the distortion is directed. This angle is measured relative to the flow direction. However, because of symmetry, this angle is either $0°$ or $90°$. In the case of scattering dichroism measurements, it is often convenient to express the results as the product, $\Delta n'' \cos 2\theta$. Since scattering dichroism is normally a positive quantity, positive values of this product imply orientation parallel to the flow direction, whereas negative values are interpreted as orientations perpendicular to the flow, and parallel to the vorticity axis.

The electric field experiments were carried out using a Kerr cell fashioned from a plastic, Delrin cup. Approximately 5 ml of sample were contained within the cup between two quartz windows. The interior walls of the cup held two parallel, stainless steel electrodes that were 1.65 mm apart and 9 mm in length. This cell was controlled in temperature to an accuracy of $0.01°C$. Electric fields up to 1000 V/mm could be applied.

Results and Discussion

Polymer Solutions Subject to Flow. In Figure 3, a SALS pattern is shown for the solution of PS/DOP. This solution was subjected to an extensional flow using the four roll mill and the extension axis was at $-45°$ to the horizontal axis of the figure. The strain rate was 3 s^{-1} and this pattern was collected after steady state had been achieved. There are two prominent features of the pattern that are evident. The primary structure of the pattern is a butterfly shape, with the "wings" of the butterfly parallel to the extension axis of the flow. This is in qualitative agreement with the predictions of the Helfand-Fredrickson model and is predicted even for Newtonian mixtures. Superimposed onto the butterfly shaped patterns are four distinct peaks that have a four fold symmetry. This additional feature is predicted by the model for the case of a symmetric extensional flow when the elasticity of the fluid (the presence of normal stresses in the second order fluid model, for example) is sufficiently large. This is a reversible phenomena, and the structure factor ultimately relaxes back towards an isotropic shape once the flow is removed.

Experiments in simple shear flow using scattering dichroism are shown for the PS/DOP solution in Figure 4. These are steady state data and are plotted as the product of the dichroism times the cosine of twice the orientation angle relative to the flow direction. The horizontal axis shows the applied stress. As discussed earlier, positive and negative values of this product are interpreted as orientations parallel and perpendicular to the flow direction, respectively. At low shear rates, this product is negative, which is in agreement with a structure factor of a butterfly shape where the wings of the butterfly are oriented parallel to the flow direction. Such an orientation is what has

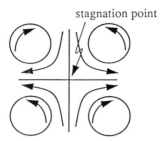

Figure 1. The four roll mill.

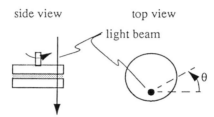

Figure 2. Schematic diagram of the parallel disk flow cell.

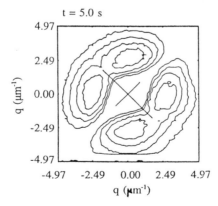

Figure 3. Small angle light scattering pattern for a PS/DOP solution subject to a four roll mill flow. The principal axis of strain was oriented at $(-45)°$ relative to the horizontal axis and a strain rate of 3 s^{-1} was applied.

been reported by researchers measuring SALS in parallel plate flows. However, at a sufficiently large shear rate, this product changes to positive values, indicating that the concentration fluctuations ultimately are deformed parallel to the flow direction. Also plotted in this figure are the shear viscosity and the birefringence. The viscosity is observed to undergo a pronounced shear thickening at the onset of the transition from perpendicular to parallel flow alignment by the concentration fluctuations. The birefringence, which is always negative, shows a marked increase in magnitude once this transition is reached. The negative sign of the birefringence for polystyrene indicates that this property is predominately a result of intrinsic, local segmental orientation.

Polymer Solutions Subject to Electric Fields. In Figure 5 two SALS patterns are shown for a solution of PS in cyclohexane. The molecular weight was 4×10^6 and the solution was at the critical concentration of 5.89 vol%. The temperature was slightly below the critical point ($T_C - T = 0.03\,°C$), so that with no field applied, this system was de-mixed and slightly turbid. The SALS pattern on the left shows the system prior to application of an electric field. Because it is in the two phase region, the solution is strongly scattering. Furthermore, the in the absence of a field, the isointensity contours are circular and isotropic. The pattern on the right, show the same system at a time of 2.5 seconds following the application of a field of 5000 V/cm directed along the horizontal axis. At this time, the SALS patterns had achieved a steady state appearance. There are two important observations: (1) the isointensity contours become highly elliptical in shape, with the axis of the ellipse oriented perpendicular to the field and (2) the amount of transmitted light and light scattered in the forward direction has increased substantially. This latter effect can be seen in the two patterns by noting the number of isointensity contours and the associated intensity values that label each contour. These findings are both qualitatively and quantitatively captured by the simple, modified Cahn-Hilliard theory discussed above.

Surfactant Solutions Subject to Flow. Experiments investigating flow induced structure in extensional flow were performed on the mixed surfactant system consisting of 40 mmol/L cetylpyridinium chloride/40 mmol/L sodium salicylate. A sequence of SALS images as functions of time is shown in Figure 6 for a strain rate of $10\ s^{-1}$. The rate of strain axis in these plots is in the horizontal direction. The initial frame shows the rest state structure factor, which is isotropic. Upon inception of flow, the structure factor first distorts in the direction of flow and develops a butterfly shaped pattern with the wings of the butterfly parallel to the flow direction. This suggests that concentration fluctuations in this system first grow in the direction perpendicular to flow, as in the case of the polymer solutions. However, at steady state, there is a transformation to a flow aligned state that is evident in the last frame showing a highly elongated structure factor that is oriented perpendicular to the principal axis of strain.

Summary

Examples of the coupling between concentration fluctuations and external fields have been described for polymeric and micellar solutions. Optical methods, such as small

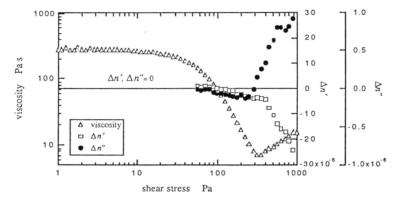

Figure 4. Dichroism (circles), viscosity (triangles), and birefringence (squares) as a function of shear stress for the PS/DOP solution subject to simple shear flow.

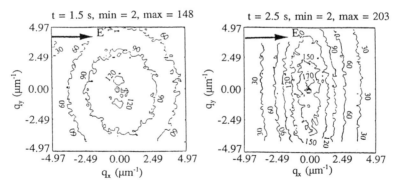

Figure 5. Small angle light scattering from the PS/CH solution. The pattern on the left shows the system in the absence of an electric field. The pattern on the right shows the effect of applying a field of 500 V/mm directed along the horizontal direction in this figure.

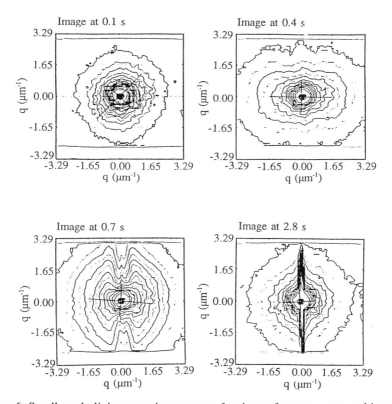

Figure 6. Small angle light scattering patterns for the surfactant system subject to an extensional flow of 10 s^{-1}. The flow was started at zero time.

angle light scattering and scattering dichroism, can effectively measure the influence of flow and electric fields on the structure factor, and reveal the distortion and orientation of fluctuations. These effects can be qualitatively modeled using a simple modification of the Cahn-Hilliard theory that is frequently used to analyze phase transition in simple mixtures and predicts that fluctuations are enhanced in the case of flow along the direction perpendicular to the principal axis of strain. In the case of electric fields, the distortion is parallel to the field, and this is accompanied by a shift in the phase boundary. Surfactant solutions where the concentration is sufficiently high to form entangled micelles also show concentration fluctuation enhancement in flow, but this process must complete with the breakage and reformation kinetics of the micelles.

Literature Cited

1. Rangel-Nafaile, C.; Metzner, A.; Wissbrun, K. *Macromolecules* **1984**, *17*, 1187.

2. Rehage, H.; Hoffmann, H. *Molecular Physics* **1991**, *74*, 933.

3. Helfand, E.; Fredrickson, G. H. *Phys. Rev. Lett.* **1989**, *62*, 2468.

4. Milner, S. T. *Phys. Rev. Lett.* **1991**, *66*, 1477.

5. Onuki, A. *Phys. Rev. Lett.* **1989**, *62*, 2472.

6. Hashimoto, T.; Kume, T. J. Phys. Soc. Japan **1992**, *61*, 1839.

7. Wu, X.-L.; Pine, D. J.; Dixon, P. K. Phys. Rev. Lett., **1991**, *66*, 2408.

8. Yanase, H.; Moldenaers, P.; Abetz, V.; van Egmond, J.; Fuller, G. G.; Mewis, J. *Rheol. Acta* **1991**, *30*, 89.

9. Wirtz, D.; Berend, K.; Fuller, G. G. *Macromolecules* **1992**, *25*, 7234.

10. Wirtz D.; Fuller, G. G. *Phys. Rev. Lett.* **1993**, *4*, 2236.

11. Goldenfeld, N. *Lectures on Phase Transitions and the Renormalization Group*; Frontiers in Physics; Addison Wesley: Reading, MA, **1992**; Vol. 85.

12. Larson, R. G. *Rheol. Acta* **1992**, *31*, 497.

13. van Egmond, J.; Fuller, G. G. *Macromolecules* **1993**, *26*, 7182.

14. Onuki, A.; Doi, M. *J. Chem. Phys.* **1986**, *85*, 1190.

RECEIVED February 25, 1995

Chapter 3

String Phase in Semidilute Polystyrene Solutions Under Steady Shear Flow

Takuji Kume and Takeji Hashimoto

Division of Polymer Chemistry, Graduate School of Engineering, Kyoto University, Kyoto 606-01, Japan

The shear-flow induced concentration fluctuations in semidilute solutions of polystyrenes with high molecular weights (~10^6) are studied as a function of shear rate, $\dot{\gamma}$, by means of light scattering and transmission optical microscopy. The solutions, which are homogeneous in the quiescent state, show enhanced concentration fluctuations parallel to the flow axis at $\dot{\gamma} > \dot{\gamma}_c$ (the critical shear rate), giving rise to unique butterfly-type scattering patterns, as reported previously. Upon further increasing $\dot{\gamma}$, we found for the first time that the solutions form long strings parallel to flow using shear microscopy. The string formation occurs in parallel to the onset of rheological anomalies and to the appearance of strong streak-like small-angle scattering at $\dot{\gamma} > \dot{\gamma}_a$, the critical shear rate above which the string and the anomalies are clearly discerned.

Shear-induced concentration fluctuations or phase-separation (1-19) have been studied quite extensively for semidilute solutions of high molecular weight polystyrenes, typically higher than 10^6, as an intriguing phenomenon concerning the nonequilibrium statistical mechanics of polymers. A polymer solution which is homogeneous and in the single phase state without shear flow exhibits strong concentration fluctuations at shear rates, $\dot{\gamma}$, larger than a critical shear rate, $\dot{\gamma}_c$. These shear-enhanced fluctuations were confirmed by measurements of changes in the turbidity or transmitted intensity (2-4,10,15,17), rheological properties (2,10,15,16) and form dichroism (10,13,15,17) with $\dot{\gamma}$.

The fluctuations induced were found to be highly anisotropic, with strong fluctuations along flow but fluctuations normal to the flow or the neutral axis of the flow being nearly identical to those at $\dot{\gamma} = 0$, giving rise to the unique butterfly-type scattering patterns (7,11,13,18,19). The anisotropic concentration fluctuations were further confirmed by a direct visualization with **shear microscopy** (18), i.e., by in-situ observation with optical microscopy. The images obtained by shear microscopy were confirmed as representing true structural entities by comparing the real light scattering patterns, simultaneously recorded by **shear light scattering**, and the Fast Fourier Transform (FFT) patterns corresponding to the microscope images.

In this paper we aim to report a novel experimental result on the formation of long strings parallel to flow at shear rates much higher than $\dot{\gamma}_c$. The string phase was directly visualized by shear microscopy and confirmed by shear light scattering which gives a streak-like scattering pattern normal to the flow direction. The string formation will be shown to be intimately related to the anomalies found in the rheological properties of these solutions.

Experimental Methods

As in previous experiments, we used semidilute polymer solutions of high molecular weight polystyrene (PS) with dioctylphthalate (DOP) (7,11,18). The PS/DOP system has a UCST-type phase diagram, and its θ temperature (T_θ) is 22 °C. We used two PS samples. PS548 is a standard sample supplied by TOSOH Co. Ltd., which has weight average molecular weight (M_w) of 5.84×10^6 and a heterogeneity index (M_w/M_n) of 1.15. PS1000 is a sample supplied by Japan Synthetic Rubber Co. Ltd., which has a molecular weight in the range $9-14 \times 10^6$. These polystyrenes were dissolved into solutions using prescribed amounts of DOP and an excess of methylene chloride, and then the PS/DOP solutions were prepared by completely evaporating the methylene chloride. The solutions used in this study are listed in Table I.

The small-angle light scattering and optical microscopy measurements under shear flow were performed by the two types of rheo-optical apparatuses which were constructed in our laboratory (20,21). We used transparent cone-and-plate type sample cells which were made of quartz. The optical setup and coordinate system in this study, shown in Figure 1, are the same as those previously used. The velocity gradient exists in the plane Oxy, where Ox is the flow direction. We send the incident beam along the Oy axis, and the scattering profiles were detected in the plane Oxz. The temperature of the sample cell was controlled with an accuracy of ±0.1°C.

The rheological measurements were performed with a mechanical spectrometer, RMS-800, and a fluid spectrometer, RFS II (both from Rheometrics, Inc.). The sample cells used also have a cone-and-plate geometry. The steady-state properties at various shear rates, $\dot{\gamma}$, were measured more than $20/\dot{\gamma}$ seconds after the specified shear rate was attained, in the cycle of increasing shear rate from a quiescent state. We confirmed that the scattering intensity at a given $\dot{\gamma}$ didn't change after shearing for the specified period of time.

Experimental Results

Figure 2 shows various properties under steady state shear flow as a function of shear rate for the 3.0 wt% DOP solution of PS548 plotted on a double logarithmic scale: η and ψ_1 are shear viscosity and coefficient of the first normal stress difference, respectively; $\mathcal{I}_{//}(\dot{\gamma})$ and $\mathcal{I}_{\perp}(\dot{\gamma})$ are the integrated scattered intensity parallel to x and z axis, respectively, and $\mathcal{I}_{//}(\dot{\gamma}=0)$ and $\mathcal{I}_{\perp}(\dot{\gamma}=0)$ are the corresponding quantities in the quiescent state. Here the quantities $\mathcal{I}_{//}$ and $\mathcal{I}_{\perp}$ were defined by

$$\mathcal{I}_{//} = \int_{q_{//S}}^{q_{//L}} I_{//}(q)dq \qquad (1)$$

Table I.   Polymers and polymer solutions used in this work

polymer	c: concentration (wt%)	c/c*[1]	cloud point[2] (°C)
PS548	3.0	20	16.3
PS548	6.0	40	13.8
PS1000	2.0	20	15.9

1) c*: overlap concentration
2) The cloud points of the solutions were determined by the same method as described previously (7).

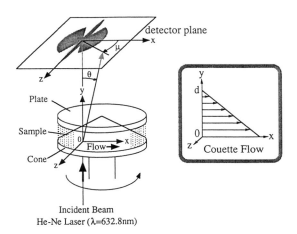

Figure 1. Cone-and-plate type sample cell to generate Couette flow, the coordinate system, and definition of the scattering angle, θ, and azimuthal angle, μ.

$$\mathcal{I}_\perp = \int_{q_{\perp S}}^{q_{\perp L}} I_\perp(q)dq \qquad (2)$$

where $I_{//}(q)$ and $I_\perp(q)$ are the scattered intensities parallel to the x and z axes, respectively, and q is the magnitude of scattering vector defined by $q = (4\pi/\lambda)\sin(\theta/2)$, with θ and λ being scattering angle and wavelength of light in the medium, respectively. The solution was sheared at 24 °C for light scattering experiments and 25 °C for the rheological experiments, 7.7 and 8.7 °C above the cloud point, respectively.

The solution, which is in a single-phase state at $\dot{\gamma} = 0$, exhibits remarkable shear-enhanced concentration fluctuations as observed in the change of light scattered intensity. At a shear rate smaller than the critical shear rate, $\dot{\gamma}_c$, the scattered intensities $\mathcal{I}_{//}(\dot{\gamma})$ and $\mathcal{I}_\perp(\dot{\gamma})$ are the same as those at $\dot{\gamma} = 0$, and hence the concentration fluctuations are not affected by shear (Regime I). However at $\dot{\gamma} > \dot{\gamma}_c$, a tremendous increase in the scattered intensities is observed, indicating the shear-enhanced concentration fluctuations reported earlier (7,11). The critical shear rate, $\dot{\gamma}_c$, for the shear enhanced concentration fluctuations is 0.2 s^{-1} for this solution. In parallel to the change in the scattered intensities $\mathcal{I}_{//}$ and $\mathcal{I}_\perp$ the rheological properties undergo remarkable change with $\dot{\gamma}$. At $\dot{\gamma} > \dot{\gamma}_c$, both η and ψ$_1$ drop with $\dot{\gamma}$, which may be approximated by the following power laws,

$$\eta \sim \dot{\gamma}^{-0.61 \pm 0.02} \quad \text{and} \quad \psi_1 \sim \dot{\gamma}^{-1.47 \pm 0.02} \qquad (3)$$

The power law behavior given by equation 3 starts to break at $\dot{\gamma} > \dot{\gamma}_a$ (Regime IV), a critical shear rate (20 s^{-1} for this solution) for the rheological anomaly as reported earlier (11): the decrease of η with $\dot{\gamma}$ starts to level off (22) and deviates from equation 3; ψ$_1$ starts to increase with $\dot{\gamma}$. This anomaly appears to occur in parallel to that in the scattered intensities $\mathcal{I}_{//}$ and $\mathcal{I}_\perp$: (i) a sharp increase in $\mathcal{I}_{//}$ and $\mathcal{I}_\perp$ with $\dot{\gamma}$ ($\dot{\gamma}_a$ appears to be an inflection point for $\mathcal{I}_{//}(\dot{\gamma})$ and $\mathcal{I}_\perp(\dot{\gamma})$) and (ii) a sharp decrease in the scattering anisotropy ($\mathcal{I}_{//}(\dot{\gamma}) - \mathcal{I}_\perp(\dot{\gamma})$) with $\dot{\gamma}$.

In our previous report (11) on 6.0 wt% solution of PS548 in DOP, we indicated that the shear-rate dependence of the properties at $\dot{\gamma}_c < \dot{\gamma} < \dot{\gamma}_a$ can be further classified into Regime II and III: in Regime II both $\mathcal{I}_{//}$ and $\mathcal{I}_\perp$ start to increase sigmoidally and reach certain saturated intensity levels, and in Regime III these intensity levels stay nearly constant. In this 3.0 wt% solution, we cannot clearly distinguish Regimes II and III in the same way as in the 6.0 wt% solution.

Figure 3 shows the change of the scattering patterns with $\dot{\gamma}$ observed in the Oxz plane. The solution examined is the 3.0 wt% PS548 in DOP solution at 25 °C. The solution exhibits no strong small angle scattering at $\dot{\gamma} < \dot{\gamma}_c$, as shown in Figure 3a, but a typical strong anisotropic scattering designated as the butterfly pattern (7,11) is exhibited at $\dot{\gamma} > \dot{\gamma}_c$, as shown in Figure 3b and 3c. The butterfly pattern has characteristics of a strong intensity parallel to the flow but a weak intensity along the neutral axis (Oz axis). The intensity along the Oz axis is as weak as that at $\dot{\gamma} = 0$, giving rise to the unique feature called the "dark streak" (11). Upon further increasing $\dot{\gamma}$ beyond $\dot{\gamma}_a$, a strong streak-like pattern appears at small-angle regions in the direction parallel to Oz axis, as seen in Figure 3d to 3g. The appearance of the streak-like scattering pattern upon increasing $\dot{\gamma}$ occurs in parallel to the onset of the rheological anomaly.

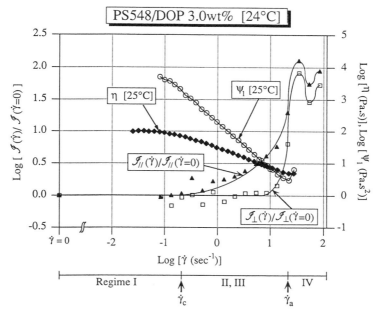

Figure 2. Shear viscosity, η, coefficient of the first normal stress difference, ψ_1, normalized integrated scattered intensity parallel and perpendicular to flow, $\mathcal{I}_{//}(\dot{\gamma})/\mathcal{I}_{//}(\dot{\gamma} = 0)$ and $\mathcal{I}_{\perp}(\dot{\gamma})/\mathcal{I}_{\perp}(\dot{\gamma} = 0)$ for the PS548/DOP 3.0 wt% solution, as a function of shear rate $\dot{\gamma}$ in steady state shear flow. $\dot{\gamma}_c$ and $\dot{\gamma}_a$ are the critical shear rates for the onset of the shear-enhanced concentration fluctuations, and of the anomalies in the rheological and scattering behaviors, respectively

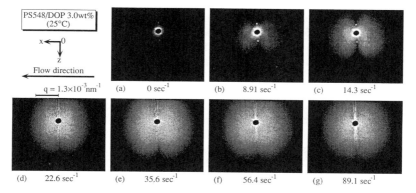

Figure 3. Steady state scattering patterns at various shear rates for the solution shown in Figure 2.

Figure 4 represents contour plots of the scattered intensity distributions at various $\dot{\gamma}$, measured by the CCD camera system previously reported (21). The features demonstrated in Figure 3 are more quantitatively shown in this figure. It is clearly shown that the intensity of the streak pattern along the neutral axis increases with $\dot{\gamma}$ beyond $\dot{\gamma}_a$, as seen in Figure 4d to 4g. It is also observed that, superimposed on the streak pattern, the butterfly pattern appears at q regions larger than those where the streak pattern appears.

Figure 5 represents a direct visualization of the structure at $\dot{\gamma} = 89.1$ s^{-1} ($> \dot{\gamma}_a$) by means of shear microscopy (a) and its FFT image (b). The image in part a (a snap shot taken with a shutter speed of 10^{-3} s) shows long strings of length about 50μm or longer oriented along flow (Ox axis). The corresponding FFT image shows a strong streak parallel to the z axis, consistent with the real scattering pattern shown in Figure 4g. The spread of the streak shown in the FFT image is approximately equal to that observed in the real light scattering pattern. Therefore the image shown in Figure 5a seems to reflect the true structural entity corresponding to the streak-like scattering. Unfortunately we could not recover the butterfly-type scattering pattern in the FFT image shown in Figure 5b from the optical image shown in Figure 5a. This may be due to the fact that the intensity of the butterfly-type scattering pattern is much weaker than that of the streak-type scattering pattern, i.e., the concentration difference between the strings and their matrix associated with the streak scattering are much greater than the concentration fluctuations within the strings associated with butterfly pattern (see discussion related to Figure 8 later) (18).

Figures 6 and 7 show formation of the string phase or the streak-like scattering pattern in other systems such as the 2.0 wt% solution of PS1000 in DOP at 25 °C and the 6.0 wt% solution of PS548, at 27 °C, respectively. The former and the latter may represent effects of molecular weight and concentration on the string phase formation, when compared with the results obtained for the 3.0 wt% solution of PS548. The streak-like scattering patterns in Figure 6b and 6c and in Figure 7 imply that the string phase is formed in these solutions as well, implying that the string phase formation may be a general phenomenon.

Discussion

The rheological anomaly as observed in the upturn of ψ_1 and η at $\dot{\gamma} > \dot{\gamma}_a$ (Figure 2) was found to be intimately related to the appearance of the streak-like scattering pattern normal to flow and along the neutral axis (Oz axis) as shown in Figures 3 and 4 (patterns d to g). A simultaneous observation under shear light scattering and shear microscopy for the same solution further clarified that this anomaly is related to the formation of the strings aligned parallel to flow (see Figures 4 and 5).

Our preliminary results indicate that the critical shear stress at the shear rate $\dot{\gamma}_a$ for the onset of the anomaly appears to increase with the concentration, e.g., 120 and 300 Pa for the 3.0 and 6.0 wt% DOP solution of PS548, respectively. Thus the higher the concentration, the harder the formation of the string phase. The string formation implies a build-up of the concentration fluctuations along the Oz axis, normal to flow. The higher the concentration, the larger the osmotic pressure and hence larger the mechanical energy required to generate the concentration fluctuations normal to flow. The critical stress for the onset of butterfly formation at $\dot{\gamma}_c$ was found (23) to be almost independent of concentration (nearly equal to 20 Pa for PS548 in the concentration range of 1.5 to 6.0 wt%).

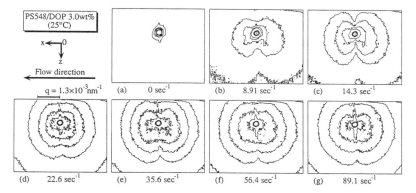

Figure 4. Contour plots of the scattered intensity distribution for the solution shown in Figure 2.

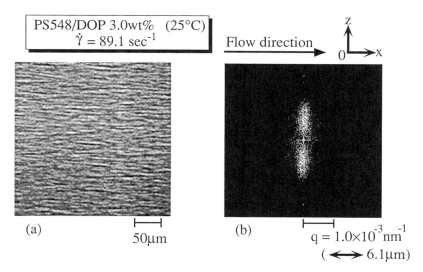

Figure 5. Microscope image of concentration fluctuations (a) and its 2D-FFT pattern (b) at $\dot{\gamma} = 89.1$ s^{-1} which correspond to Figure 3g and Figure 4g. The microscope image is a snap shot taken at a shutter speed of 10^{-3} s.

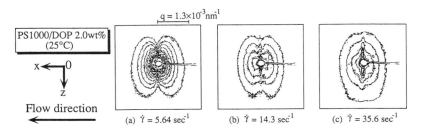

Figure 6. Contour plots of the scattered intensity distribution for the PS1000/DOP 2.0 wt% solution at $\dot{\gamma}$ = 5.64 (a), 14.3 (b) and 35.6 s^{-1} (c) in the steady state.

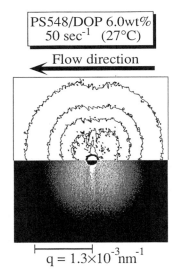

Figure 7. Steady-state scattering pattern (lower half) and contour plot of the intensity distribution (upper half) for the PS548/DOP 6.0 wt% solution at $\dot{\gamma}$ = 50 s^{-1}.

The fact that the streak-like scattering appears at small scattering angles and the butterfly appears at large scattering angles in Figure 4g suggests that the strings are larger than the concentration fluctuations responsible for the butterfly pattern. Moreover the micrograph in Figure 5a shows the structure uniformly composed of the strings but not a two-phase structure composed of the strings and the structure responsible for butterfly pattern. Thus the results in Figure 4g and 5a may suggest the model shown in Figure 8 for a self-assembly at $\dot{\gamma} > \dot{\gamma}_a$.

This model consists of the long strings oriented parallel to flow. The strings, S, have polymer concentration higher than that of the matrix, M. These concentration fluctuations give rise to the strong streak-like scattering along the Oz axis. The strings also may have internal concentration fluctuations along their axes as schematically illustrated by regions A and B having a higher and a lower concentration, respectively. These internal concentration fluctuations are expected to be responsible for the butterfly pattern. Moreover the length scale for the internal fluctuations, Λ, is shorter than the average length, L, of the strings, which account for the fact that butterfly patterns appear at larger scattering angles than the streak-like scattering patterns.

In order to better demonstrate the unique scattering pattern composed of a superposition of the butterfly-type and streak-like scattering patterns, we present here scattering patterns and optical micrographs obtained simultaneously from a very different system in Figure 9. The system to be considered here is a DOP solution of an 80/20 wt%/wt% (77/23 vol%/vol%) mixture of polystyrene (PS)/polybutadiene (PB). The polymer concentration, c, is 3.3 wt% which is about twice as large as the overlap concentration, c*. The PB has a weight-average molecular weight of $M_B = 3.13 \times 10^5$ and PS has $M_S = 2.14 \times 10^5$. Figure 9 shows the steady state structure (a) and the scattering pattern from the solution sheared at $\dot{\gamma} = 4$ s^{-1} (b) and at 10 °C below the cloud temperature $T_{cl} = 54$ °C of the quiescent solution. It also shows the transient change of the structure (c and e) and scattering pattern (d and f) after cessation of the steady shear.

This solution was shown (24) to form highly elongated strings along flow (Figure 9a) and thus to exhibit a very sharp streak-like scattering pattern normal to it (Figure 9b) under steady shear flow. The strings are rich in PB and dispersed in a matrix rich in PS. DOP is a neutral solvent for both phases. In this system the strings have a uniform polymer concentration along their axes and hence do not exhibit the butterfly pattern in steady state as in the case of PS/DOP system. This is because the strong shear flow stabilizes the strings against their intrinsic surface tension instabilities (25).

However, when the shear flow is ceased, surface tension instabilities occur, and the strings are broken into a series of droplets rich in PB in a matrix rich in PS (as seen Figure 9c). The droplets keep a memory of the string so that they have a one-dimensional spatial correlation along the string, just like a series of droplets A in the matrix M as schematized in Figure 8. This nematic alignment of the droplets gives rise to both the horizontal sharp streak-like scattering and the diffuse butterfly-type scattering oriented along flow as shown in Figure 9d. The butterfly pattern is a consequence of the concentration fluctuations of PB along the original string with a characteristic wavelength of the order of interdroplet distance, Λ (see Figure 8) and the streak pattern is a result of the spatial correlation of the concentration fluctuations with correlation length, L (see Figure 8).

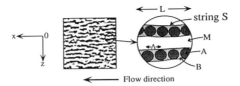

Figure 8. Sketch of concentration fluctuations in polymer solutions at $\dot{\gamma} > \dot{\gamma}_a$. Λ and L are a length scale of the internal fluctuations and an average length of the strings, respectively. Strings, S, have polymer concentrations higher than that of matrix, M. Regions, A, in the strings, S, have higher polymer concentrations than region B.

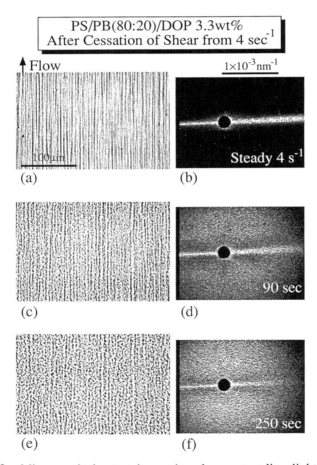

Figure 9. Microscopic images (a, c, e) and corresponding light scattering patterns (b, d, f) for the PS/PB(80:20)/DOP 3.3 wt% solution at DT = 10°C. (a) and (b) were obtained under steady state shear flow at 4 s^{-1}. (c) to (f) were obtained after cessation of shear.

As time elapses after cessation of flow, the droplets grow as a consequence of diffusion-coalescence processes and hence Λ is expected to increase. At the same time the nematic alignment of the droplets tends to be destroyed, giving rise to a decrease of L with time. The former would shift the butterfly pattern toward smaller scattering angles and increase its scattered intensity with time as seen in Figures 9d and 9f. The latter would suppress the intensity of the butterfly pattern and make the lateral spread of the streak smaller with time as seen again in Figures 9d and 9f. The pattern in Figure 9f is quite analogous to the pattern observed for the PS/DOP system under steady shear. Thus the real-space structure of the PS/DOP system is expected to be analogous to that shown in Figure 9e. Although the structure in Figure 9e was obtained after cessation of shear in the PS/PB/DOP system, it exists in a steady state under strong shear flow of $\dot{\gamma} > \dot{\gamma}_a$ in the PS/DOP system.

Summary

The self-assembly of the PS/DOP systems under steady shear flow is summarized in Figure 10 in which the change of the light scattering pattern and real-space structure with $\dot{\gamma}$ are schematically illustrated in the upper and lower halves, respectively. Upon imposing shear flow on the single-phase PS/DOP solutions in the quiescent state (Figure 10a) strong concentration fluctuations are induced at $\dot{\gamma} > \dot{\gamma}_c$, as shown in Figure 10b to 10d in the order of increasing $\dot{\gamma}$. At $\dot{\gamma} < \dot{\gamma}_c$ (in Regime I) the solution remains homogeneous presumably, because $\dot{\gamma}_c$ is smaller than the relaxation rate for the reptation process (26). Under this condition shear does not strongly affect molecular orientation and hence the elastic effect (5,6,8,27) is not significant. At $\dot{\gamma} > \dot{\gamma}_c$, however, the elastic effect is significant, which enhances the concentration fluctuations along flow (7,11). The shear-enhanced concentration fluctuations may be expressed by a superposition of Fourier modes with wave vectors **k**.

At relatively small $\dot{\gamma}$ above $\dot{\gamma}_c$ (i.e., in Regime II) the wave vectors **k** spread over a narrow angular range around the flow direction (i.e., the fluctuations are enhanced along flow) and hence the scattering pattern spreads over a narrow azimuthal angle, μ, around $\mu = 0°$ (i.e., the wing of the butterfly is narrow) as shown in Figure 10b. See Figure 1 for the definition of μ. Upon further increasing $\dot{\gamma}$ (in Regime III), the wave vectors **k** spread over a wider angular range around the flow direction and hence the butterfly wings became wider, giving rise to a thinner dark streak normal to flow as shown in Figure 10c.

With increasing $\dot{\gamma}$ from Regime III to IV the wave vectors **k** spread even to the direction normal to flow and a remarkable change in self-assembly must occur to result in formation of the string phase at $\dot{\gamma} > \dot{\gamma}_a$ in Regime IV. The region rich in polymers (arrays of higher concentration blobs, A, in Figure 8) become more and more rich in polymer with increasing $\dot{\gamma}$ from Region III to IV, through the effect of polymer chains being squeezed from polymer-poor regions (region M in Figure 8) to polymer-rich regions, due to the elastic effect (6,27). At $\dot{\gamma} > \dot{\gamma}_a$, the blobs, A, are interconnected to form strings. This string formation should be responsible for shear thickening as observed in Figure 2.

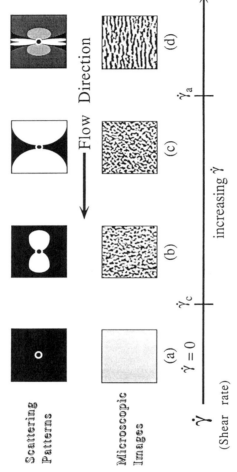

Figure 10. Schematic representation of the small-angle light scattering patterns (upper halves) and concentration fluctuations observed under shear microscopy (lower halves) at quiescent state and under steady shear flow.

Acknowledgments

This work was supported in part by a Grant-in-aid for Scientific Research (05453149) and by a Grant-in-aid for Encouragement of Young Scientists-A (00062753), both from the Ministry of Education, Science and Culture, Japan.

Literature Cited

1. Larson, R. G. *Rheol. Acta* **1992**, *31*, 497.
2. Ver Strate, G.; Philippoff, W. *J. Polym. Sci., Polym. Lett. Ed.* **1974**, *12*, 267.
3. Wolf, B. A. *Makromol. Chem.* **1980**, *1*, 231.
4. Rangel-Nafaile, C.; Metzner, A. B.; Wissbrun, K. F. *Macromolecules* **1984**, *17*, 1187.
5. Helfand, E.; Fredrickson, G. H. *Phys. Rev. Lett.* **1989**, *62*, 2468.
6. Onuki, A. *Phys. Rev. Lett.* **1989**, *62*, 2472, *J. Phys. Soc. Japan* **1990**, *59*, 3427.
7. Hashimoto, T.; Fujioka, K. *J. Phys. Soc. Japan* **1991**, *60*, 356.
8. Milner, S. T.; *Phys. Rev. Lett.* **1991**, *66*, 1477.
9. Wu, X.-L.; Pine, D. J.; Dixon, P. K. *Phys. Rev. Lett.* **1991**, *66*, 2408.
10. Yanase, H.; Moldenaers, P.; Mewis, J.; Abetz, V.; van Egmond, J.; Fuller, G. G. *Rheol. Acta* **1991**, *30*, 89.
11. Hashimoto, T.; Kume, T. *J. Phys. Soc. Japan* **1992**, *61*, 1839.
12. Dixon, P. K.; Pine, D. J.; Wu, X.-L. *Phys. Rev. Lett.* **1992**, *68*, 2239.
13. van Egmond, J. W.; Werner, D. E.; Fuller, G. G. *J. Chem. Phys.* **1992**, *96*, 7742.
14. Mavrantzas, V. G.; Beris, A. N. *Phys. Rev. Lett.* **1992**, *69*, 273.
15. Moldenaers, P.; Yanase, H.; Mewis, J.; Fuller, G. G.; Lee, C.-S.; Magda, J. J. *Rheol. Acta*, **1993**, *32*, 1.
16. Magda, J. J.; Lee, C. S.; Muller, S. J.; Larson, R. G. *Macromolecules* **1993**, *26*, 1696.
17. van Egmond, J. W.; Fuller, G. G. *Macromolecules* **1993**, *26*, 7182.
18. Moses,E.; Kume, T.; Hashimoto, T. *Phys. Rev. Lett.* **1994**, *72*, 2037.
19. Boué, F.; Lindner, P.; *Europhys. Lett.* **1994**, *25*, 421.
20. Hashimoto, T.; Takebe, T.; Suehiro, S. *Polym. J.* **1986**, *18*, 123.
21. Kume, T.; Asakawa, K.; Moses, E.; Matsuzaka, K.; Hashimoto, T. *Acta Polymer.* in press.
22. In ref. 11 for 6.0 wt% DOP solution of the same sample, we stated that there is no anomaly in η even in Regime IV (though there is anomaly in ψ_1), because the first overshoot of shear stress decays to the level expected from the $\dot{\gamma}$-dependence of η in the lower $\dot{\gamma}$ regime, if the solution is kept under shear. However we recently found that this solution exhibits at least two overshoots on its approach toward steady state, if the solution is kept under shear even longer and that the steady values are larger than those reported previously. Therefore this solution was found to exhibit the anomaly in both η and ψ_1, as reported in this work and Yanase et. al. (*10*).
23. unpublished data.
24. Hashimoto, T.; Matsuzaka, K.; Moses, E.; Onuki, A. *Phys. Rev. Lett.*, in press.
25. Tomotika, S.; *Proc. Royal Soc. London, Ser. A* **1935**, *150*, 322; **1936**, *153*, 302.
26. Doi, M.; Edwards, S. F. *The Theory of Polymer Dynamics*; Oxford Univ. Press, 1986.
27. Doi, M.; Onuki, A. *J. Phys. II France* **1992**, *2*, 1631.

RECEIVED January 25, 1995

Chapter 4

Small Angle Neutron Scattering from Sheared Semidilute Solutions: Butterfly Effect

François Boué[1] and Peter Lindner[2]

[1]Laboratoire Léon Brillouin, Centre d'Etudes Saclay, F-91191 Gif-sur-Yvette, France
[2]Institut Laue-Langevin, BP 156, F-38042 Grenoble Cédex 9, France

An anisotropy of the two-dimensional SANS pattern is observed on a semidilute polymer solution under shear, with an increase of the low q scattering in direction parallel to the flow. The corresponding double winged shape of the pattern resembles the "butterfly" shapes found for the interchain scattering in deformed gels and rubbers. Spontaneous fluctuation as well as frozen fluctuation models may both be compared with the results.

We show here what can be revealed using SANS from the structure of a polymer semidilute solution under shear. Using this technique, shear effects were investigated up to now in only two situations : dilute solutions, revealing the deformation of the single chain (1), and solutions close to phase separation, leading to an isotropic enhancement of the scattering (first results by Hammouda et al. display no anisotropy of the scattering (2)). As reported in other contributions of this Symposium book, most of the scattering observations on semidilute solution have been done using light scattering (3-6). The striking effect is a strong increase of the intensity, present along the flow direction for all shear rates. In this paper, we show, on the same polymer/solvent system, that SANS observations are complementary to light scattering results. They involve a larger momentum transfer range and allow progressive monitoring of the increase of the signal close to the phase separation line or in theta regime (as for light) as well as, at higher temperatures, in the good solvent regime.

Formerly, in the same q range, the effect of a different deformation, namely stretching, was investigated on polymer networks permeated by a mobile labeled species, solvent (7,8) or small chains (9-13). Let us recall now the main results : measurements at a progressive increase of the elongation ratio show a progressive increase of scattering in direction parallel to the elongation (as well as any direction corresponding to an increase in the dimensions of the sample) (7-13). This low contribution displays a Lorentzian variation $1/(1+q^2\xi_s^2)$ for $q < 1/\xi_s$ and a crossover for $q > \xi_s$ to a function which is independent of λ and depends only on the network. This function appears on a wider domain as λ is increased, because the lower limit , $1/\xi_s$, decreases progressively (ξ_s increases with λ). In other words, as the deformation is increased, the intensity collapses on a single curve, which we call the "limit curve", within a domain $1/\xi_s < q < 1/\xi_{iso}$ of increasing width. It was therefore proposed that such a mechanism of "descreening" was revealing a "superstructure" characterized by the "limit curve" which depends on the network.

A similar increase of the scattering was observed for short mobile labeled chains embedded in a melt of long entangled chains, which is a kind of temporary network (14,15). An entangled semidilute solution is, from the same point of view, a kind of temporary gel, and could exhibit also, under deformation, a "superstructure". This could explain the increase of the scattering under shear (4), but its origin should be made more precise. On the other hand, other explanations have been proposed as a result from the coupling between the polymer concentration fluctuations and shear flow through the concentration dependent viscosity (16). This was extended to entanglement effects (17). In the following we mainly present our first data, as a basis for discussion, which is here very brief.

Phase Diagram and Scattering Geometry

Figure 1 shows the schematic temperature-concentration phase diagram of polymer solutions (18) for the example of polystyrene / dioctyl phthalate (PS/ DOP), a system which has been studied by the previously cited experiments as well as here. Indicated in Figure 1 are the different domains investigated in the different works (2-6).

As for the quiescent state, theory and experiment agree to describe the scattering of the semidilute solutions (due to polymer concentration fluctuations) by a cross-over from an Ornstein-Zernike law for $\xi q \ll 1$

$$d\Sigma/d\Omega\,(q) = (d\Sigma/d\Omega)_{q=0} / (1+\xi^2 q^2) \tag{1}$$

towards a power law

$$d\Sigma/d\Omega\,(q) \sim q^{-D_f} \tag{2}$$

for $\xi q > 1$, where $D_f \sim 5/3$ in the good solvent regime (high temperature in PS/ DOP, which is a upper critical solution temperature (UCST)-system) and $D_f \sim 2$ in the theta domain (lower temperature, above the phase separation line). A further decrease of the temperature will lead to phase separation, accompanied by a strong enhancement of the scattering at low q. ξ is the correlation length and is identical to the "blob" size in a good solvent (19) above which excluded volume correlations are screened out. The temperature during the SANS experiment has been varied in order to move from good solvent to theta conditions (see line 1 in Figure 1). Therefore we should be able to test how elastic deformation effects (which we assume to be dominant in good solvent) couple with phase separation effects when approaching the phase separation line.

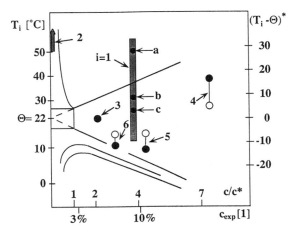

Figure 1. Schematic temperature-concentration diagram for the different shear experiments (numbered i= 1 to 6) with polystyrene done at different temperatures T_i and with different molecular masses M_i. The abscissae are given in reduced concentrations c/c* as well as in real numbers for the present study. The ordinate is in real temperature units [°C] for the open symbols and in equivalent temperature with respect to our experiment $(T_i - \Theta)^* = [(T_i - \Theta) \cdot \sqrt{(M_i/M_1)}]$ for the hatched symbols. i=1: results of this paper, i=2: d-PS in good solvent (oligostyrene + toluene) (1), i=3(2), i=4:(3,4), i=5:(5), i=6:(6) (i=1, 3- 5: solvent is DOP).

In Figure 2 we have sketched for comparison different Couette-geometries of light scattering and SANS experiments so far used. In our study, the incoming primary neutron beam hits the sample normal to the axis of rotation of the outer cylinder and is thus *perpendicular* to the flow $\bar{v}$ (x-direction) and *parallel* to the shear gradient $\dot{\gamma} = \bar{v}/d$ (y-direction). Differences with light scattering geometries become evident from Figure 2. Moreover, cone-plate geometries are used as well (3,4,5).

Experimental Details

A solution made of fully deuterated, linear polystyrene (d-PS, synthesized by anionic polymerization of styrene-d8 (Fluka) (20)) in (protonated) dioctyl phthalate (DOP, Aldrich) has been prepared, for which the theta point is at $T \approx 22$ °C (21). Polystyrene with a molecular mass of M_w=514 000 g/mole with a polydispersity $M_w/M_n = 1.12$ at a concentration of c=9% (w/w) has been used.

From the R_g/M_w relationship (22) for amorphous polystyrene $[R_g^2/M_w]^{0.5} = 0.275$, we get as an estimate for the radius of gyration (under theta conditions) $R_g \approx 200$ Å for the studied polymer sample. With the usual definition of the concentration, c*, where single chains start to overlap (19), $c^* = 3 M_w/(4\pi R_g^3 N_L)$ and N_L being Avogadro's number, this entanglement concentration is equal to $c^* \approx 2.5 \% < c$ in our case.

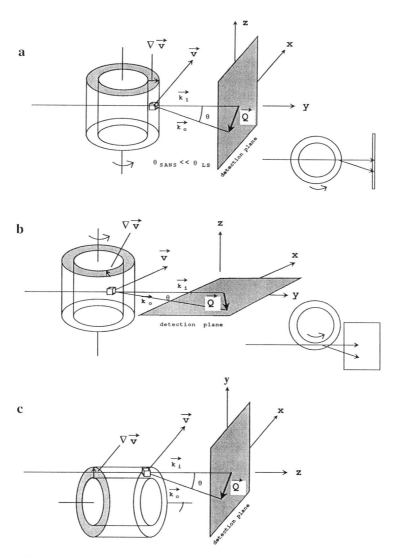

Figure 2. Sketch of various scattering experiments using Couette geometry.
a) SANS shear apparatus (26) as used on PAXY at LLB: The sample is confined in the d=0.5 mm gap between two concentric quartz cylinders, with the outer one rotating at a constant speed. The inner static cylinder is under temperature control. b) light scattering geometry, reference (5). c) light scattering, reference (3,4).

Mechanical measurements of the dynamic viscosity $\eta(\omega)$, using a Weissenberg Rheogoniometer (Carrimed) in cone-plate configuration under temperature control (23) gave access to the terminal relaxation time τ_c of the system d-PS/ DOP at c=9% by fitting the Carreau viscosity equation (24). At temperatures T= 15 °C, 22 °C and 30 °C we obtained τ_c = 0.172, 0.0787 and 0.0375 s respectively.

Small angle neutron scattering experiments have been performed at the instrument PAXY of the Laboratoire Léon Brillouin, Saclay, France. Spectra were recorded on a position sensitive multidetector with 128 x 128 cells (0.5 cm² each). With a neutron wavelength of λ = 15 Å and a sample to detector distance of L = 3.23 m, the effective range of momentum transfer, q = $(4\pi/\lambda)$ • sin $(\theta/2)$, (θ = scattering angle) was $0.53 \cdot 10^{-2}$ nm^{-1} ≤ q ≤ 0.46 nm^{-1}. Measuring times for the SANS spectra were typically of the order of 2 hours. The neutron scattering intensity of the raw data is normalized with the scattering of a 1 mm water sample as standard reference and subsequently corrected for sample container and background scattering. Subtraction of the normalized solvent intensity yields the coherent differential scattering cross section of the polymer as a function of momentum transfer, $d\Sigma/d\Omega$ [cm^{-1}] = f(q).

Results

The series of temperature dependent scattering experiments at rest has been performed by using standard sample holders under temperature control, the sample being confined between two quartz plates at a distance of 1 mm (25). By fitting an Ornstein-Zernike law (equation 1) we obtain the correlation length ξ varying with temperature between 62 Å (T=12 °C) and 39 Å (T=55 °C). This behavior at rest is a reasonable result for a binary system, approaching the phase boundary with decreasing temperature. However, the quiescent system at T=13 °C is still in the one phase region; the sample was optically transparent after the run at this temperature.

For the shear experiments the Couette type shear apparatus (26) of the Institut Laue-Langevin has been used at PAXY. Figure 3 presents the two dimensional patterns of the (uncorrected) scattering, measured at rest and at a temperature of T = 25 °C (Figure 3a) and in laminar shear flow at three different combinations of shear rate, $\dot{\gamma}$, with temperature T,: $\dot{\gamma}$=150 s^{-1} at T=25 °C (Figure 3b), $\dot{\gamma}$=300 s^{-1} at T=25 °C (Figure 3c) and $\dot{\gamma}$=150 s^{-1} at T=15 °C (Figure 3d). At rest (Figure 3a), an azimuthally isotropic scattering pattern is observed. As a general feature under shear (Figure 3b, c, d), the patterns are slightly anisotropic at large q, with a long axis perpendicular to the flow direction. At low q, closer to the beamstop, a different behavior is observed for Figures 3b and c: a pronounced increase parallel to the flow direction leads to a double winged shape of the pattern which resembles closely the "butterfly patterns" observed in deformed gels and rubbers (7-15). On the butterfly effect as observed in Figure 3b and 3c, a strong low q scattering seems to be superimposed when approaching the phase boundary by lowering the temperature (Figure 3d, T=15 °C). Whether or not the butterfly effect (anisotropy parallel to the flow) is maintained under these conditions is not clear from the figure and is the subject of ongoing experiments. However, when shearing the solution at T=15 °C just above ≈ 150 s^{-1}, a shear induced turbidity was visible by eye.

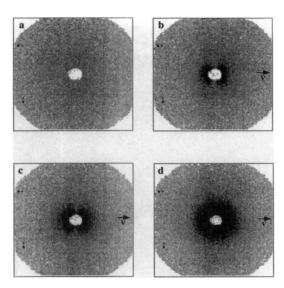

Figure 3. Two dimensional patterns of the uncorrected scattering measured at rest and in laminar shear flow at different combinations of shear rate $\dot{\gamma}$ and temperature T.
The grey scale levels are incremented linearly from 0 to 30. a) at rest ($\dot{\gamma}=0$) and at a temperature of T=25 °C. b) $\dot{\gamma}=150$ s^{-1} at T=25 °C. c) $\dot{\gamma}=300$ s^{-1} at T=25 °C. d) $\dot{\gamma}=150$ s^{-1} at T=15 °C.

Effect of temperature. We turn now to the description of the radial intensity distribution along the preferential directions. Radial regrouping of the anisotropic spectra (see Figures 3 b-d) with a narrow sector mask of 15 degrees in parallel direction (\\) and 30 degrees in perpendicular direction (⊥) with respect to flow lead to the corresponding distribution functions $(d\Sigma/d\Omega)_{\backslash\backslash} = f(q)$ and $(d\Sigma/d\Omega)_\perp = f(q)$.

Figure 4 shows the radial distribution function $(d\Sigma/d\Omega)_{iso}=f(q)$ of the isotropic scattering for three temperatures (T=25 °C, 50 °C and 15 °C) at rest (either measured in the Couette cell or in standard sample holders) together with the intensities obtained from sheared samples at the same temperatures and at shear rates of $\dot{\gamma}=150$ s^{-1}, $\dot{\gamma}=1000$ s^{-1} and $\dot{\gamma}=150$ s^{-1}, respectively.

For the three temperatures, $(d\Sigma/d\Omega)_\perp$ remains identical within the statistical error to the isotropic scattering at rest. Such lack of shear deformation in perpendicular direction (z-direction in Figure 2) is observed with sheared dilute polymer solutions (1) as well as with melts (27) and is a sensible result because the macroscopic deformation tensor has a zero component along this z-direction.

In the large q-regime, however, we observe for all three temperatures lower intensities $(d\Sigma/d\Omega)_{\backslash\backslash}$ than in the perpendicular direction. The latter overlaps with the isotropic scattering, $(d\Sigma/d\Omega)_{iso}$. An asymptotic behavior, characterized by a power law of q^{-2} is observed, at least for temperatures around the theta point, as better revealed by complementary SANS measurements at even larger q (data not shown here).

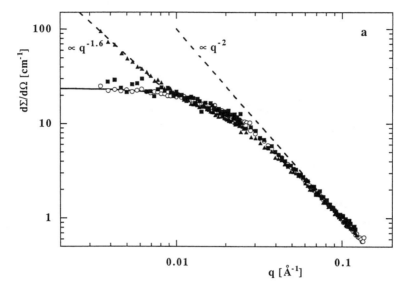

Figure 4. Log-log representation of the coherent differential scattering cross section $d\Sigma/d\Omega$ [cm^{-1}] as function of momentum transfer, q, of the polymer in semidilute solution at rest and under shear (d-PS/ DOP at c=9%) for three representative temperatures. a. T=25 °C, b. T=50 °C, c. T=15 °C. Open circles: data acquired at rest. Filled squares: data under shear and in perpendicular direction. Filled triangles: data under shear and in parallel direction. The dashed lines (~q^{-2}) in a. and c. indicate the asymptotic slope in the large q-limit, expected for theta conditions. The full line in a. and c. is the Ornstein-Zernike function (equation 1).

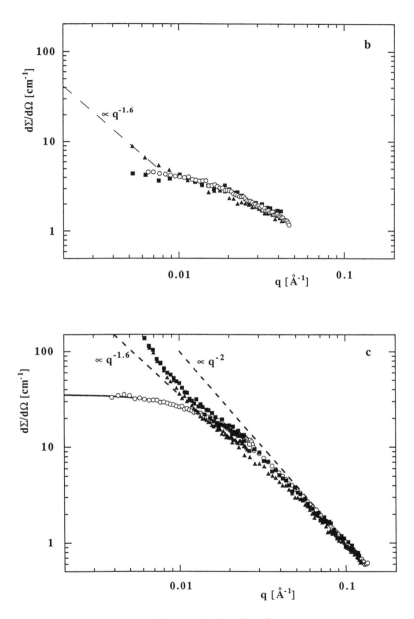

Figure 4. *Continued*

On the contrary, a pronounced increase is observed for $(d\Sigma/d\Omega)_\parallel$ at the three temperatures. For T=25 °C the increase can be approximated by a straight line with a slope of $\alpha \sim -1.6$. For T=50 °C the effect is smoother but indicates a similar increase.

For T=15 °C the anisotropy at intermediate q values is of the same type but more pronounced. This result is to be expected because of the fast increase of the characteristic relaxation time with decreasing temperature. Given the relaxation times of the system d-PS/ DOP, as determined by dynamic viscosity measurements (see **Experimental Details**) which agree with an Arrhenius type behavior, we can estimate the Weissenberg numbers Wi = $\dot{\gamma} \cdot \tau_c$. We find for $\dot{\gamma}$=150 s^{-1} : Wi=26 at T=15 °C and Wi=9 at T=25 °C; for $\dot{\gamma}$=1000 s^{-1} the extrapolation gives Wi=6 at T= 50 °C. These values are definitely larger than 1 ("strong shear" regime (*16*)).

In contrast to the results obtained at higher temperatures, the low q scattering for T=15 °C increases for both directions parallel and perpendicular with a steeper power law. The apparent exponent is close to 3. Finally, even in this case at the lowest measured q values, we note that the scattering in parallel direction becomes slighty larger than in perpendicular direction.

Effect of shear rate. Figure 5 shows the scattering in parallel direction for two sets of shear rates at a temperature of T=22 °C, which is the theta temperature for the system PS/ DOP. Figure 5 a shows data in the "weak shear" regime (Wi ≤ 1) at $\dot{\gamma}$=6.8 s^{-1} (Wi=0.5), $\dot{\gamma}$=12.5 s^{-1} (Wi=1) and $\dot{\gamma}$=25 s^{-1} (Wi=2) together with the results of the isotropic solution at rest. A noticeable difference (increase of scattering at low q) appears as soon as Wi is of the order of one or larger. Figure 5b deals with the "strong shear" regime (Wi >>1) at $\dot{\gamma}$=150 s^{-1} (Wi=12), $\dot{\gamma}$=300 s^{-1} (Wi=24) and $\dot{\gamma}$=450 s^{-1} (Wi=36) together with the results of the isotropic solution at rest. With increasing shear gradient, the scattering at low q strongly increases, the slope changes from ≈ 1.6 to ≈ 2.5.

Discussion

The experimental data for the sheared semidilute system show two regimes:
- on the large q side of Figure 4 we see that $(d\Sigma/d\Omega)_\parallel / (d\Sigma/d\Omega)_\perp$ < 1, which corresponds to the behavior observed for the shear deformation of a single chain in solution at intermediate q values (*25*). Hence, in the large q-limit ($\xi q > 1$) we expect to explore here the single chain conformation.
- on the contrary, for q --> 0 , on the low-q side (Figure 4) we do not observe a single chain behavior where the intensity at q --> 0 levels at the value proportional to the molecular mass of the chain in both directions parallel and perpendicular. As already mentioned, $(d\Sigma/d\Omega)_\parallel / (d\Sigma/d\Omega)_\perp$ > 1 in the sheared semidilute solution at low q. This is an essential feature of the "butterfly effect" (*7-15*).

The increase of intensity at q --> 0 in direction parallel to the flow is qualitatively consistent with two different theoretical pictures. The first model considers a spontaneous enhancement of the polymer concentration fluctuations due to the following coupling (*16,17*): the deformation depends on the viscosity which is itself dependent on the concentration. A region in space which is less concentrated in polymer at a given time (due to thermal fluctuations) will be less viscous. In comparison to the more concentrated and thus more viscous regions, stress will produce a stronger deformation in the low viscosity regions: calculations show that the global deformation energy is lowered (*16*) compared to a homogeneous deformation.

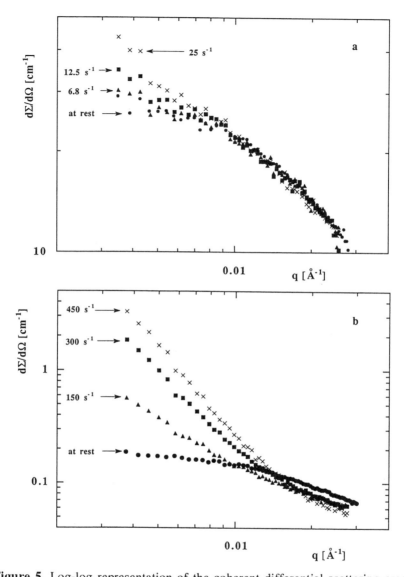

Figure 5. Log-log representation of the coherent differential scattering cross section in parallel direction, $(d\Sigma/d\Omega)$, as parallel function of momentum transfer q of the sheared semidilute polymer solution (d-PS/ DOP at c = 9%) at T=22 °C.
a. weak shear regime: at rest (circles), $\dot{\gamma}=6.8$ s^{-1} (triangles), $\dot{\gamma}=12.5$ s^{-1} (squares), $\dot{\gamma}=25$ s^{-1} (crosses).
b. strong shear regime: at rest (circles), $\dot{\gamma}=150$ s^{-1} (triangles), $\dot{\gamma}=300$ s^{-1} (squares), $\dot{\gamma}=450$ s^{-1} (crosses). The aspect of the curves at large q is irrelevant with respect to the discussion since the data are not yet corrected for background in this Figure.

The same reasoning has been developed for rubbers and gels at equilibrium deformation replacing the viscosity by the elastic modulus (28). For these materials the predicted Lorentzian shape of the "limit curve" disagreed with observations, as well as the variation of the zero q limit intensity and correlation length with deformation. This could be due to the range of validity of reference (28) which is limited to deformation ratios lower than the experimental ones. In both models the treatment of the linear hydrodynamics or linear elasticity lead at rest to a simple Lorentzian type behavior, which can be fitted to the scattering of isotropic semidilute solutions (at rest). The same treatment for sheared solutions predicts also a Lorentzian shape, which is not observed in our case. The theory given in reference (16) assumes Rouse dynamics and reduced shear rates $\dot{\gamma}\tau_c = 1$. In the case of most of our measurements (except Figure 5a) the reduced shear rate is larger ($\dot{\gamma}\tau \geq 5$).

As a matter of fact, given the experimental concentration the chains may be entangled. Hence, the Rouse dynamics (non-entangled chains) could be replaced by a reptation picture (17). In this case an enhancement of the fluctuations should vanish at large length scales, where the thermal fluctuations become slower than chain reptation. The result is a maximum of the scattering function at low momentum transfers in a range accessible to light scattering but not SANS. It seems to us that the model is valid in its principle for theta and good solvent regimes investigated here. This model predicts at $q\xi > 1$ a slope of q^{-2}. The observed apparent slope varies with shear rate between ≈ 1.6 and ≈ 2.5 in our experiment in the "strong shear" regime. Given the limited q-range in our SANS experiment we cannot exclude a superposition of a "classical" semi-dilute solution scattering (large q) and scattering of enhanced thermal fluctuations, occurring at $q\xi < 1$ and disconnected from the former. In such a case the low q range of our experiment would correspond to a cross over region with an ill-defined slope. We can only say that our measurements in parallel direction, $(d\Sigma/d\Omega)_\|$, would lie in the range of q larger than the correlation length and that the observed slope of $\alpha \sim -1.6$ is not directly predicted.

The second model considers disinterpenetration of an assembly of regions of higher concentrations and regions of lower concentrations (29). The deformation in the parallel direction acts as a dilution of the high concentration regions (harder) inside the low concentration regions (softer). These regions may be highly "interpenetrated" in the quiescent state, i.e. their screening length (observable using scattering) is much smaller than their maximum size. Dilution, as equivalent for instance to shear in parallel direction, will increase the "visible" size of the clusters. The typical situation considered in (29) are percolation clusters of regions of high crosslinking ratio after random crosslinking of a semidilute solution, which is well suited to the description of the "butterfly effect" in uniaxially deformed swollen gels (7,8).

However, the system investigated here is a semidilute solution, entangled but not crosslinked. This is similar to the case where free mobile labelled chains are embedded in a stretched matrix : butterfly effects are observed when the matrix is crosslinked, as well as when it is a melt of long chains, which are just entangled. In stretched gels, a power law $q^{-1.6}$ is predicted (29). Apparent slopes observed in practice are 1.6 in some cases (7,8) and larger values 2 to 2.5 for other gels and networks with small chains(13). Similar values are found in this paper. Thus a model of disinterpenetration of frozen heterogeneities, which is also proposed to explain light scattering results (4), agrees also with our findings within the experimental accuracy. An extension of these first results will be reported elsewhere.

Conclusion

In summary, for semidilute solutions under shearing, an excess of intensity appears along the direction of the flow when the shear rate $\dot{\gamma}$ is of the order of the terminal time or higher, i.e. when the Weissenberg number $\dot{\gamma}\cdot\tau \geq 1$. At large q, one observes the usual scattering of a deformed polymer chain. Going from large q to low q, the crossover to the low q excess of intensity is sharp, for not too high shear. Both, the excess intensity and the correlation length increase with shear rate. Moreover, the shape of the curve changes. For Weissenberg numbers close to one, one tends to a power law with an exponent 1.6 and for large shear rates the exponent increases to 2.5. The unscreening of preexisting frozen concentrations fluctuations can explain both the increase in correlation and the apparent power laws, with no precise prediction at this stage. The Helfand-Fredrickson-Milner model (16,17) can also predict an increase of the correlation, but an available predicted shape in q to fit the experimental one in the SANS regime remains to be achieved. Finally, when the solution is in the bad solvent regime (low temperature) the effect is of a different nature: it is dominated by an increase of intensity along all directions characteristic of shear-induced demixing.

Acknowledgments

Grant of beamtime at the Laboratoire Léon-Brillouin, Saclay is gratefully acknowledged. We acknowledge help by A. Magnin, Laboratoire de Rhéologie, St. Martin d'Hères, who performed dynamic viscosity measurements on the system d-PS/DOP.

Literature Cited

1. Lindner, P.; Oberthür, R. *Coll. & Polym. Sci.* **1988**, *266*, 886.
2. Hammouda, B.; Nakatani, A. I.; Waldow, D. A.; Han C. C. *Macromolecules* **1992**, *25*, 2903.
3. Hashimoto, T.; Fujioka, K. *J. Phys. Soc. Japan* **1991**, *60*, 356.
4. Hashimoto, T.; Kume, T. *J. Phys. Soc. Japan* **1992**, *61*, 1839.
5. van Egmond, J. W.; Werner, D. E.; Fuller, G. G. *J. Chem. Phys.* **1992**, *96*, 7742.
6. Wu, X. L.; Pine, D. J.; Dixon, P. K. *Phys. Rev. Lett.* **1991**, *66*, 2408.
7. Mendès, E. Ph.D thesis Université Louis Pasteur Strasbourg, **1992**.
8. Mendès, E.; Bastide, J.; Buzier, M.; Boué, F.; Lindner, P. *Phys. Rev. Lett.* **1991**, *66*, 1595.
9. Oeser, R.; Picot, C.; Herz, J. *Springer proceedings in physics* **1988**, *29*, 104.
10. Oeser, R. Ph.D thesis University of Mainz, **1992**.
11. Zielinski, F. Ph.D thesis Université Paris VI Pierre et Marie Curie, **1992**.
12. Zielinski, F.; Buzier, M.; Lartigue, C.; Bastide, J.; Boué, F. *Prog. Coll. Pol. Sci.* **1992**, *90*, 115.
13. Ramzi, A. Ph.D thesis Université Paris VI and submitted papers, **1994**.

14   Bastide, J.; Boué, F.; Buzier, M. *Springer proceedings in physics* **1988**, *29*, 112.
15   Boué, F.; Bastide, J.; Buzier, M.; Herz, J.; Vilgis, T. A. *Coll. Polym. Sci.* **1991**, *269*, 195.
16   Helfand, E.; Fredrickson, G. H. *Phys. Rev. Lett.* **1989**, *62*, 2468.
17   Milner, S. *Phys. Rev. Lett.* **1991**, *66*, 1477.
18   Daoud, M.; Jannink, G. *J. Phys. (Paris)* **1976**, *37*, 97.
19   deGennes, P. G. *Scaling Concepts in Polymer Physics* ; Cornell University Press, Ithaca, N.Y., **1979**.
20   Zimmer, G. Ph.D thesis University of Mainz, **1985**.
21   Park, J. O.; Berry , G. C. *Macromolecules* **1989**, *22*, 3022.
22   Cotton, J. P.; Decker, D.; Benoit, H.; Farnoux, B.; Higgins, J ; Jannink, G.; Ober, R.; Picot, C.; desCloizeaux, J. ; *Macromolecules* **1974**, *7*, 863.
23   Magnin, A.; Boué, F.; Lindner, P. to be published.
24   Bird, R. B.; Armstrong, R. C.; Ossager, O. *Dynamics of Polymeric Liquids* ; John Wiley: New York, **1977**; Vol. 1, p. 210.
25   Boué, F.; Lindner, P. *Europhys. Lett.* **1994**, *25 (6)*, 421.
26   Lindner, P.; Oberthür, R.C. *Rev. Phys. Appl.* **1984**, *19*, 759.
27   Boué, F. *Adv.Pol.Sci.* **1987**, *82*, 47.
28   Rabin, Y.; Bruinsma, R. *Europhys. Lett.*, **1992**, *20*, 79.
29   Bastide, J.; Leibler, L.; Prost, J. *Macromolecules* **1990**; *23*, 1821.

RECEIVED February 25, 1995

Chapter 5

Structural Evolution and Viscous Dissipation During Spinodal Demixing of a Biopolymeric Solution

A. Emanuele[1] and M. B. Palma-Vittorelli[1,2]

[1]Department of Physics and Gruppo Nazionale di Struttura della Materia-Istituto Nazaionale di Fisica della Materia
and [2]Consiglio Nazionale delle Ricerche, Institute for Interdisciplinary Application of Physics, University of Palermo, Via Archirafi 36, I 90123 Palermo, Italy

> We study the time-resolved behavior of shear viscosity and of its shear dependence, in the course of spinodal demixing of agarose solutions in pure water as well as in presence of compatible cosolutes. We also report partial viscoelasticity data. At the beginning of the ripening stage of demixing we observe a large surge of shear viscosity followed by a decay towards normal values. The time of appearance of the surge exhibits a critical behavior and its shear dependence follows an inverse power law. The exponent has an exceptionally high value, which is independent of quenching depth, polymer concentration and solvent perturbation. Scaling properties of the viscosity peak and light scattering data show that the viscosity surge is due to the onset of a loose, non-permanent structure, characterized by a relaxation time of the order of 100 s. The solution remains nevertheless in the liquid state and all its measurable properties still remain essentially unchanged. A much longer time must elapse before the measurable start of the gelation process. Low shear slightly counteracts the occurrence of this structure and eventually causes its deformation and breaking. Sufficiently high shear prevents its occurrence. Co-solutes are found to affect the spinodal temperature but not the critical and scaling behaviors.

The effect of shear on demixing (*1-7*) and, conversely, the rheological behavior of polymer solutions during spinodal demixing (*8-22*) have drawn a marked interest, particularly with reference to scaling and to critical and universal behaviors. These complementary aspects are jointly considered in the present experimental study of the rheological response of a biopolymeric solution to shear perturbation. The system is agarose in aqueous solvent, and in presence of compatible cosolutes. This system is studied in the course of spinodal demixing. Our results show that shear perturbs the developing structure severely. The observed shear dependence and the critical and scaling properties of the viscous response allow identifying unsuspected features of structural and dynamic changes due to the demixing process.

Agarose is an essentially uncharged polysaccharide. Its aqueous solutions can form thermoreversible gels upon cooling, providing an interesting example of self-assembly of biologically significant supramolecular order (23-24). Gelation has been related to a conformational (2 coils) → (1 double helix) transition, necessary for crosslinking (25-27). As for many other gel-forming biopolymers, agarose gelation occurs even at very low concentrations (24, 28-29). Previous work at our laboratory (30-36) has elucidated the self-assembly process and allowed determining the phase diagram shown in Figure 1. Experiments have unambiguously shown that in the region of low and moderate concentrations (region 2 in Figure 1), the thermodynamic phase transition of spinodal demixing precedes and promotes the topological phase transition of gelation (30-36). The present experiments concern this region. The symmetry break of the initially homogeneous sol due to demixing provides regions where the agarose concentration is sufficiently high for the occurrence of double helices and gelation (30, 31, 37). These solute-solute correlations lower considerably the concentration threshold expected for random crosslinking (38-41). Time resolved experiments have shown that, after a delay time, depending upon quenching depth and polymer concentration, the pattern of domains (monitored by the low-angle light scattering) stabilizes to a great extent while the sample is still macroscopically liquid (32). Stabilization is reflected by the accompanying onset of a much longer characteristic time in the noise

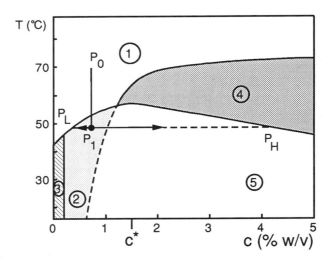

Figure 1 - Quantitative phase diagram of agarose water systems. ① Thermodynamically stable sol. ② Quenching from P_0 to P_1 causes spinodal demixing followed by macroscopic gelation. The latter occurs in the canvas of polymer-rich regions such as P_H. ③ As in ②, but polymer-rich regions are disconnected. A mesoscopic gel is formed. ④ Direct gelation. ⑤ Demixing-promoted and direct gelation compete kinetically.

of scattered light (*32*). In the conditions of our experiments, the timescale change was from less than 1 s to the order of 100 s. Concurrently with these changes, a surge and decay of shear viscosity is observed (*15-17*). Only later, and well separated in time, gelation is observed to start: this is revealed by a decrease in the number of polymer coils observed to diffuse freely, an increase of the optical rotation signal, reflecting a coil-helix transition (necessary for crosslinking), and an increase of viscoelasticity, reflecting the occurrence of crosslinks (*31*). At very low concentrations (region 1 of Figure 1) demixing is still observed, but gelation occurs only within disconnected regions, freely drifting in the system which remains macroscopically liquid (*36*).

The present experimental study of shear-viscosity, along with partial viscoelasticity data also presented, opens a view on structural and dynamic changes occurring in the system at the intermediate stage, when the linear regime of spinodal demixing is completed and gelation has not yet started. Experiments concern the time-resolved behavior of shear viscosity during spinodal demixing and the dependence of features of the viscosity surge upon the rate of shear, quenching depth and polymer concentration. We observe a critical behavior and unexpected scaling properties of shear viscosity, that we discuss along with the observed viscoelastic behavior.

Experiments were also performed in solvent perturbed by co-solutes. Use of small amounts of compatible cosolutes allows more information to be obtained, since they are known to modulate solvent-induced interactions (*37, 42-44*). In our system, experiments in perturbed solvent had already shown that solvent-induced interactions contribute to determining the line of thermodynamic instability of the solution (*35, 42*), and enter in the (2 coils) ⇌ (1 double helix) conformational equilibrium (*27*). Qualitative evidence for the role of solvent in the kinetics of stabilization of the domain pattern occurring prior to gelation was also available (*35*). For the present experiments, we have used two different classes of cosolutes: alcohols (ethanol, ETOH and tert-butanol, TBOH) and trimethylamine N-oxide (TMAO). In the following we report and compare results and scaling behavior obtained in aqueous agarose solutions with and without co-solute perturbation.

EXPERIMENTAL

Viscosity measurements were performed using a computer-interfaced Couette-type viscometer (Contraves Low-Shear 30). Available (constant) shear-rates, $\dot{\gamma}$, are in the range $1.6 \times 10^{-2} - 1.2 \times 10^2$ s^{-1}. An oscillating mode was used for viscoelasticity measurements, with an oscillation frequency $\omega = 0.1395$ s^{-1}, an oscillation amplitude of 0.7 degrees and a strain amplitude 0.15. An argon laser (488 nm) was used for light-scattering experiments (elastic light scattering, ELS, and dynamic light scattering, DLS). Data were automatically collected at different angles with a Brookhaven BI2030AT correlator and BI200SM goniometer. Agarose (Seakem HGT(P) from FMC Bioproducts) was dissolved at 100 °C, for 20 min, filtered at T > 80 °C through 0.22 μm filters and directly transferred to pre-thermostatted cells. Analysis grade absolute EtOH and TBOH (from Merck), and di-hydrated TMAO (from Sigma Chemical Co.) were used without further purification. The required amount was added to the solution before filtration. Temperature control was within ± 0.2 °C. Quenching depths (region 2 of Figure 1) not exceeding 5 °C allowed sufficiently slow kinetics for time-resolved studies.

Measurements of different quantities were simultaneously performed on different aliquots of the same solution, quenched at the same temperature. Experiments were repeated for various concentrations and quenching temperatures. In one aliquot, the growth of scattered light intensity was recorded at different angles, in the absence of shear. As already reported (*15, 30, 31, 36*), unmistakable "signatures" of spinodal demixing (not always observed in real system) are neatly observed in our case. These are the initial exponential growth of scattered light and linear Cahn plots (*45*). The rate of growth and the time-dependent structure function, S(q), showed a rather broad maximum around a value q_M of the wave vector, determined by quenching depth, polymer concentration and presence of co-solutes, as already observed. For each experimental condition, we calculated the characteristic length $L_M = 2\pi/q_M$ of the developing pattern of polymer-rich and solvent-rich domains. In our conditions, L_M was of the order of few microns. The mean size of the developing polymer-rich domains also measured by DLS, was in the range of 10^{-1} μm (*36*).

On another aliquot, a constant shear was applied immediately after quenching, and shear viscosity was recorded. After a waiting time, which depends upon quenching conditions, a large shear-dependent and well reproducible surge of viscosity (corresponding to an increase by 2 or 3 orders of magnitude) was observed (*15-17*). It appeared after the end of the initial (linear) stage of demixing, concurrently with the stabilization of the low angle light scattering ring (*32*) mentioned in the Introduction. At decreasing quenching depths, the peak value of the viscosity surge decreased and the waiting time necessary for its appearance increased, as shown in Figure 2a. In the presence of cosolute perturbation, the behavior of viscosity had similar trends (Figure

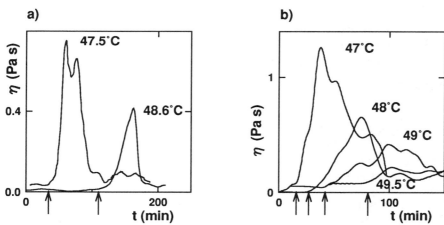

Figure 2 - Typical shear-viscosity curves, for agarose concentration c=0.5% w/v, in different solvents and different quenching temperatures (indicated on each curve), at the same shear-rate $\dot{\gamma} = 1.9 \times 10^{-1}$ s^{-1}. a): in pure H_2O. b): in H_2O+0.36% TMAO mol. fraction. For clarity, some of the curves have been multiplied by suitable factors, K: a) at 48.6°C, K=2; b) 47°C, K=0.1; 49.5°C, K=5. Delay time between quenching and viscosity surging is indicated by arrows.

2b). The time-resolved behaviors of shear-viscosity and viscoelasticity are compared in Figure 3, for a typical case. Optical rotation (O.R.) data from ref. 31, monitoring the (2 coils) → (1 double helix) transition, are also reported. We see that at the beginning of the viscosity surge, no detectable O.R. signal has appeared, yet the viscoelastic response is already essentially elastic (Figure 3a). However, both G' and G" values are much lower than those observed at the end of the experiment (about 4% for G', Figure 3b). We also see that a time one order of magnitude larger than that necessary for the appearance of the viscosity surge must elapse before the appearance of a detectable O.R. signal, while a sizeable increase of G' has already occurred. From Figure 3a and comparison with ref. 33, we further see that the behavior of G' (and, in particular, its plateau) parallels that of the turbidity signal which monitors the onset and maturing of spinodal demixing. We note that at the end of the experiment (5001 min in the case of the figure) G' and G" have very slow rates of growth, and have not necessarily reached their equilibrium values. G' values larger by about one order of magnitude are typically observed in completely formed gels at this agarose concentration. These data concurrently show that the viscosity surge must be related to the ripening stage of spinodal demixing. They also show that some relevant structural changes have occurred in the system, long before the experimental detectability of crosslinking. In order to clarify whether such changes occur in the solution even in the absence of shear or are shear-promoted, some of the quenching experiments were repeated three times at same quenching depth, polymer concentration and rate of shear, as follows: i) Shear was applied immediately after quenching, as described above. The observed viscosity surge is shown in Figure 4, curve a. ii) Shear was applied at the time corresponding to the occurrence of a stable

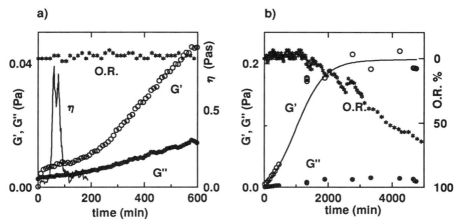

Figure 3. - Time evolution of G' (○) and G" (●) (storage and loss shear moduli, respectively) in a typical quenching experiment. Frequency $\omega = 0.1395$ s^{-1}, strain 0.15, angular amplitude 0.7°. Agarose concentration c = 0.5% w/v, in pure water, T=47.5 °C. Quenching is performed at t=0. a) Time evolution in the first 600 min. Shear viscosity, η (–), and optical rotation (✶) (from ref. 31) are also shown. b) As in a), on the 0-5,000 min scale. Note that the η peak is here omitted, as a consequence of scale compression. Note also the change of scale for G' and G".

low-angle scattering ring and to the beginning of the increase of viscosity in the previous experiment. A viscosity surge, similar to the previous one but somewhat more intense, was observed with no further delay time, as shown in Figure 4, curve b. iii) Shear was applied at the time approximately corresponding to the end of the decay of viscosity in the first experiment. A very high viscosity was recorded, followed by a rapid decay, as shown in Figure 4, curve c. Comparison of these results shows that the structural changes causing the viscosity surge occur in the system after quenching, both in the presence of low shear and in its absence. The presence of shear causes a depression of the peak value reached by η (Figure 4), suggesting its perturbing effect on the formation and stabilization of the (still unknown) structure,

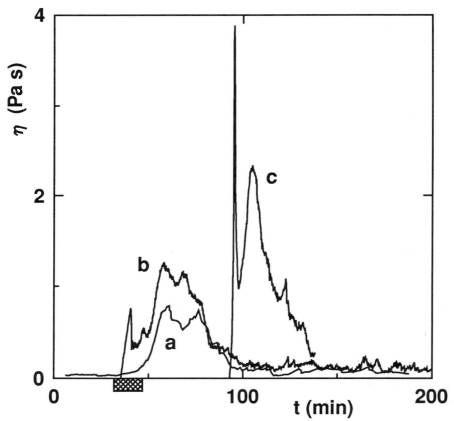

Figure 4. - Viscosity surge observed under the following experimental procedures:
a) Shear is applied immediately after quenching (t=0).
b) Shear is applied at the time corresponding to the beginning of the surge in curve a.
c) Shear is applied when the decay of the viscosity surge of curve a is completed. Stabilization of forward light-scattering pattern is marked by the box. Experimental conditions as in Figure 3.

as we shall further discuss in the following. It is worth noting that the observed peak values (a few Pa·s) are still very low in comparison with those (virtually infinite, actually non-measurable) of completely formed gels.

The rest of our experiments were done at constant shear rate, $\dot{\gamma}$, applied immediately after quenching. Delay times for the appearance of the viscosity surge and viscosity peak values were studied at several polymer concentrations, quenching temperatures and shear values, and also in the presence of cosolute perturbation. The dependence of the delay time on quenching depth followed the critical behavior shown in Figures 5a and 5b. Best fittings of experimental points in terms of the critical law $\Delta t = \text{const.} \times [(T_{sp}-T)/T_{sp}]^p$ allowed determining the critical temperature and exponent corresponding to each condition. The exponent p = -1.3 was common to all agarose concentrations (*16, 17, 30, 34*) and solvent perturbations, as illustrated in Figures 5c and 5d. Checks in conditions where other methods were also available (*35, 42*) showed the coincidence of this critical temperature with the spinodal temperature, T_{sp}. Sizeable shifts of T_{sp} due to co-solutes are shown in Figure 5b. The T_{sp} values decreased in presence of ETOH and TBOH, and increased in presence of TMAO. As a consequence of the T_{sp} shifts, L_M values (independently measured by light scattering) were also affected.

In experiments performed at different shear rates, the viscosity peak showed a strong inverse dependence upon $\dot{\gamma}$ while the delay time was not affected. In Figure 6 the peak values $\Delta\eta/\eta$ observed at selected polymer concentrations and quenching depths are plotted vs $\dot{\gamma}^{-1}$. In all cases, satisfactory fits were obtained with inverse power laws with exponent $\phi = 1.86 \pm 0.06$, in the entire range $0.016 < \dot{\gamma} < 0.2$ s^{-1}. At higher shear rates the amplitude of the peak was below detectability. A similar shear-dependence was also observed in presence of cosolutes, as shown in Figure 7.

DISCUSSION AND CONCLUSIONS

The large viscosity surge is related to spinodal demixing. This is shown by the critical behavior of its delay time after quenching, related to a critical temperature coincident with the spinodal temperature determined by other techniques (*30,31*). The time of appearance of the viscosity surge coincides with the end of the linear regime of spinodal demixing which is clearly observed in our system (*15,30,31,36*). At this time, the sample remains macroscopically liquid and gelation has not yet started. Double helices, if any, are below detectability. Also, within experimental uncertainties, all polymer coils are still freely diffusing (*31*). However, the low-angle light scattering pattern is stabilized and the mechanical response has become prevailingly elastic (Figure 3). Values of η (peak) and of the plateau reached by G' (Figure 3) are much larger than in the sol and much smaller than in the gel. All this suggests tracing the observed viscosity behavior to the onset and subsequent distortion and breaking of a very loose structure, whose evolution is disturbed by shear (Figure 4). Information on this structure is provided by the exceptional shear-dependence of the viscosity peak. The very high value of the related exponent requires the existence of a relevant relaxation time, τ, and associated length scale, l, such that the shear deformation satisfy $\dot{\gamma}\tau > 1$ in all experimental conditions, notwithstanding the low values of applied shear. We can assume that the relation $\tau = (k_BT)^{-1} 6\pi\eta l^3$ holds, at least

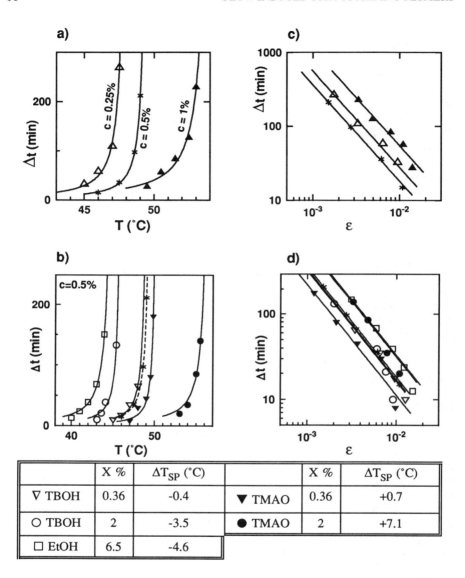

	X %	ΔT_{SP} (°C)		X %	ΔT_{SP} (°C)
▽ TBOH	0.36	-0.4	▼ TMAO	0.36	+0.7
○ TBOH	2	-3.5	● TMAO	2	+7.1
□ EtOH	6.5	-4.6			

Figure 5. - Delay-time between quenching and viscosity surge (as indicated by the arrows in Figure 2), vs quenching temperature, for different agarose concentrations a) in pure water, and b) in perturbed solvents. c) and d): same as Figure 5a, 5b, on a log-log scale, vs $\varepsilon = (T_{sp}-T)/T_{sp}$. Values of T_{sp} are obtained by best-fittings of data in a) and b) to a critical law $\Delta t=[(T_{sp}-T)/T_{sp}]^p$. Slopes of the straight lines in the log-log plots are $p = -1.3$ for all cases. Shifts of T_{sp} caused by co-solutes are also given. Co-solute concentrations are given in the table, in percent of molar fraction.

approximately (15-17). Among the relevant length scales are the molecular size (of either coils or double helices), the thermodynamic correlation length $\xi = \xi_0 \epsilon^{-\nu}$ (where ϵ is the quenching depth $\epsilon = (T_{sp} - T)/T_{sp}$), and the size of domains, which would all correspond to $\dot{\gamma}\tau \ll 1$. Therefore, we expect the shear to cause negligible deformation of polymer conformation, of thermal fluctuations and of polymer-rich domains generated by spinodal demixing. The remaining relevant length is the mean inter-domain distance, L_M. The associated relaxation time $\tau_M = (k_B T)^{-1} 6\pi\eta L^3_M$ is of the order of 100 s, similar to the long characteristic time which appears at this very stage in the noise of scattered light (32). This τ_M value gives $\dot{\gamma}\tau_M > 1$ in all our experimental conditions. It is therefore of interest to plot all $\Delta\eta/\eta$ data vs. $\dot{\gamma}\tau_M$, which measures the deformation experienced by the overall distribution of domains throughout the sample. This plot of $\Delta\eta/\eta$ is shown in log-log form in Figure 8 where L_M values come from independent static light scattering experiments, performed in each experimental condition. As shown in Figure 8a, all data corresponding to

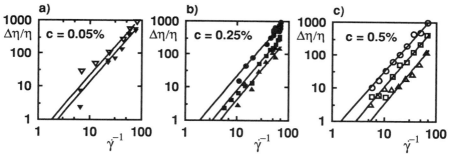

Figure 6. - Log-log plots of viscosity peak values $\Delta\eta/\eta$, vs inverse shear-rate $\dot{\gamma}^{-1}$ for agarose-water solutions. a) c=0.05%, b) c=0.25%, c) c=0.5%. Agarose concentrations are in weight per volume. Quenching temperatures are: ▽ T=36°C; ▼ T=37°C; ● T=45°C; ■ T=46°C; ▲ T=47°C; ○ T=46°C; □ T=47.5°C; △ T=48.6°C. Slopes of best fitting straight lines are all within the range 1.86 ± 0.1.

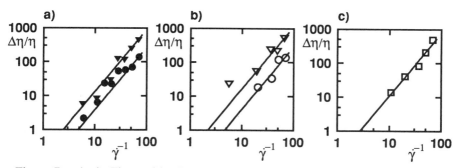

Figure 7. - As in Figure 6 in the presence of cosolutes. a) TMAO, ▼ = 0.36%, ● = 2.0% b) TBOH, ▽ = 0.36%, ○ = 2.0% c) EtOH □ = 6.5%. Agarose concentration c=0.5% w/v, quenching depth $\Delta T=4°C$. Slopes of best fitting straight lines are again within the range 1.86 ± 0.1.

different shear rates, different polymer concentration and quenching depths fall on one and the same master line, corresponding to $\Delta\eta/\eta$ = const. $\dot{\gamma}\tau_M^{-1.97}$. A similar plot of data obtained at a specific polymer concentration in cosolute-perturbed solutions is shown in Figure 8b. We see that data points lie on the same straight line, as those obtained with pure water. We note that a slight shift of data points towards lower $\Delta\eta/\eta$ (or $\dot{\gamma}\tau_M$) values visible in the figure, might be an instrumental artifact due to a 10% uncertainty in L_M values. Such error might cause a horizontal shift of all data points pertaining to a given sample and quenching condition. The size of this error is shown in Figure 8b. We conclude that the appropriate scaling factor of shear rate is the relaxation time, τ_M, associated with the dynamic response of the overall interdomain structure. Accordingly, the observed viscosity peak must be due to distortion and breaking of some sort of links among polymer-rich domains generated by spinodal demixing.

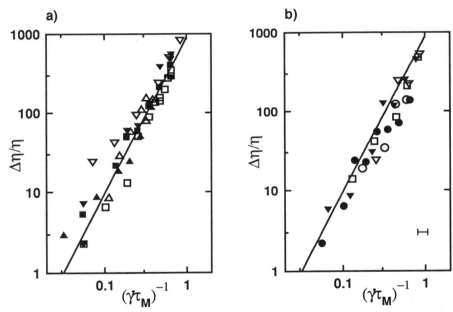

Figure 8. - Log-log plot of the viscosity peak values, $\Delta\eta/\eta$, vs a scaled shear, $(\dot{\gamma}\tau_M)^{-1}$. Here τ_M is the relaxation time associated with the interdomain structure, $\tau_M = (k_B T)^{-1} 6\pi\eta L_M^3$, and L_M is independently measured by light scattering in each experimental condition.
a): data from Figure 6, for different agarose concentrations and quenching depth, in pure water. Best fitting straight line is indicated. ▽ c=0.05%, T=36°C; ▼ c=0.05%, T=37°C; ■ c=0.25%, T=46°C; ▲ c=0.25% T=47°C; □ c=0.5%, T=47.5°C; △ c=0.5%T=48.6°C.
b): data from Figure 7, agarose concentration c=0.5% w/v, quenching depth $\Delta T=4°C$, different co-solutes. Straight line is the same as in the a) panel. ▼ TMAO 0.36%, ● TMAO 2.0% ;▽ TBOH 0.36%, ○ TMAO 2.0%; EtOH □ = 6.5%.

As to the nature of these interdomain links, values of η (peak) and G' indicate that they must be much weaker (or much fewer) than those pertaining to the gel. At first sight, it seems unlikely that they are due to a number (even if undetectably small) of double helices, since formation of the latter in polymer-poor regions is unfavored. However, one cannot totally disregard the possibility that some sort of mutual tethering of polymer-rich regions via single or double (*25, 46*) polymer strands takes place through the operation of complex and perhaps competing mechanisms. The latter might include those involved in the ripening of demixing (*47*), the drifting of polymer-rich domains [clearly evidenced by experiments (*36*) and statistical simulation (*48, 49*)], and their occasional contacts, also evidenced by simulations (*48, 49*). The contribution of solvent-induced forces to such mechanisms may be expected (*37*). Contacts among polymer-rich domains are favored by the elongated shape of the latter, evidenced by recent depolarized light scattering experiments (Bulone, D.; San Biagio, P.L. *Biophys. J.*, in press).

Solvent perturbations by cosolutes cause marked shifts of spinodal temperature. These effects are easily understood at a plain thermodynamic level, in terms of the mean-field Flory-Huggins approach (*38, 50*) and ascribed to changes of the Flory-Huggins χ parameter (*37, 42, 51, 52*). This makes them similar to other effects of critical temperature shifts, such as those due to dilution by impurities in ferromagnets (*53*). The similarity, however, is limited to thermodynamics, since in the present case the microscopic and statistical mechanical picture is much richer. This is a consequence of the dynamics of solvent, solute and co-solute system, closely related to the existence and non-additive modulability of solvent-induced interactions among solutes (*37,42-44*), and to the apparently stochastic, yet non-zero-average nature of such interactions (*37,42-44,54*). Correspondingly, as shown by the opposite shifts caused by two geometrically similar cosolutes (TMAO and TBOH), the observed effects are closely linked to specific microscopic features of the perturbers (Bulone, D. et al., to be published; Noto, R. et al., to be published). This can be exploited for understanding the underlying microscopic mechanism (*42-44*). Two features of the reported effects of cosolutes are worth noting. One is that the scaling behavior remains unchanged and the scale invariance of the shear dependence remains valid. The other is that the large shifts of T_{sp} imply that at equal temperatures and polymer concentrations, phenomena reported here may or may not occur, depending on the presence and type of co-solutes. If they occur, the system exhibit different time scales and interdomain distances, and rheological behavior. Such differences are expected (*33,42,43*) to cause structural differences in gels, which may have extended implications for biology and for applications.

In conclusion, our results show that in aqueous agarose solutions, at the end of the linear stage of spinodal demixing, a loose structure is formed. This structure is not permanent. It leaves the system in the liquid state, and its relaxation time is of the order of 100 s. Following its onset, all properties of the solution remain essentially unchanged, for a long time before the measurable start of the gelation process. Low shear slightly counteracts the occurrence of this structure and eventually causes its deformation and breaking. High shear totally prevents its occurrence. The related rheological behavior of our biopolymeric solutions as affected by spinodal demixing and by co-solute perturbations of solvent, is the main point of interest of the present work. However, understanding the precise molecular nature of the constraints responsible for this structure and the kinetics of their onset and later evolution is likely to bring into the picture kinetic processes, not yet taken into consideration by current theories of gelation (*39,40,55*).

ACKNOWLEDGMENTS

We are grateful to M.U. Palma for helpful discussions. We also thank P.L. San Biagio and D. Bulone for friendly help and exchange of data, and M. Lapis and G. Giacomazza, of the Institute for Interdisciplinary Applications of Physics of C. N. R., Palermo, for technical help. Discussions with A. Onuki, J.V. Sengers and R. Boscaino are also acknowledged. This work was done within the Network "Dynamics of Protein Structures" of the European Community "Human Capital and Mobility" program, and was partially supported by MURST (national and local funds) and by CRRNSM.

LITERATURE CITED

1. Hashimoto, T. et al. In *Dynamics and Patterns in Complex Fluids*; Onuki, A.; Kawasaki, K., Eds.; Springer: Berlin, 1990; pp 86-99.
2. Perrot, F.; Baumbergen, T.; Chan, C. K.; Beysens, D. *Physica A* **1991**, *172*, 87.
3. Perrot, F.; Chan, C. K.; Beysens, D. *Europhys. Lett.* **1989**, *9*, 65.
4. Baumbergen, T,; Perrot, F.; Beysens, D. *Phase Transitions* **1991**, *31*, 251.
5. Moses, E.; Kume, T.; Hashimoto, T. *Phys. Rev. Lett.* **1994**, *72*, 2037.
6. Nakatani, A. I.; Johnsonbaugh, D.; Han, C. C., present volume.
7. Hashimoto, T.; Matsuzaka, K.; Kume, T.; Moses, E., present volume.
8. Doi, M; Harden, J. L. ; Ohta, T. *Macromolecules* **1993**, *26*, 4935.
9. Ohta, T; Enamoto, Y; Harden, J.L.; Doi, M. *Macromolecules* **1993**, 26, 4928.
10. Ohta, T.; Nozaki, H; Doi, M. *J. Chem. Phys.* **1990**, *93*, 2664.
11. Onuki, A. *J. Phys. : Condensed Matter* **1994**, *6 A*, 193.
12. Onuki, A. *J. Phys. Soc. Jpn.* **1990**, *59*, 3423 and refs. therein.
13. Onuki, A. *Phys. Rev.* **1987**, *35*, 5149. *Int. J. Thermophysics* **1989**, *10*, 293.
14. Dhont, J. K. C.; Duyndam, A. F. H.; Ackerson, B. J. *Physica A* **1992**, *189*, 532.
15. Emanuele, A.; Palma-Vittorelli, M. B. *Phys. Rev. Lett.* **1992**, *69*, 81.
16. Emanuele, A.; Palma-Vittorelli, M. B. *Int. J. Thermophysics* **1995**.
17. Emanuele, A. Ph. D. Thesis, Palermo, Italy, 1992.
18. Krall, A. H.; Sengers, J. V.; Hamano, K. *Phys. Rev. E* **1993**, *48*, 357.
19. Krall, A. H.; Sengers, J. V.; Hamano, K. *Phys. Rev. Lett.* **1992**, *69*, 1963.
20. Hamano, K.; Yamashita, S.; Sengers, J. V. *Phys. Rev. Lett.* **1992**, *68*, 3579.
21. Axelos, M. A. V.; Kolb, M. *Phys. Rev. Lett.* **1990**, *64*, 1457.
22. Kolb, M.; Axelos, M. A. V. In *Correlations and connectivity;* Stanley, H. E.; Ostrowsky, N., Eds.; NATO ASI E; Kluwer: The Netherlands, 1990, Vol. 188; pp 255-261.
23. Suggett, A. In *Water: a Comprehensive Treatise;* Franks, F., Ed.; Plenum Press: New York, NY, 1975, Vol. 4; pp 519-567.

24. *Functional Properties of Food Macromolecules*; .Mitchell, J. R.; Ledward, D. A., Eds.; Elsevier: London, 1985.
25. Arnott, S.; Fulner, A.; Scott, W.E.; Dea, I. C. M.; Moorhouse, R.; Rees, D.A *J. Mol. Biol.* **1974**, *90*, 269.
26. Indovina, P.L.; Tettamanti, E.; Micciancio-Giammarinaro, M.S.; Palma, M.U. *J. Chem. Phys.* **1979**, *70*, 2841.
27. Vento, G.; Palma, M.U.; Indovina, P.L. *J. Chem. Phys.* **1979**, *70*, 2848.
28. *Physical Networks;* Burchard, W.; Ross-Murphy, S. B., Eds.; Elsevier: London, 1988.
29. Clark, A.H.; Ross-Murphy, S.B. *Adv. Polym. Sci.* **1987**, *83*, 57.
30. San Biagio, P. L.; Bulone, D.; Palma-Vittorelli, M. B.; Palma, M. U. *Food Hydrocolloids* **1995**.
31. Emanuele, A.; Di Stefano, L.; Giacomazza, D.; Trapanese, M.; Palma-Vittorelli, M.B.; Palma, M.U. *Biopolymers* **1991**, *31*, 859.
32. San Biagio, P.L.; Madonia, F., Newman, J.; Palma, M.U. *Biopolymers* **1986**, *25*, 2250.
33. Leone, M.; Sciortino, F.; Migliore, M.; Fornili, S.L.; Palma-Vittorelli, M.B. *Biopolymers* **1987**, *26*, 743.
34. San Biagio, P.L.; Bulone, D.; Emanuele, A.; Madonia, F.; Di Stefano, L.; Giacomazza, D.; Trapanese, M.; Palma-Vittorelli, M.B.; Palma, M.U. *Makromol. Chem. Macromol. Symp.* **1990**, *40*, 33.
35. San Biagio, P.L.; Newman, J.; Madonia, F.; Palma, M.U. *Chem. Phys. Lett.* **1989**, *154*, 477.
36. Bulone, D.; San Biagio, P.L. *Chem. Phys. Lett.* **1991**, *179*, 339.
37. Palma, M. U.; San Biagio, P. L.; Bulone, D.; Palma-Vittorelli, M. B. In *Hydrogen Bond Networks*; Bellissent-Funel M. C.; Dore, J. C., Eds.; Nato ASI C; Kluwer: The Netherlands, 1994, Vol 435; pp. 457-479.
38. de Gennes, P.G. *Scaling Concepts in Polymer Physics*; Cornell Univ. Press: Ithaca, N.Y., 1979.
39. Coniglio, A.; Stanley, H.E.; Klein, W. *Phys. Rev. B* **1982**, *25*, 6805.
40. Stauffer, D.; Coniglio, A.; Adam, M. *Adv. Polym. Sci.* **1982**, *44*, 103.
41. Lironis, G; Heermann, D.W.; Binder K. *J. Phys. A: Math. Gen.* **1990**, *23*, L329.
42. Bulone, D.; San Biagio, P.L.; Palma-Vittorelli, M.B.; Palma, M.U. *J. Mol. Liquids* **1993**, *58*, 129.
43. Palma-Vittorelli, M.B.; Bulone, D.; San Biagio, P.L.; Palma, M.U. In *Water-Biomolecule Interactions*; Palma, M. U.; Palma-Vittorelli, M. B.; Parak, F., Eds.; Conference Proc. Series of the Italian Physical Society; Compositori: Bologna, 1993; Vol. 43, pp 253-260.
44. Palma-Vittorelli, M.B. In *Hydrogen Bond Networks;* Bellissent-Funel M. C.; Dore, J. C., Eds.; Nato ASI C; Kluwer: The Netherlands, 1994, Vol 435; pp. 540-541.
45. Cahn, J. W. *J. Chem. Phys.*, **1965**, *42*, 93.
46. Foord, S.A.; Atkins, E.D.T. *Biopolymers* **1989**, *28*, 1345.
47. Siggia, E.D. *Phys. Rev. A* **1979**, *20*, 595.
48. Brugé, F.; Martorana, V.; Fornili, S. L.; Palma-Vittorelli M.B. *Int. J. Quantum Chem.* **1992**, *42*, 1171.

49. Brugé, F.; Martorana, V.; Fornili, S. L.; Palma-Vittorelli M.B. *Makromol. Chem. Macromol. Symp.* **1991**, *45*, 43.
50. Kurata, M. *Thermodynamics of Polymer Solutions*; Harwood Acad. Publ.: Chur, N.Y., 1982.
51. Bulone, D.; San Biagio, P.L.; Palma-Vittorelli, M.B.; Palma, M.U. *Science* **1993**, *259*, 1335.
52. San Biagio, P.L.; Palma M.U. *Biophys. J.* **1991**, *60*, 508.
53. Neda Z. *J. Phys. 1 (France)*, **1994**, *4*, 175.
54. Brugé, F.; Fornili, S.L.; Palma-Vittorelli, M.B. *J. Chem. Phys.* **1994**, *101*, 2407.
55. Stauffer, D.; Aharony, A. *Introduction to Percolation Theory*; Taylor and Francis: Bristol, U.K., 1991.

RECEIVED January 25, 1995

Chapter 6

Flow-Induced Structuring and Conformational Rearrangements in Flexible and Semiflexible Polymer Solutions

A. J. McHugh, A. Immaneni, and B. J. Edwards

Department of Chemical Engineering, University of Illinois, Urbana, IL 61801

This article presents a discussion of two examples of the rheo-optics of flow-induced structure formation in solutions of flexible and semi-rigid chain systems. Polystyrene solutions show shear-thickening, turbidity, and associated dichroism maximization during flow which can be related to the formation of micron-sized aggregates. Hydroxypropylcellulose in acetic acid represents a semi-rigid system which can undergo a conformational change under shear flow which is distinct from the overall molecular ordering. Such changes can be conveniently monitored in terms of the circular birefringence.

The molecular ordering and supermolecular structuring that can be induced by flow in otherwise isotropic, polymer solutions and blends have been extensively studied using optical and rheo-optical techniques. Supermolecular structure formation is generally accompanied by significant increases in the solution turbidity and small angle light scattering (1,2,3,4,5,6), as well as measurable increases in the viscosity with shear rate (i.e., shear-thickening) (3,7,8). Our recent studies with solutions of an optically active, semi-rigid system (hydroxypropylcellulose (HPC) in acetic acid), demonstrate that flow can also induce intramolecular, conformational changes, which appear as measurable changes in the optical rotatory power (i.e., circular birefringence) (9,10).

Since most of the above-noted studies have been extensively reviewed elsewhere (11,12) (or will be the subjects of further discussion in this symposium text), this chapter will simply focus on a discussion of two examples from our studies of flexible and semi-rigid systems which illustrate the above phenomena. The principal technique we use is modulated polarimetry in combination with in-situ measurement of the shear stress/viscosity behavior.

Rheo-optical Preliminaries

The flow cell used in our studies is a small gap (0.5 mm) Couette device in which the outer cylinder is rotated while torque is monitored at the stationary inner cylinder. The optical properties of interest are simultaneously measured in the flow/shear plane by passing a laser beam (λ = 632.8 nm) down the fluid gap (d = 7.745 cm). The optical train consists of a laser beam which passes through a polarizer aligned with its transmission axis at 90° relative to the flow direction, followed by a photoelastic modulator (PEM) oriented at 45°, before entering the flow cell. Depending on the properties to be measured, additional optical elements are placed on either side of the flow cell. The PEM produces a sinusoidal dependence in the polarization of the light, δ_{pem} = A sin(ωt). The configurations used in our studies are listed in Table 1. The leading letter indicates the property whose signal would normally be dominant in the particular configuration. The first two harmonics of the output intensity ($I_\omega, I_{2\omega}$) are registered by lock-in amplifiers and the dc component of the signal, I_{dc}, is isolated using a low-pass filter. The digitized signals are transferred to a computer by the lock-ins, where ratios of the two harmonic intensities with respect to the dc intensity are formed as follows:

$$R_{1\omega} = \frac{I_{1\omega}}{2I_{dc}J_1(A)} \quad , \quad R_{2\omega} = \frac{I_{2\omega}}{2I_{dc}J_2(A)} \quad . \tag{1}$$

The J_i (i = 1,2) are Bessel functions of the first kind of order i and A is the adjustable amplitude of the photoelastic modulator which is set such that $J_o(A) = 0$. In the most general case (i.e. one in which the fluid exhibits circular as well as linear optical properties), these ratios will be functions of all the optical properties, i.e., linear dichroism ($\Delta n''$), linear birefringence ($\Delta n'$), their respective orientation angles (χ_d, χ_b), circular birefringence ($\Delta n'_c$) and circular dichroism ($\Delta n''_c$). For the wavelength used in our studies, the dichroism is due to scattering as opposed to absorption.

Output ratios for each configuration listed in Table 1 can be determined using the Jones calculus. This results in a set of complex, non-linear, coupled equations which are functions of the optical properties (*9,10,12*). These can be simplified depending on the relative magnitudes of the optical properties one is measuring. However, care needs to be taken, particularly in the analysis of circular properties, since terms which might otherwise be thought negligible, can have important consequences for the interpretation of the dynamical behavior (*9,10*).

Flexible Chain Behavior

Our rheo-optical studies of flexible-chain systems have focused on dilute polystyrene solutions (c/c* ~ 0.1-1.1) in decalin, with fractionated polymer molecular weights ranging from 4.3 x 10^5 to 17.9 x 10^6. The viscosity and linear optical properties have been monitored at 25 °C over the range of shear rates (500 s^{-1} to 8000 s^{-1}) where both shear-thinning and shear-thickening occur. Figure 1 shows an example which illustrates the steady-state viscosity and dichroism behavior of the lowest molecular weight solutions that exhibit the phenomenon of shear-thickening (MW = 1.54 x 10^6, denoted PS2). At the lowest concentration, the scattering dichroism rises with shear

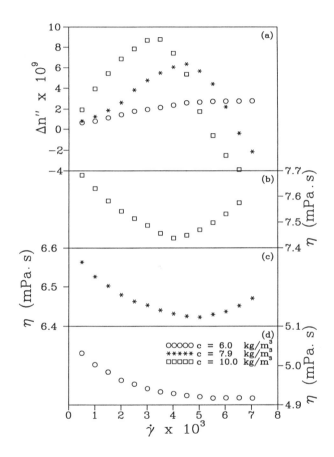

Figure 1. - Shear rate dependence of the steady-state, linear dichroism (a), and viscosity (b,c, and d) of 1.54×10^6 MW (PS2) polystyrene/decalin solutions. Reproduced with permission from reference 8.

rate, eventually reaching a plateau, while the viscosity shows the usual shear-thinning pattern, dropping to a Newtonian plateau at more or less the same shear rate where the dichroism saturates. At higher concentrations, a dramatically different pattern emerges in which the dichroism rises more rapidly to a maximum, while the viscosity simultaneously drops to a minimum. This is followed by a region of shear-thickening in which the viscosity continuously rises with shear rate, while the dichroism continuously decreases and eventually becomes negative. The region of positive, steady-state dichroism extends for as much as 2000 s^{-1} beyond the maximum point (i.e. well into the shear-thickening region), and, within this region, no instabilities occur in the stress(i.e. the flow remains one-dimensional, laminar shear). Within the shear-thickening region, the dichroism orientation angle continuously drops to a constant, low value, indicative of near-alignment with the flow axis (8). On the other hand, for these same conditions, the steady-state, linear birefringence (Figure 2) exhibits a steady, monotonic increase with shear rate, independent of whether the solution shear-thickens or shear-thins. Both the shape and magnitude of the shear rate dependence of the birefringence are characteristic of the behavior of stretched, flexible-chain molecules in a non-matching solvent (13).

Analysis using the anomalous diffraction approximation (ADA) suggests that the magnitude and shear rate dependence of the steady-state dichroism can be interpreted in terms of the formation and growth of micron-sized, spheroidal-shaped aggregates (13,14,15). A similar proposal for the formation of spheroidal, micron-sized aggregates has been used to explain the magneto-optical behavior of polystyrene solutions (16). Since, in our measurements, the transient optical ratios and mechanical signals did not exhibit measurable time constants (i.e. they both mapped with the Couette motor profiles during start up and cessation of flow) (8), one cannot deduce from them information about the kinetics of the structure formation. However, the dc intensity (I_{dc}), which drops out of the analysis on forming the harmonic ratios, does contain useful information. Figure 3 shows a comparison of the transmittance, T, (ratio of the transmitted intensity to that of the quiescent, isotropic solution) as a function of time and shear rate for two of the higher concentration solutions of the MW = 6.8 x 10^6 (PS7) fraction. These data typify the behavior seen in all the solutions studied (17). In this case, the lower concentration solution corresponds to the range where increasing shear-thickening and dichroism maximization occurred with increasing concentration (i.e, $c \leq 2$ kg/m^3), such as illustrated in Figure 1. However, for this same molecular weight, the solutions exhibited a diminution in shear-thickening (and the associated dichroism maximum) with further increases in concentration, eventually exhibiting simple shear-thinning (and dichroism saturation) behavior at the highest concentration ($c = 3$ kg/m^3) (i.e. the inverse of the pattern shown in Figure 1). At high shear rates, both solutions exhibit a sharp decrease in transmittance during flow (Figure 3). We attribute this to thermal gradients due to viscous dissipation which cause deflection of the light into the absorbing walls of the flow cell (8). In all cases, due to the relatively low solution viscosities, these thermal gradients rapidly dissipate on flow cessation, leaving a reduced transmittance due to scattering from the structures which have formed during flow. The transmittance of the 2 kg/m^3 solution for the two lowest shear rates, corresponding to the shear-thinning region, exhibits a continuous drop during flow, reaching a saturation value just prior to flow cessation. After flow cessation, the intensity eventually returns to the initial

Table 1. Configurations Used in Rheo-optical Experiments

B1	B2	B3
Polarizer at 90°	Polarizer at 90°	Polarizer at 90°
PEM at 45°	PEM at 45°	PEM at 45°
Retarder at 0°	Sample	Sample
Sample	Polarizer at -45°	Retarder at 0°
Retarder at 0°		Polarizer at -45°
Polarizer at -45°		

D1	D2
Polarizer at 90°	Polarizer at 90°
PEM at 45°	PEM at 45°
Retarder at 0°	Retarder at 45°
Sample	Sample

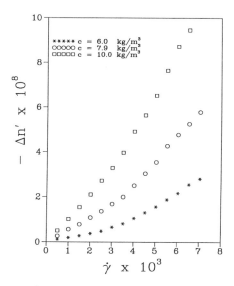

Figure 2. - Shear rate dependence of linear birefringence of PS2 solutions. The 6.0 kg/m^3 solution exhibited simple shear-thinning, the other two exhibited shear-thickening. Reproduced with permission from reference 8.

quiescent value on a time scale (not shown) of about 45 s. At the three highest shear rates where shear-thickening occurs, the transmittance on flow cessation has a much lower value and returns to the quiescent value on a much longer time scale (~130 s, not shown). In the case of the 3 kg/m^3 solution, which exhibited shear thinning at all shear rates, the magnitude of the transmittance on flow cessation is much higher than that for the shear-thickening solutions, and the return to the quiescent transmittance level occurs on a smaller time scale (~45 s).

The turbidity on flow cessation (i.e., $\tau = \ln(T) / d$ where d is the optical path length) is shown in Figure 4 as a function of the steady-state shear rate for the PS2 and PS7 solutions. In both cases the lowest concentration corresponds to the pure shear-thinning region, while the higher concentrations are those at which shear-thickening occurred, beginning at the shear rates indicated by the arrows. These data further illustrate the sharp increase that occurs in the turbidity in the shear-thickening region and lend consistency to the interpretation of flow-induced, supermolecular structure formation mentioned above.

Semi-rigid Chain Behavior

Hydroxypropylcellulose (HPC) in acetic acid (AA) represents an example of a semi-flexible system which forms a cholesteric liquid crystalline phase at high concentrations. In this case, the combination of chain rigidity and optical activity, (due either to the presence of chiral centers in the molecule or intramolecular hydrogen bonding leading to a helical conformational (*18,19*)) plays a critical role in the phase behavior. Rheo-optics presents an excellent tool for probing both orientational and conformational changes which can be induced in such systems in the isotropic, pretransitional state. In our studies we have developed a technique to monitor changes in the circular birefringence in the shear/flow plane of HPC/AA solutions in the dilute to semi-dilute region (2.5, 5.0, 7.5, and 10.0 wt% polymer). The initial portion of our work (*9*) involved measurements of the linear dichroism and birefringence, the orientation angles of each, the circular birefringence and circular dichroism, and the *in-situ* shear stress over a wide range of shear rates using the small gap Couette device. Both the linear birefringence and dichroism exhibited dynamical behavior characteristic of rigid rods (i.e., monotonic increases with shear rate to eventual plateaus and, in the case of the dichroism, a corresponding, monotonic decrease in the orientation angle which saturated at a low value at high shear rates for all concentrations). On the other hand, for the three highest concentration solutions, measurable changes in the circular birefringence occurred after a minimal shear rate had been reached. At a given shear rate these changes occurred after the linear properties had saturated, i.e. steady-state, orientational equilibrium had been reached. Moreover, start-up and relaxation times for the circular birefringence were found to be much larger than those of the linear optical properties, hence the change in optical rotatory power could not be attributed to molecular orientation induced by the shear. Start-up times decreased with increasing shear rate and increased with increasing concentration. The shear stress/shear rate behavior of the solutions exhibited simple power law behavior for the three highest concentration solutions, including the shear rate ranges where changes in optical rotatory power occurred. The transient rheological response mapped closely with that of the linear optical properties at

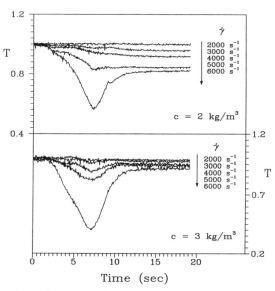

Figure 3. - Time dependent transmittance for two PS7 solutions for various shear rates. Flow start-up occurs during the first 2 s and cessation occurs between 7 and 9 s of elapsed time.

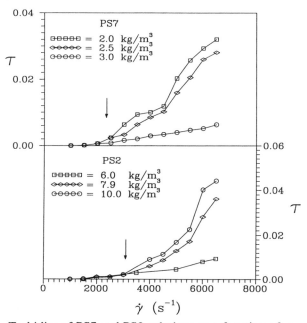

Figure 4. - Turbidity of PS7 and PS2 solutions as a function of steady-state shear rate. Arrows denote the onset of shear-thickening.

concurrent shear rates. Hence the rheology, or at least the shear stress, of these solutions is apparently governed by the orientational mechanism as opposed to that which leads to the alteration of the optical rotatory power of the solutions.

We have carried out a more extensive investigation of the optical and rheo-optical behavior of these same HPC/AA solutions and have developed techniques whereby we can measure, essentially independently, the optical rotatory power of the flowing solutions under steady-state and transient conditions. Since details of the optical alignment and analysis procedures are given elsewhere (10), we simply present here a summary of the major observations and results. We begin with measurements of the quiescent rotatory power of HPC/AA solutions which can be done very accurately using modulated polarimetry. This is followed with a discussion of the change in circular birefringence under flow.

Quiescent Circular Birefringence

The cell used for measuring the rotatory power of the quiescent HPC/AA solutions consisted of a thermostatted (solution temperature 25 °C ± 0.2 °C) aluminum cylinder with an inside diameter of 4 cm and a length of 20 cm. The cylinder is attached to a bottom plate which contains a window seat and has an adjustable top plate, containing an additional window seat. This allows the pathlength through the solution to be varied from 0 to 20 cm ±1 mm. The optical windows are devoid of measurable residual birefringence. Similar to the Couette device, the entire apparatus is anodized black to eliminate polarization changes due to reflection.

The optical configuration used for quiescent measurements of the circular birefringence is B2 (Table 1). In the absence of circular dichroism, the following expressions relating the circular birefringence, ξ (= $2\pi(d/\lambda)\Delta n'_c$), to the harmonic ratios hold:

$$R_{1\omega}^{B2} = 0, \quad R_{2\omega}^{B2} = -\sin(\xi). \tag{2}$$

At higher concentrations, finite linear birefringence may be present due to orientational fluctuations or orientation of surface layers near the optical windows. However, in this case, it can be shown (10) that the effect of the linear birefringence can be eliminated by rotating the solution cell to null the first harmonic signal. At that point, equations 2 again hold. Acetic acid solutions ranging from 1.0 wt% to 40.0 wt% of HPC were used and measurements of the harmonic intensities were made for three or four pathlengths at each concentration. Interpretable measurements were only possible for concentrations less than the lyotropic transition, 25.8 wt%. The measured value of ξ (in radians) was divided by the pathlength (and converted to degrees) to give the optical rotatory power in units of degrees per centimeter of pathlength. Using this method, the lower limit on measurable optical rotary power was about 0.005°/cm.

Values of the circular birefringence per unit of solution path length for concentrations between 1.0 and 25.0 wt% are shown in Figure 5. The linear dependence with concentration demonstrates that the specific rotation is constant over the concentration range studied. From this we conclude that intermolecular interactions are negligible in the isotropic state and that there are no observable pretransitional effects, almost up to the critical concentration for formation of the biphasic state. These data also provide the quiescent value of ξ for the solutions used in our flow experiments.

Flow Experiments

The Couette flow cell described earlier was used to evaluate the optical properties and viscosity as functions of shear rate. In addition to the steady state value, the start-up time (time required for the optical property to reach steady state) and the relaxation time (time required for the steady state value to drop by half after the cessation of flow) were also evaluated. Four concentrations of HPC in AA (2.5, 5, 7.5 and 10 wt%) were used in these experiments. Under all conditions studied, changes in the solution transmittance were negligible, indicating little or no supermolecular aggregation was occurring. The dynamics one sees in semi-rigid systems under flow (i.e. orientational or configurational) will, however, depend on which optical property dominates in the particular configuration being used. In the following discussions we will illustrate this with data obtained on the 10 wt% solution.

It was observed that the response of linear birefringence of the 10 wt% solution (measured using configuration B1) at a shear rate of 7.5 s^{-1} was controlled by a single time constant, and it was that of the linear birefringence (10). At this concentration, the shear rate is well into the region where measurable changes in the circular properties and their associated long time constants have begun to occur ($\dot{\gamma}_{cr} \sim 1$ s^{-1} (9)). This behavior can be quantitatively understood by considering the following, simplified form of the general expressions for the signal ratios in configuration B1(10):

$$R_{1\omega}^{B1} = \frac{-\delta \cos(2\chi_b) \sin(z)}{z} + \frac{2\delta\xi \sin(2\chi_b) \sin^2(z/2)}{z^2},$$

$$R_{2\omega}^{B1} = \frac{\delta \sin(2\chi_b) \sin(z)}{z} + \frac{2\delta\xi \cos(2\chi_b) \sin^2(z/2)}{z^2}.$$

(3)

In these expressions, $z = (\delta + \xi)^{1/2}$, and δ (= $2\pi(d/\lambda)\Delta n'$) is the linear birefringence. Terms involving the linear dichroism have been neglected since they are at least an order of magnitude smaller than the terms containing linear birefringence for these conditions (9). In the initial portion of the transient, signal response corresponding to the low shear rate region, δ will be very small and ξ will be on the same order or larger. However, if ξ remains at its quiescent value (~0.05 radians from Figure 5), the transient response rapidly becomes dominated by the overall chain orientation, i.e., δ. In the region above the critical shear rate, the circular birefringence does begin to change, however, its dynamics are not seen since δ is now much greater than ξ. Hence the start up and relaxation times are those associated with the overall orientation process. On the other hand, a separation of the signals associated with the orientation and conformational changes which are occurring can be seen in the response of the first harmonic ratio using configuration D1, as shown in Figure 6. Here data for the first few seconds of the start-up period have been deleted to allow more accurate analysis of the long time behavior, and the relaxational behavior has also been deleted to expand the abscissa for clarity. For this configuration, the following, approximate expression for the first harmonic ratio holds (*10*):

$$R_{1\omega}^{D1} \approx \frac{\kappa}{\delta}[\sin(\delta)\sin(2\chi_d) + \frac{\xi}{\delta}(1-\cos(\delta))\cos(2\chi_d)] \qquad (4)$$

where κ ($= 2\pi(d/\lambda)\Delta n''$) is the linear dichroism. For all shear rates beyond the critical value, the steady-state magnitude of the first harmonic ratio is very close to zero and is reflected in a small value of χ_d. If on the time scale of the linear birefringence start-up (12 s), ξ maintains the small negative value associated with the quiescent state, then as χ_d drops rapidly to near zero after flow inception, the first term in equation 4 will become very small, leaving the negative second term. This is reflected during the first 12 s of flow. At longer times, if ξ begins to decrease in absolute value, the intensity ratio will increase in positive character, eventually reaching a steady-state value after an elapsed time of t_c, corresponding to the time required for start-up of ξ.

Separation of the two dynamical processes (overall orientation and conformational change), particularly the time constants during relaxation, can be more clearly seen in the B2 configuration. Figure 7 shows, again for the same solution and shear rate conditions, the response of the second harmonic ratio in the B2 configuration. Under these conditions, the intensity ratio will be given by (*10*):

$$R_{2\omega}^{B2} = -\frac{\xi\sin z}{z} + \frac{\delta^2\sin(4\chi_b)\sin^2(z/2)}{z^2}. \qquad (5)$$

After 12 s of flow, δ reaches its steady-state value, at which point the intensity maximizes since the first term in equation 5 will be positive and thus add to the second term. As ξ begins to decrease in absolute value, the first term decreases, leading to the overshoot shown at $t_b = 12$ s prior to the achievement of steady state after $t_c = 80$ s. On cessation of flow, the second term in equation 5 drops rapidly due to its quadratic dependence on δ, leaving ξ as the sole relaxing quantity. Hence, immediately after flow cessation, the magnitude of the signal drops to a value lower than before flow inception, before gradually regaining its quiescent value as the circular birefringence relaxes to its pre-flow value. From these experiments we found that the time required for the circular birefringence to return to its quiescent value appears to be constant at about 95 s, independent of shear rate and concentration.

The transient dynamics of the process associated with the loss of the circular birefringence are best seen using a variation of configuration B2 in which the analyzer is rotated from its initial configuration. Once the linear properties have reached steady-state, the second harmonic intensity can be nulled by orienting the analyzer at an angle α. Under these conditions, it can be shown that the following will hold during flow start-up (*10*):

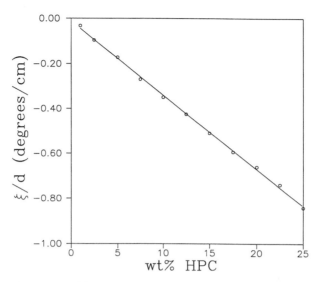

Figure 5. - Circular birefringence of quiescent HPC/AA solutions versus weight percent of polymer.

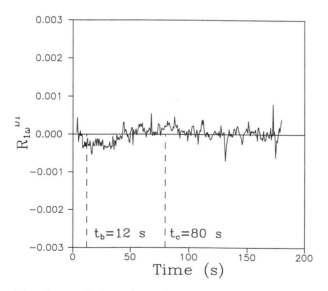

Figure 6. - First harmonic intensity ratio in configuration D1 versus time for 10 wt% HPC/AA solution at 7.55 s^{-1}. Start-up times for linear and circular birefringence are noted.

$$\frac{dR_{2\omega}^{B2}}{dt} = (C_1 \xi + C_2)\frac{d\xi}{dt} \tag{6}$$

where $C_1 = 2 \cos(2\alpha) \sin^2(\delta/2)/\delta^2$ and $C_2 = \sin(2\alpha) \sin(\delta)/\delta$. Thus, once δ reaches steady state, the start-up times and consequently the steady state value of ξ can be more accurately determined. Figure 8 shows the transient behavior of ξ so obtained. The start up time for ξ is very clear in this plot, and, for comparison, the start up time for δ is indicated as well. Once δ reaches steady state (about 12 seconds in the plot) the measurements of ξ become accurate. Extrapolation of the ξ curve back to the ordinate yields a value of $\xi_o = -0.05$ radians, which is close to that obtained in the quiescent experiments. This indicates that, above the critical shear rate, the optical rotatory power of the HPC molecules begins to change shortly after flow inception. Figure 9 shows the steady state values of ξ as a function of shear rate, calculated using the above described technique. Once past the critical shear rate, all the three higher concentration solutions display decreasing circular birefringence. These data suggest that the HPC molecules most probably untwist or uncoil under shear above a critical shear rate. This effect would probably have been noticed in the 2.5% solution had we been able to obtain measurements at high enough shear rates. These results give credence to the conclusion that the conformational twisting-untwisting is an intramolecular process and the optical rotation relaxes at a rate quite different from that of the linear properties.

Rheological Measurements

In-situ shear stress measurements were also made in our Couette viscometer. However, the torque transducer used for these measurements had a limited range, so that only a small portion of the shear rate range investigated optically could be examined rheologically. We extended these results using a Weissenberg R21 rheogoniometer to make measurements of the shear stress and the first normal-stress difference over the entire range of shear rates. Rheological experiments were performed at room temperature on the 5.0, 7.5, and 10.0% solutions using a cone (angle = 2.01°) and plate geometry with a diameter of 10.0 cm. The measured values of the shear viscosity were found to agree within 2% of the values determined using the Couette viscometer over the same shear-rate range (9). Power-law behavior was observed over the entire range of shear rates studied optically, with power-law indices close to 1. For all three solutions, the first normal-stress difference was negligibly small over the entire shear rate range, indicating the absence of elasticity in the system. Thus, for the conditions studied, the HPC solutions behave as Newtonian fluids. Moreover, as noted in the previous study, the transient response of the shear stress is essentially that of the linear birefringence. All of this implies that the rheological response of the system is governed by the molecular orientational mechanism, as opposed to the conformational rearrangement mechanism.

On the other hand, further useful information can be derived from the stress-optic behavior. Figure 10 shows the stress-optical ratios (S.O.R.) for the three HPC solutions. The S.O.R. is defined in terms of the shear stress, τ, and the linear birefringence in the usual fashion.

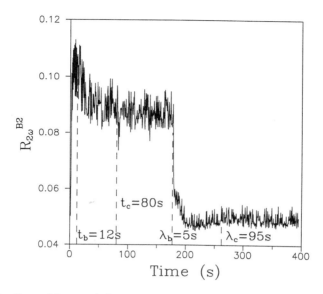

Figure 7. - Second harmonic intensity ratio in configuration B2 versus time for the same solution and conditions of Figure 6. Start-up and relaxation times are noted.

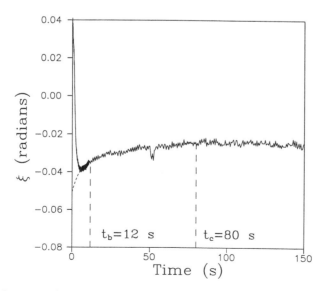

Figure 8. - Transient behavior of the optical rotation for the same solution and conditions of Figure 6. Optical configuration used was B2, with analyzer set at $\alpha = -41.77°$.

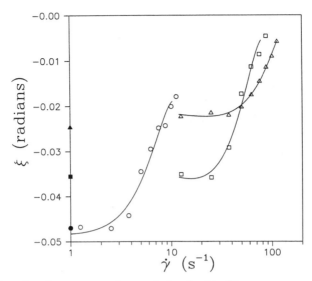

Figure 9. - Steady-state retardance of circular birefringence versus shear rate for the 10.0% (O), 7.5% (□), and 5.0% (△) HPC/AA solutions. The filled in symbols on the ordinate indicate the quiescent values.

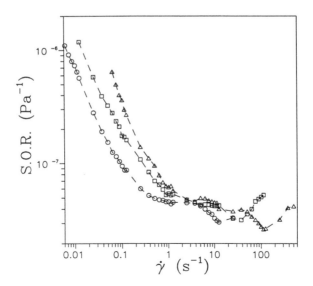

Figure 10. - Stress-optical ratio versus shear rate for HPC/AA solutions. Symbols are defined in Figure 9.

$$S.O.R. = \frac{\Delta n' \sin(2\chi_b)}{2\tau} \quad (7)$$

In the low shear rate region the S.O.R. decreases with shear rate in a more-or-less power law fashion. Mead and Larson (20) reported similar behavior for solutions of poly(γ-benzyl-L-glutamate) in m-cresol and attributed this phenomenon to the molecular weight dispersity. However, in contrast to their results, our S.O.R. measurements decrease with concentration at constant shear rate. The plateau region, where the S.O.R. is essentially constant with shear rate and concentration, corresponds to classical stress-optical behavior. Moreover, in this region, all molecules are affected by the shear and have achieved their maximum overall orientation. It is in this shear rate region that the changes in circular birefringence, discussed earlier, begin to occur. This offers further evidence that the dynamics of the latter process are separate from those of simple molecular orientation.

Conclusions

The examples discussed in this article illustrate the dynamic and steady-state rheo-optical behavior one sees in systems which exhibit supermolecular structure formation or conformational rearrangement under flow. In the first case, dilute solutions of flexible polystyrene molecules may exhibit the phenomenon of shear-thickening at high shear rates. Shear-thickening is accompanied by a maximum in the linear dichroism of the solution as well as increases in the turbidity which persist for longer times. These patterns can be understood in terms of the shear-induced formation of micron-sized molecular aggregates during flow. These aggregates undoubtedly serve as precursors for the more massive phase separation which has been visually observed in capillary flows (21). On the other hand, the linear birefringence of the solutions is dominated by molecular stretching and chain orientation which behave relatively independently of the large scale structuring. We believe the phenomena illustrated here arise from essentially the same physical processes seen in other light scattering and rheo-optical experiments on more concentrated systems (see reviews and references mentioned in the introduction).

Based on the results and discussion presented above, we believe that semi-rigid, helical systems, such as HPC/AA, can undergo independent processes of overall chain orientation and molecular conformational rearrangement, depending on the shear rate and concentration. Our results indicate that, below a critical shear rate, the rheooptical behavior of initially isotropic solutions is dominated by the linear birefringence and dichroism, suggesting that, while the overall molecular orientation increases with increased shear rate, the internal conformation of the HPC molecule remains essentially constant. Above a critical shear rate, the molecule begins to untwist from its quiescent and low shear conformation as noted by the decrease in its circular birefringence. The birefringence continues to rise in this shear rate range, indicating that the molecules untwist and form a more extended conformation. The time constant for the untwisting process is much larger than that for the molecular orientation mechanism. On cessation of flow, the molecules rapidly randomize their orientation and slowly retwist to their quiescent value, in a reversible fashion.

Acknowledgments

Portions of this work have been supported under a grant from the National Science Foundation (DMR 89-20538), administered through the Materials Research Laboratory at the University of Illinois. One of us (AJM) wishes to thank Dr. G. H. Meeten for pointing out the necessary conceptual correction to our interpretation of the ADA calculations.

Literature Cited

1. Rehage, H.; Wunderlich, I.; Hoffmann, H. *Prog. Coll. Polym. Sci.* **1986**, *72*, 51.
2. Mani, S.; Malone, M. F.; Winter, H. H. *Macromolecules* **1992**, *25*, 5671.
3. Yanase, H.; Moldenaers, P.; Mewis, J.; Abetz, V.; van Egmond, J.; Fuller, G. G. *Rheol. Acta* **1991**, *30*, 89.
4. van Egmond, J. W.; Werner, D. E.; Fuller, G. G. *J. Chem. Phys.* **1992**, *96*, 7742.
5. Wu, X.-L.; Pine, D. J.; Dixon, P. K. *Phys. Rev. Lett.* **1991**, *66*, 2408.
6. Hashimoto, T.; Fujioka, K. *J. Phys. Soc. Japan* **1991**, *60*, 356.
7. Vrahopoulou, E. P.; McHugh, A. J. *J. Non-Newtonian Mech.* **1987**, *25*, 157.
8. Kishbaugh, A. J.; McHugh, A. J. *Rheol. Acta.* **1993**, *32*, 9.
9. Edwards, B. J.; McHugh, A. J. *J. Rheol.* **1993**, *37*, 743.
10. Edwards, B. J.; McHugh, A. J.; Immaneni, A. *J.Rheol.* (in press).
11. Larson, R. G. *Rheol. Acta* **1992**, *31*, 497.
12. McHugh, A. J.; Edwards, B. J. in *Rheo-optical Techniques for the Characterization of Multiphase Polymer Systems*, Søndergaard, K.; Lyngaae-Jørgensen, J. Eds.; Technomic Publishing: Lancaster, PA, 1995; Chap. 5.
13. Kishbaugh, A. J.; McHugh, A. J. *Rheol. Acta.* **1993**, *32*, 115.
14. Our analysis attributes the scattering to optically isotropic spheroids. Meeten (*15*) has recently pointed out that ADA theory requires inherent optical anisotropy of the scattering center. This could be accommodated by assuming that the overall anisotropy is related to the inherent anisotropy of the partially ordered chain segments in the aggregate. A similar intrinsic diamagnetic susceptibility anisotropy was inferred for the micron-sized entities related to the magneto-optical birefringence behavior noted in ref. 16.
15. Meeten, G. H. *Rheol. Acta.* (submitted) and personal communication.
16. Meeten, G. H. *Polymer* **1974**, *15*, 187.
17. Kishbaugh, A .J. *Ph.D. Thesis*, University of Illinois: Urbana, IL, 1992.
18. Samuels, R. J. *J. Polym. Sci. Part A-2* **1969**, *7*, 197.
19. Ritcey, A. M.; Gray, D. K. *Biopolym.* **1988**, 27, 479.
20. Mead, D. W.; Larson, R. G. *Macromolecules* **1990**, *23*, 2524.
21. McHugh, A. J.; Spevacek, J. A. *J. Polym. Sci.: Part B:, Polym. Phys.* **1991**, *29*, 969.

RECEIVED January 25, 1995

Chapter 7

Changes of Macromolecular Chain Conformations Induced by Shear Flow

M. Zisenis[1], B. Prötzl[2], and J. Springer[2]

[1]Max Planck Institute for Colloid and Interface Research, 14513 Teltow, Germany
[2]Division of Macromolecular Chemistry, Institute of Technical Chemistry, Technical University of Berlin, 10623 Berlin, Germany

From light scattering measurements on polymer solutions subject to shear flow, the molecular orientation and all three main radii of the ellipsoidal gyration space of the polymer coils are determined by both variation of the wavelength of the incident light and the detection position. The observed coil deformability is much weaker than predicted by theory, while the shear induced orientation proves to be a complicated function of solution properties. On the basis of a series of investigations the specific contributions of hydrodynamic and thermodynamic interactions are discussed.

Subject to shear flow, a coiled macromolecule in solution will be both oriented and deformed if the acting shear force is strong enough. The knowledge of the exact behavior of the single polymer molecule in flow is a focus of scientific interest because new insights into polymer physics and an understanding of the complex viscous flow of polymer solutions are expected.

In principle, flow light scattering (FLS) and flow birefringence (FBR) are related methods for the direct determination of the molecular orientation. FBR, however, is restricted to polymer-solvent systems that show low depolarization in the quiescent solution, i.e. to systems that have no or low scattering intensity contrast. For FLS, exactly the opposite condition is required. Additionally, FBR yields no information about the molecular conformation except for an estimation of the change of the mean square end-to-end-distance, $<R^2>$, with flow, based on several assumptions (1). Basically, only scattering methods can give evidence of the molecular size and shape of macromolecules, and only FLS is able to give detailed information about each of the main gyration radii of a macromolecule in flow, with almost no assumptions.

Theory of Light Scattering on Macromolecules in Sheared Solution

The dimensions of macromolecules are usually characterized by the radius of gyration $<r_{gyr}^2>^{1/2}$ which can be determined by scattering methods. From a number of statistical calculations and Monte Carlo simulations (2-6), it is well understood that a random coil is not spherical, but has an ellipsoidal shape characterized by three main coil radii in the ratio of 1 to 1.73 to 3.74 (6). In quiescent solution, however, no information about the coil main radii is yielded because, due to the Brownian rotation of the molecules and the comparatively large experimental scattering volume, only an averaged spherical shape, the gyration space of the molecules, is observed.

Predictions about the orientation and deformation of polymer molecules in solution are made by the dynamic polymer models of Kuhn (elastic dumbbell model), Rouse (spring-bead model) and Zimm (spring-bead model including hydrodynamic interaction of the beads). These models are widely accepted, though important properties of polymer solutions such as the shear thinning behavior are not explained: While orientation causes a viscosity drop, the molecular extension raises the solution viscosity. These two effects are predicted to compensate exactly, in obvious contradiction to reality.

It was some decades ago, that Peterlin, Heller and Nakagaki (7-9) recognized that FLS should be a useful tool to investigate the conformation of macromolecules in shear flow, and developed methods to evaluate FLS measurements. Using these procedures it is possible to determine the orientation angle of polymer molecules in shear flow as well as all three independent main radii of the molecular gyration space, as a function of shear rate.

Orientation in Shear Flow. The above theories predict orientation and deformation of the macromolecules when the hydrodynamic shear force overcomes the Brownian motion, i.e. when the reduced shear rate, β, described by the molecular relaxation time τ and the shear rate $\dot{\gamma} = dv_x/dy$, becomes $\beta = \tau\dot{\gamma} \geq 1$. Taking the rheologically relevant rotatory relaxation time τ_r, the reduced shear rate for macromolecules in dilute solution is given by

$$\beta = \tau_r \dot{\gamma} = ([\eta] \eta_s M / RT) \dot{\gamma} \tag{1}$$

where $[\eta]$ is the intrinsic viscosity, η_s the solvent viscosity, M the molar mass, R the gas constant, and T the absolute temperature.

In principle, in a weak flow orientation and deformation are expected to occur in the flow plane only, which is defined by the direction of flow velocity, v_x, and shear rate, $\dot{\gamma}$. Both effects cause a change of the molecular gyration space from the spherical shape at rest to a shear dependent anisotropic form as shown in Figure 1: With increasing shear rate (I-IV) the main radius r_A grows while r_B diminishes, and r_C remains constant because no velocity gradient acts in that direction (10). In the given designation the rectangular coordinates A, B, C refer to the gyration space which is placed within the external coordinates x, y, z. The directions C and z are identical and normal to the flow plane which is defined by x and y. The angle between the longest main radius, r_A, and the direction of flow (x) is called the orientation angle, χ. In general, the variation of χ with shear rate is described by

$$\chi = 0.5 \arctan(m/\beta) \tag{2}$$

where the orientation resistance parameter, m, characterizes the influence of thermodynamic and hydrodynamic interactions in the given polymer-solvent system, and of coil conformation effects as well. The parameter m is numerically calculable for the orientation of the diffusion, stress and gyration tensors of polymers in dilute solution in the presence of steady shear flow (11,12). From Kuhn, m = 1 was obtained for the elastic dumbbell, and in agreement with FBR results, m = 2.5 and m = 4.83 were calculated for the stress tensor relating to the spring-bead models without (Rouse) and with strong (Zimm) hydrodynamic interaction, respectively. For the orientation of the gyration tensor, the limiting values of m = 1.75 (Rouse) and m = 2.55 (Zimm) were obtained (Bossart, J., Öttinger, H. C., ETH Zürich, preprint 1994). Finally, the gyration tensor will be the quantity corresponding to the gyrational dimensions observed by light scattering techniques.

Deformation in Shear Flow. For LS, determination of molecular dimensions is usually done using the Zimm approximation of the Debye scattering function:

$$P(q) = 1 - q^2 \langle S^2 \rangle + \ldots \tag{3}$$

for particles of any shape, with $q = (4\pi/\lambda) \cdot \sin(\Theta/2)$, the absolute value of the scattering vector $\vec{q}$, and $\langle S^2 \rangle$ being the mean square correlation length.

$$\langle S^2 \rangle = \frac{1}{2N^2} \sum_{k}^{N} \sum_{l}^{N} \langle r_{kl,q}^2 \rangle \tag{4}$$

In equation 4, N is the number of chain segments (k,l) and $\langle r_{kl,q}^2 \rangle$ represents the mean square projection of the segment distance vector $\vec{r}_{kl}$ on $\vec{q}$.

In quiescent solution the observable correlation length, $\langle S^2 \rangle^{1/2}$, is constant in all directions and the mean square radius of gyration, $\langle r_{gyr}^2 \rangle = 3\langle S^2 \rangle$, is determinable by use of the classical Zimm method (plotting P^{-1} vs. $\sin^2(\Theta/2)$, and extrapolation to zero concentration and $q \to 0$) where q changes through the variation of the detection angle, Θ, in the Zimm plane. The three main radii of the oriented ellipsoidal gyration space, however, cannot be determined simply by variation of the angle in the Zimm plane and evaluation as usual, because at each detection position a different correlation length determines the local scattering intensity. This leads to strongly curved

Zimm plots, and to rather uncertain extrapolations. Therefore, for the determination of the radius of gyration of oriented polymer coils in shear flow, all three main correlation lengths $<S_A^2>^{1/2}$, $<S_B^2>^{1/2}$ and $<S_C^2>^{1/2}$ of the ellipsoidal gyration space must be known since

$$<r_{gyr}^2> = <S_A^2> + <S_B^2> + <S_C^2> \tag{5}$$

According to Peterlin (7), the determination of the correlation lengths can be performed on the basis of the Zimm scattering function by variation of the incident wavelength while the detector position remains constant. Peterlin showed that the angular dissymmetry and the wavelength dependence of q are equivalent, and that one is permitted to determine the radius of gyration by measuring the scattering intensity as a function of the scattering angle (Θ) at a given wavelength or by measuring the scattering intensity as a function of wavelength at a constant angle of observation. Evaluation of LS data corresponding to equation 3 then gives the mean square correlation length, $<S^2>$, of the gyration space in the direction of the detection position.

It can be shown (7) that each value of $<S^2>$ is related to certain contributions of the three main correlation lengths according to:

$$<S^2> = <S_A^2>(\cos^2\frac{\Theta}{2}\cos^2\omega') + <S_B^2>(\cos^2\frac{\Theta}{2}\sin^2\omega') + <S_C^2>(\sin^2\frac{\Theta}{2}) \tag{6}$$

where $\omega' = \omega + \chi$ is the degree of orientation in respect to the detection position ω in the x-y plane. The theoretical predictions (13,14) for the deformation ratio of polymers in shear flow are:

$$\delta^2 = <r_{gyr}^2>/<r_{gyr,0}^2> \approx <R_e^2>/<R_{e,0}^2> = 1 + \frac{2}{3m}\beta^2 \tag{7}$$

where m is the previously mentioned orientation resistance parameter, $<r_{gyr,0}^2>$ is the mean square radius of gyration at rest, and $<R_e^2>$ and $<R_{e,0}^2>$ are the mean square end-to-end-distances in flow and at rest, respectively. In the regime of small deformations, the coil is assumed to be still Gaussian, i.e. the deformation ratios of $<R_e^2>$ and $<r_{gyr}^2>$ are approximately equal (14).

Fundamentals of Flow Light Scattering Measurements

Experimental Setup. The laser light scattering measurements are performed by use of a Springer-Wölfle rheophotometer (15, 16, Figure 2) consisting of a cylindrical shear cell with rotating inner cylinder (Searle-type arrangement). The small gap (width=1mm) between the inner and outer cylinders is traversed by a vertically polarized Ar-ion-laser beam parallel to the rotor axis, i.e. in the z or C direction. The laser light intensity is highly stable and focused by an achromatic lens to a diameter of approximately 0.5 mm. The irradiated power usually is about 100 mW. The scattered light leaves the shear cell through a spherical lens that is fixed onto the wall of the outer cylinder,

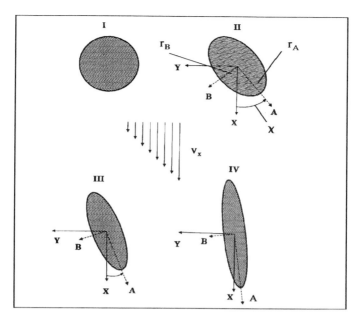

Figure 1. Theoretical estimation of the gyration space radii r_A and r_B of coil molecules with increasing shear flow (I-IV). v_x = flow velocity, χ = orientation angle.

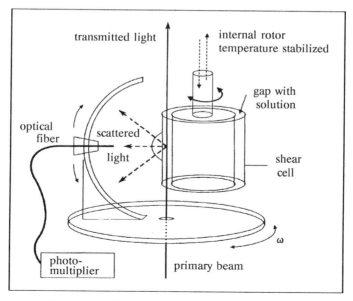

Figure 2. Scheme of the experimental setup.

and is detected by an optical fiber which can be moved horizontally and vertically around the scattering center.

In principle, the LS photometer not only allows lateral detection of the scattered light in one plane as usual, but in a hemi-spherical geometry. Consequently, various Zimm planes, i.e. planes which contain the primary beam, are accessible. This important difference to all comparable photometers used before (17) opens the access to the three dimensional LS-detection. The Zimm plane passing through $\omega = 90°$ complies with the conventially used detection geometry. Furthermore, the vertical polarization of the primary beam is always related to this plane. When changing the position of the Zimm plane, the polarization is steadily adjusted by rotating the laser's polarizer.

Through the Ar-ion-laser, a small spectrum of 5 lines (458 nm, 476 nm, 488 nm, 496 nm, and 514 nm) is available for λ. As the intensity of all of these wavelengths is quite different, after performing static LS on quiescent solutions the intensities were normalized to the results obtained for $\lambda = 514$ nm.

In order to investigate the shear dependent behavior of macromolecules under a variety of shear processes, the drive of the rotor is made by a step motor combined with computer driven electronic gearing. This arrangement enables time dependent measurements at constant shear rates (0.003 s$^{-1} \leq \dot\gamma \leq 3000$ s^{-1}) as well as stepwise or continuous increases or decreases of shear rate. Additionally, the rotor is temperature stabilized inside to an accuracy of 0.1 °C.

Experimental Preconditions. The LS measurements for the determination of molecular orientation and deformation of macromolecules in shear flow are restricted to some optical and rheological conditions. At first, the molecules must be large enough to enable multiple scattering and hence interference of the scattered light emitted by one molecule. This generates the characteristic intensity distribution in the flow plane from which the orientation is determined.

Generally (18), this condition will be met when $q <r_{gyr}^2>^{1/2} \geq 1$. For incident light of $\lambda = 514$ nm, $\theta = 90°$ (flow plane) and $n_s = 1.5$, i.e. when the refractive index of the solution, n_s, matches the refractive index of the glass outer cylinder, the minimum value of $<r_{gyr}^2>^{1/2} = 40$ nm. With this limiting quantity, the minimum required molar mass of polystyrene is about $M = 2 \cdot 10^6$ g/mol in a θ-solvent (trans-decalin) and $M = 7.5 \cdot 10^5$ g/mol in the very good solvent benzene (19).

From the rheological point of view, a sufficiently high molar mass and/or solvent viscosity is required to reach adequate values of β (see equation 1). The solution viscosity (η) and density (ρ) furthermore determine the range of shear rates accessible to the investigation, because after reaching a certain critical shear rate, $\dot\gamma_{crit}$, the stable laminar flow is superimposed by ring shaped vortices which change the orientational behavior. FLS is very sensitive to the appearance of this so-called Taylor vortex flow (14,15), which is due to centrifugal forces caused by the rotor. Through

$$\dot{\gamma}_{crit} = [41.2 \, (r_a/r_i - 1)^{3/2} \cdot r_i^2 \eta/\rho] \cdot r_a^2/r_a^2 - r_i^2 \qquad (8)$$

the critical shear rate can be calculated from η, ρ, and the inner (r_i) and outer (r_a) radii of the cylindrical gap. However, η is very often dependent on shear rate and concentration, while ρ is quasi independent of both parameters in dilute solution.

The concentration itself plays a further important role: In order to obtain molecular dimensions, the experiments have to be performed on dilute solutions and the concentration dependent results have to be extrapolated to zero concentration to match the theoretically treated state of the single unperturbed molecule. Therefore, a sufficiently high refractive index increment is required to obtain concentration dependent scattering.

Determination of Orientation. The theoretical scattering behavior of a polymer solution subject to shear flow was first shown by Peterlin (7,8). Numerical evaluation of his scattering functions based on the elastic dumbbell model and the Rouse model reveal a characteristic change of the scattered light distribution in the flow plane: In certain angular regions, the scattered intensity is increased, and at other angles decreasing intensity is predicted with increasing shear rate. Later, Nakagaki and Heller (9) calculated similar curves for rigid ellipsoids in shear flow. In both cases one gets sinusoidal curves showing a distinct maximum between $\omega = 45°$ and $\omega = 90°$. With increasing β, the location of the maximum (ω_{max}), which is predicted to coincide with the direction of the minor axis of the oriented coil (r_B), moves towards 90° while the amplitude increases.

With increasing shear rate, ω_{max} approaches 90° while the longitudinal axis (r_A), which is perpendicular to r_B in the flow plane, moves toward the direction of flow. Finally, the orientation angle is obtained from the maximum of the scattered light distribution according to

$$\chi = 90° - \omega_{max} \qquad (9)$$

In practice, the scattered light is detected in the flow plane in the range of $40° \leq \omega \leq 140°$ with respect to the flow direction, at first while the solution is at rest (I_0) and afterwards as a function of shear rate (I_β). For a higher sensitivity, the intensity distribution is then plotted in terms of the relative scattering intensity, $I_{rel} = (I_\beta - I_0)/I_0$, versus the detection angle, ω.

For comparison to theory, the orientation resistance parameter, m, is finally calculated from the corresponding χ and β data according to equation 2.

Determination of Deformation. For the determination of the molecular deformation according to Peterlin (7), three detector positions [ω, θ] (I: [140°,90°], II: [85°,90°], and III: [90°,120°]), were chosen. Though it should be irrelevant which (non-coplanar) positions are taken, the preferred positions were where the scattering vector points approximately at the gyration main axes. As a consequence, each of the three positions is dominated by one of the gyration radii. At each position the scattered intensity was measured at rest and under shear with all the wavelengths available by the Ar-ion-laser. The narrow range 458 nm $\leq \lambda \leq$ 514 nm causes uncertainty in the following Zimm extrapolation of the inverse reduced scattering intensity ($R_{(q)}$) over q^2 to $q^2 \to 0$. Therefore, the zero angle scattering in quiescent solution was determined for all wavelengths by the usual angle variation in the Zimm plane passing through $\omega = 90°$.

Since the osmotic pressure virial expansion is limited to two coefficients, which is reasonable in a near theta solvent, one can assume

$$Kc/R_{(q \to 0)} = 1/M_w + 2 A_2 c = \text{const.} \quad (10)$$

assuming the scattering constant, K, and the second virial coefficient, A_2, to be unaffected by the applied shear flow. Comparison of the results for $Kc/R_{(q \to 0)}$ obtained by angular variation on solutions at rest and under shear prove that this quantity remains constant within experimental accuracy.

From $Kc/R_{(q)}$ vs. q^2, obtained by wavelength variation, the quantity $<S^2>$ is finally determined for the three detection positions, and solving the set of three equations in the form of equation 6, the main correlation lengths $<S_A^2>$, $<S_B^2>$, and $<S_C^2>$ follow.

Measurements on High Molar Mass Polystyrenes

Until now a number of measurements on high molar mass polystyrene standards (10^6 g/mol $\leq M_w \leq 2 \cdot 10^7$ g/mol), most of them on the sample PS10 (M_w = 10.3·10^6 g/mol), were performed in a variety of good to poor solvents (*16,19*). The nominal M_w of the samples was confirmed within 10% by static LS.

Sample Preparation. For the preparation of samples for the FLS measurements all the solvents were cleared from dust by passing them through 0.2 μm filters. The solutions usually were also filtered (0.45 μm filters) or, when their viscosity was to high, exposed to centrifugation at 3000G for one hour. In some viscous solvents (dioctylphthalate = DOP and diethylphthalate = DEP), where stirring could damage the polymer molecules, they were softly shaken at increased temperature of about 50-55 °C for 2-4 weeks (especially in DOP) to ensure complete dissolution. To check for effects on FLS results due to shear degradation, the constancy of molar mass was checked by optical and viscometrical characterization before and after the scattering experiment.

Molecular Orientation. The number of FLS experiments (*16,19-21*) performed on polystyrene in different solvents prove conclusively that the shear induced orientation of macromolecules in solution is strongly dependent on the acting shear force. Moreover, for high molar mass polymers the orientational behavior clearly can be correlated with the viscometric behavior of the solutions. From the following series of measurements, conclusions relating to the shear dependent variation of χ, m, and $[\eta]$ are drawn, which are schematically illustrated in Figures 3 and 4:

Curve discussion. In general, the orientational behavior is divided into two phases, the onset of orientation (phase I) and the orientation dominated region (phase II). The critical shear rate, β_{crit}, relating to the transition from phase I to phase II, which is evident in m, compares quite well to the rate at which the shear thinning behavior of the solutions turns to the viscosity plateau.

In contradiction to theoretical predictions, in many measurements the data obtained in the orientation onset region (see m-curves in Figure 4) follow curves starting at the origin, describable by polynomials in the form of $m = k_1\beta \pm k_2\beta^2 \mp k_3\beta^3$. In poor solvents ($k_1, k_2 > 0$; $k_3 < 0$) this behavior is very pronounced and the observed orientation angle passes a maximum (curve a). With increasing solvent power β_{crit} also increases, while the onset behavior ($k_1, k_3 > 0$; $k_2 < 0$) becomes less pronounced (curve b). Finally, the curvature becomes rather hyperbolic and apparently a linear extrapolation of the data from regime II to the ordinate is revealed (curve c). The latter case, indeed, results in $\chi = 45°$ for $\beta \to 0$, as predicted by the basic theories.

Therefore, the intercept at zero shear rate, m_0, obtained by linear extrapolation of the data in regime II, is the quantity comparable to the theoretical values, i.e. $m_0 \equiv m_{(theory)}$. Furthermore, m_0 and the slope $dm/d\beta$ determined from region II are the characteristic numbers describing the influence of solution properties on the orientational behavior. In Figure 5, the observed variation of m_0 is shown as a function of the Mark-Houwink exponent, a, which generally describes the solution properties.

Discussion of Solution Properties. Basically, the Mark-Houwink exponent, a, represents the influence of both hydrodynamic (HI) and thermodynamic interactions on the solution properties of the macromolecule. In practice, the contribution of thermodynamic interaction is expressed by means of the excluded volume (EV) parameter, B, of the well-known Stockmayer-Fixman two-parameter-theory. The contribution of the HI can be described (22) through the interaction parameter $h^* = 8.87 \cdot 10^{-2} \cdot (a^{-2} - 1.10)$. For flexible polymer molecules, the Mark-Houwink exponent is between the limits of $a = 0.5$ at θ-temperature and $a = 0.775$ in very good solution, i.e. $0.25 \geq h^* \geq 0.05$. The theoretically lowest value $h^* = 0$ corresponds to $a = 1$ and the flexible polymer coil in the state of the Rouse theory.

Theory as well as the evaluation of the rheooptical measurements prove that m_0 declines with increasing solvent power. In theory, this decline solely is attributed to the variation of HI between the limits given by Zimm and Rouse. The examination of m_0 as a function of h^* and B, however, underlines the importance of the EV effect. The influence of HI is predominant only in

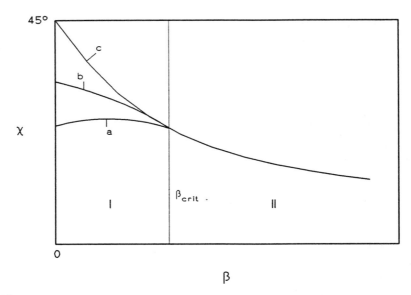

Figure 3. Schematic drawing of the variation of the orientation angle, χ, with increasing reduced shear rate β. a, b, and c = poor; half-good; good solvent, respectively.

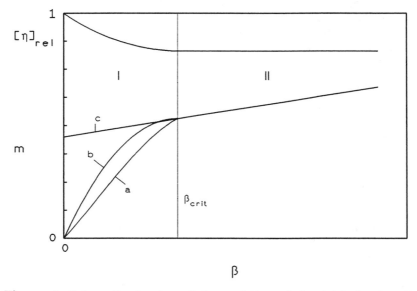

Figure 4. Schematic drawing of the variation of the intrinsic viscosity, $[\eta]$, and the orientation resistance parameter, m, with β. a, b, and c = poor; half-good; good solvent, respectively.

rather poor solvents ($0.5 \leq a \leq 0.65$) and with the assumptions $h_\theta^* = 0.25$ and $m_{0\theta} = 2.5$ (extrapolated value), the linear relation

$$m_0 = 2.5 - 2.25 \cdot (0.25 - h^*) \qquad (11)$$

is obtained for polystyrene (19). In extremely good solvents ($0.73 \leq a \leq 0.775$) the EV is predominant and HI is rather negligible, and the relation

$$m_0 = 1 + 0.04 \cdot (20 - B/10^{-28} cm^3 mol^2 g^{-2}) \qquad (12)$$

is revealed when the limiting values ($a \to 0.775$) $B = 20 \cdot 10^{-28}$ cm^3 mol^2 g^{-2} (calculated for PS10 in toluene) and $m_0 = 1$ are assumed. In the intermediate range, the orientation is both affected by HI and EV.

Change of Chain Conformations. From measurements at the three chosen detector positions (I: [140°,90°], II: [85°,90°], and III: [90°,120°]), the quantities $<S_A^2>$, $<S_B^2>$, $<S_C^2>$ and finally $<r_{gyr}^2>$ are calculated. In Figure 6 the interesting results for PS10 in the poor solvent DOP ($a = 0.54$) at 25 °C (16) are shown as an example. As predicted by theory, the longest correlation length $<S_A^2>^{1/2}$ increases with increasing shear rate while the short one in the flow plane ($<S_B^2>^{1/2}$) decreases and the third one ($<S_C^2>^{1/2}$), which is perpendicular to the flow plane, remains nearly constant.

Since completely undeformable molecules would also show a similar behavior for the three main radii, only the resulting mean radius of gyration indicates an actual deformation, if it exceeds its value at rest. For this purpose, in Figure 7 the gyration radius extension ratio $\delta^2 = <r_{gyr}^2>/<r_{gyr,0}^2>$ is plotted for some concentrations of PS10 in DOP as a function of shear rate. Here $<r_{gyr}^2>$ is the value determined at shear rate β, and $<r_{gyr,0}^2>$ is the value measured at rest ($\beta = 0$). The comparison to the theoretical results of the Kuhn, Rouse and Zimm theories (lines in Figure 7) reveals a large discrepancy between theoretical and experimental deformation ratio.

A striking feature of the experiment is the strong dependence on concentration. With $c = 2.85$ g/l, which is about 80% of the overlap concentration, the Zimm deformation ratio region is nearly reached, while the extrapolation to zero concentration yields:

$$<r_{gyr}^2>/<r_{gyr,0}^2> = 1 + (0.017 \pm 0.002) \cdot \beta^{1.40 \pm 0.05} \qquad (13)$$

Especially at higher shear rates, the data is far from the theoretical predictions. For the same polystyrene sample in the good solvent DEP ($a = 0.71$) the evaluation results in a stronger deformation than in DOP, described by

$$<r_{gyr}^2>/<r_{gyr,0}^2> = 1 + (0.032 \pm 0.004) \cdot \beta^{1.20 \pm 0.05} \qquad (14)$$

Generally, the observed deformability of the large polystyrene coils in shear flow is clearly weaker than theoretically predicted. Nevertheless, a scaling law in the form of $\delta^2 = 1 + A \cdot \beta^B$ is experimentally confirmed, but with the shear rate dependence smaller than β^2. Comparable results have been

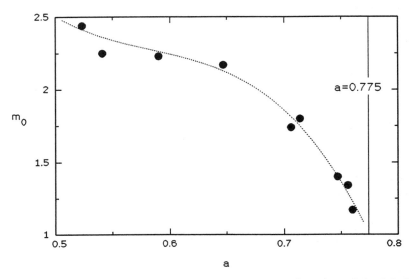

Figure 5. Orientation resistance parameter m_0 as a function of the Mark-Houwink exponent a.

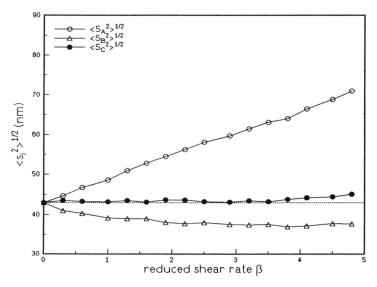

Figure 6. Main correlation lengths of PS10 in DOP at 25 °C, changing with shear rate (Adapted from ref. *16*.).

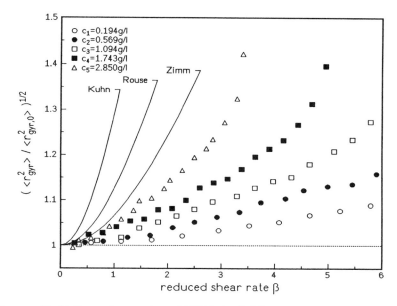

Figure 7. Deformation ratio of PS10 in DOP at 25 °C, changing with shear rate and in comparison to theoretical results (Adapted from ref. *16*.).

reported for LS-experiments on polystyrene in steady uniaxial elongational flow (*23*) recently.

Finally, it is obvious that FLS-experiments on polymer solutions subject to shear flow yield a variety of interesting results: The orientation angle and the orientation resistance parameter, as well as the main gyration axes and the change of macromolecular conformation are directly determinable from measurements on dilute solutions. The sensitivity of the FLS technique furthermore allows the investigation of secondary flow effects (*20*) and of shear induced aggregation and phase separation processes (*16*) that have not been discussed in this contribution.

Acknowledgments. The presented investigation is part of the special research program Sfb 335, 'Anisotropic Fluids', kindly supported by the Technical University of Berlin and the Deutsche Forschungsgemeinschaft.

Literature Cited

1. Peterlin, A. *J. Chem. Phys.* **1963**, *39*, 224
2. Kuhn, W. *Kolloid-Z.* **1934**, *68*, 2
3. Solc, K.; Stockmayer, W. H. *J. Chem. Phys.* **1971**, *54*, 2756
4. Solc, K. *J. Chem. Phys.* **1971**, *55*, 335
5. Bruns, W. *Coll. Polym. Sci.* **1976**, *254*, 325
6. Bruns, W. *Makromol. Chem., Theory Simul.* **1992**, *1*, 287

7. Peterlin, A. *J. Polym. Sci.* **1957**, *23*, 189
8. Peterlin, A; Heller, W.; Nakagaki, M. *J. Chem. Phys.* **1958**, *28*, 470
9. Nakagaki. M.; Heller, W. *J. Chem. Phys.* **1976**, *64*, 3797
10. Kuhn, W.; Kuhn, H. *Helv. Chim. Acta* **1944**, *28*, 97
11. Öttinger, H. C. *Phys. Review A* **1990**, *41*, 4413
12. Zylka, W.; Öttinger, H. C. *Macromolecules* **1991**, *24*, 484
13. Peterlin, A. *J. Chem. Phys.* **1963**, *39*, 224
14. Cottrell, F. R.; Merrill, E. W.; Smith, K. A. *J. Polym. Sci. A-2* **1969**, *7*, 1415
15. Wölfle, A.; Springer, J. *Coll. Polym. Sci.* **1984**, *262*, 876
16. Link, A.; Springer, J. *Macromolecules* **1993**, *26*, 464
17. Dupuis, D.; Layec, Y.; Wolff, C. In *Optical Properties of Polymers*; Meeten, G. H., Ed.; Elsevier: London, UK, 1986; pp 101-166
18. Nicholson, L. K.; Higgins, J. S.; Hayter, J. B. *Macromolecules* **1981**, *14*, 836
19. Zisenis, M. *Streulichtmessungen zum Studium der Orientierung von Hoch-Polymeren durch Scherströmung*; Verlag Köster: Berlin, D, 1994
20. Stock, H.; Zisenis, M.; Cleschinsky, D.; Springer, J. *Rheol. Acta* **1992**, *31*, 274
21. Zisenis, M.; Springer, J. *Polymer* **1994**, *35*, 3156
22. Hsu, Y. T.; Schümmer, P. *Rheol. Acta* **1983**, *22*, 12
23. Menasveta, M. J.; Hoagland, D. A. *Macromolecules* **1991**, *24*, 3427

RECEIVED February 25, 1995

Polymer Blends and Homopolymers

Chapter 8

Effect of Flow on Polymer–Polymer Miscibility

M. L. Fernandez and J. S. Higgins

Chemical Engineering Department, Imperial College of Science, Technology, and Medicine, London SW7 2BY, England

The effect of shear flow on the miscibility of three model polymer blends has been studied by one-dimensional and two-dimensional light scattering, differential scanning calorimetry and in-situ optical microscopy. Shear was applied between parallel plates and the range of shear rates used was from approximately $0.5 \, s^{-1}$ to $15 \, s^{-1}$. Shear was observed to induce both miscibility and phase separation depending on the temperature and shear rate used. Under certain conditions, a large wave-like pattern was observed by in-situ microscopy which gave rise to elongated streaks in the two-dimensional light scattering patterns. These "waves" appeared to be independent of phase separation and instead seem to be consistent with the presence of fluid instabilities.

The literature concerning the study of polymer polymer miscibility has become very large in recent years (*1, 2*). This is for reasons both of academic and of industrial significance. Blends of two or more polymers offer opportunities to produce in a mixture properties not available in a single species, but the underlying thermodynamics rest on a very delicate balance of terms difficult to predict from simple theory. Both prediction and experimental studies of miscibility under equilibrium conditions have been reported extensively (*3 - 7*), while the process of spinodal decomposition which follows when a blend is changed from the one to the two phase region of the phase diagram (usually by a sudden change in temperature) has also received considerable attention from theorists (*5, 8, 9*) and experimentalists (*10 - 13*). Given that in practice all such mixtures will be subject to shear and or extensional flow during processing the extension of these studies to the investigation of the interaction of rheology with thermodynamics was both inevitable and logical. In fact the first reported results concentrated on the effects of flow on polymer solutions (*14 - 16*) and it is relatively recently that data have been obtained on binary polymer blends. All the reported observations confirm that, if the thermodynamics of equilibrium systems was already complex and difficult to predict, the situation is even more confusing when the rheological history of the sample is involved. There are reports of both shear induced mixing and of shear induced demixing, sometimes in the same system (*15, 17 - 20, 22 - 28, 30, 31*). Recent theoretical approaches suggest that this latter observation could be a common occurrence, depending on the rheological behavior of the mixed system (*31,32*).

We have undertaken a systematic study of the effect of shear flow on the miscibility limits of a number of high molecular weight polymer blends exhibiting lower critical solution temperatures. Using a simple parallel plate shear cell and light scattering, as well as direct observation, we have observed large shifts in the cloud point curves as blends are heated *while shear is applied*. We have observed complex behavior of the cloud points, and have been able to show, using ancillary measurements of glass transition temperatures on quenched samples, that these effects are the results of changes in the miscibility of the blends and not artifacts of the scattering techniques. Much of the data has already been reported in the literature (*24, 30, 31, 33 - 35*) and we take the opportunity here to summarize the complex behavior observed and discuss the conclusions reached.

Experimental

The materials used in the present study were ethyl vinyl acetate copolymer (EVA) 45% by weight in vinyl acetate, solution chlorinated polyethylene (SCPE1 and SCPE2) 55% and 57% by weight in chlorine respectively, poly (butyl acrylate) (PBA), polystyrene (PS1 and PS2) and poly (vinyl methyl ether) (PVME1 and PVME2). Four different blends were prepared in the form of thin films by solvent casting and the film thicknesses were of the order of 0.15 - 0.30 mm. The characteristics of the blends and the individual polymers are shown in Table I. Details on the preparation procedure can be found elsewhere (*28, 30*).

Shear was applied between parallel glass plates. The bottom plate remained fixed while the top plate rotated at a user controlled velocity. The top plate was a disk, 5 cm in diameter with a notch on the side. The top plate was placed inside a gear with a pin that fits inside the notch in the plate. The gear was connected through a series of rods and a gear box to a motor which can impart angular velocities to the top plate from 0.01 rad s^{-1} to 0.47 rad s^{-1}. For the measurements reported here, the range of angular velocities chosen is from approximately 0.08 rad s^{-1} to 0.26 rad s^{-1}. The assembly formed by the thin film "sandwiched" between the parallel plates was kept inside a copper block which could be heated, cooled or kept at a constant temperature. The copper block had two holes, one at the bottom to allow light to impinge on the sample and one on the top to allow the scattered radiation to pass through. In a rotating parallel plate shear device the shear rate varies across the radius, it is zero at the center and maximum at the edge of the plate. In our set-up, the laser beam impinged on the sample at an intermediate radial position (11.5 mm from the center) and, except where otherwise stated, the shear rates quoted here are from the position of the laser beam. The entire assembly was mounted on a freely rotating shaft with an extended arm which rested against a calibrated leaf spring. Measurements of torque were possible by measuring the displacement of the arm by means of a displacement transducer and this signal was recorded by a computer. Further details on the experimental set-up can be found elsewhere (28, 30). Four types of measurements were carried out.

1.- One-dimensional light scattering: The sample assembly was placed in the path of a 5 mW He-Ne Aerotech laser beam. The scattered radiation was detected by a linear array of 32 photodiodes which covered an angular range of 64° and which were oriented in the direction perpendicular to the applied shear flow. The readings of each of the photodiodes were collected by a computer as a function of time with the help of an A/D converter card. The signals of the photodiodes were normalised by accounting for the different response of each of the photodiodes, which had been calibrated by measuring the scattering from an isotropic scatterer.

2.- Two-dimensional light scattering: The light source was the same as in the case of the one-dimensional light scattering measurements but in this case, instead of using the photodiode array, the scattered radiation was projected on a screen and photographed with a Nikon camera. In this way, two-dimensional light scattering patterns were obtained, examples of which will be shown in this paper (see Figure 3). The white regions in the patterns correspond to regions where the scattered intensity is maximum and the black regions correspond to areas where there is little or no scattering.

Table I. Characteristics of the polymers and blends used in this work

Code	Solvent	Polymer	% in blend (by weight)	M_w (kg/mol)	M_n (kg/mol)	T_{cl}[‡] (°C)	T_g[†] (°C)
1	THF[1]	EVA	40	106	38.7	118	-26
		SCPE1	60	332	62.7		47
2	MEK[2]	PBA	50	75.5	27.4	109	-44
		SCPE2	50	274	44.3		46
3a	toluene	PS1	30	330	128	94	107
		PVME1	70	89	~ 20		-30
3b	toluene	PS2	30	96	92	131	107
		PVME2	70	95	38		-28

[1] tetrahydrofuran
[2] methyl ethyl ketone
† glass transition temperature
‡ cloud point temperature for the blend in the quiescent state.

3.- In-situ video microscopy: The sample was illuminated by a halogen lamp with the help of a collector lens. The image was magnified with a x20 Nikon objective which was attached to a JVC CCD video camera. In this way, it was possible to observe in-situ, the structure of the blend under shear flow. The real-time image was collected in a video tape, stills of which will be shown in this paper (see Figure 4).

4.- Differential scanning calorimetry (DSC): DSC measurements were carried out on a Perkin Elmer DSC2 instrument and the heating rate used was 20°C/min. Measurements were carried out on quenched samples. After a light-scattering or a video microscopy measurement the sample assembly formed by the film and the top and bottom glass plates could be quickly released from the instrument and quenched in iced water. The glass plates were subsequently removed by placing them over liquid nitrogen (the samples were not immersed in liquid nitrogen directly to prevent the glass plates from shattering). The required amount of material was removed from the appropriate regions in the sample and encapsulated in DSC pans ready for measurement.

One-dimensional Light Scattering Measurements.

These measurements were carried out in order to determine the phase behavior of the model polymer blends in Table I with and without shear (*28*). The procedure used was as follows: The samples were equilibrated at a temperature inside the one-phase region of the phase diagram. The temperature was then increased at a rate of 1°C/min and shear was applied to the sample shortly after starting to increase the temperature (the shear rate was kept constant throughout the experiment). In this way "cloud point curves" were obtained and the cloud point temperature was taken at the point at which the scattered intensity started to deviate from the value in the one-phase region. The same procedure was followed to measure the cloud point in the quiescent state but in this case no shear was applied to the sample. All the blends studied show LCST (lower critical solution temperature) phase behavior, i.e. they phase separate on heating. The quiescent cloud point temperatures for the systems and compositions given here have been shown in Table I and, as mentioned in the text, have been measured by one-dimensional light scattering.

For all the blends studied it was observed that the effect of shear was to shift the cloud point temperature. Figure 1 shows the observed shift in the cloud point $(T(\dot{\gamma}) - T(0))$ as a function of the applied shear rate for Blends 1, 2, 3a and 3b where $T(\dot{\gamma})$ is the cloud point for the system at a shear rate $\dot{\gamma}$ and $T(0)$ is the quiescent cloud point. In all cases the shift in the cloud point is negative indicating that the effect of shear is to extend the two-phase region by shifting the coexistence curve towards lower temperatures and therefore to induce demixing. The magnitude of the shift was found to depend on the blend type and composition. For Blends 1 and 2, the compositions shown in Figure 1 and Table I are those for which the shift in the cloud point was found to be maximum. Blend 3a was not studied as a function of the composition. It might be of significance that (at least for the systems studied here) the larger the mismatch in the viscosities of the blend components, the larger the shear-induced demixing effect. For Blend 1, the ratio of viscosities $\eta_{SCPE1} / \eta_{EVA} = 13$ and the maximum shift observed in the cloud point was $T(\dot{\gamma}) - T(0) \sim 5°C$. For Blend 2, $\eta_{SCPE2} / \eta_{PBA}$

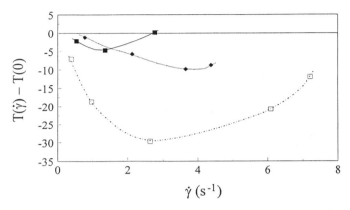

Figure 1.- Shift in the cloud point ($T(\dot{\gamma}) - T(0)$) as a function of the applied shear rate for Blends 1 (■), 2 (♦) and 3 (□) in Table I. (Adapted from ref. 28).

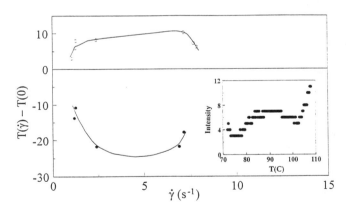

Figure 2.- Shift in the cloud point ($T(\dot{\gamma}) - T(0)$) for Blend 3a obtained starting the shear at low temperatures (ca. 70°C) (●) and starting the shear at high temperatures (ca. 90°C) (○). Inset: Scattered intensity as a function of temperature. (Adapted from refs. 30 and 31).

= 64000 and the maximum shift observed was $T(\dot{\gamma}) - T(0) \sim 10°C$. Finally, for Blend 3a which is the system for which the viscosities of the two components are more mismatched, $\eta_{PS1} / \eta_{PVME1} = \infty$, the shift $T(\dot{\gamma}) - T(0) \sim 30°C$. In all cases the viscosity has been measured at 95°C and at a shear rate $\dot{\gamma} = 0.8$ s^{-1}. The viscosity of PS1 has been taken as ∞ because 95°C is below the glass transition temperature of polystyrene. The temperature (95°C) and the shear rate (0.8 s^{-1}) for the viscosity measurements reported in Table I were chosen because they fell within the range of conditions used in the cloud point measurements of Figure 1. The viscosity however changes with temperature and therefore the polymer viscosities at each of the cloud points in Figure I will be different to those at 95°C although the temperatures covered in the cloud point measurements were not much higher than 95°C and therefore the viscosities will not change very much. In addition, when the polymer blends phase separate they will not form two phases of pure components but two phases rich in different components and therefore the viscosities of the two phases will be different to those from the pure polymers. The relationship between the viscosities of individual polymers and their phase behavior under shear is therefore complicated. Although the argument that we have presented on the relationship between the maximum shift in the cloud point under shear and the viscosities of the individual components is simplistic, the general trend of viscosity ratios is still expected to be similar to that shown in Table I and the observation that we have made could be a real effect. Clearly this point would require more measurements.

The effect of blend composition was studied for Blends 1 and 2 up to higher shear rates than those reported here (*28*). For the case of Blend 1 and for the lower shear rates, shear induced demixing was observed as shown in Figure 1. As the shear rate was increased beyond the range shown in the figure, there was a cross-over to shear induced mixing. The shear rate at which the effect of shear crossed from inducing demixing to induce mixing was similar for all the compositions investigated. This could be a significant fact or a coincidence which we cannot explain at present. Blend 2 seemed to follow the same trend as Blend 1 although the range of shear rates investigated was not large enough to observe the cross-over to shear induced mixing. The effect of blend composition was not studied for Blend 3a.

Later measurements on Blend 3a (*30, 31*) carried out over a wider temperature range showed a more complicated phase behavior. Measurements were carried out as explained earlier. Homogeneous films (in the one-phase region) were prepared and placed inside the shear apparatus. The temperature was ramped at 1°C/min and the shear was started shortly after starting increasing the temperature. When shear was started at low temperatures (ca. 70°C) shear-induced demixing similar to that in Figure 1 was observed. If the shear was however started from higher temperatures (ca. 90°C) but still in the one-phase region, shear-induced demixing was not observed and shear-induced mixing was observed instead, for the same range of shear rates. For each shear rate there were therefore two cloud point temperatures as shown in Figure 2, depending on whether the shear was started from low or from high temperatures. In the figure, the full circles correspond to the shift in the cloud point determined at low temperatures (shear-induced demixing) and the open circles correspond to the cloud point determined at higher temperatures (shear - induced mixing). In a few cases it was possible to detect both cloud points in a single measurement and an example is shown in the inset of Figure 2. The inset represents a plot of scattered intensity as a function of temperature. The scattered intensity increased from ~ 75°C (first cloud point temperature) to ~ 90°C, it then decreased until ~ 100°C and then the intensity increased for a second time from approximately 103°C (second cloud point temperature). These results were also confirmed visually by observing that, as the temperature was increased at a constant shear rate, the samples were first clear, then cloudy (after the first cloud point), then clear and then cloudy again (after the second cloud point).

These results seem to agree very well with the theoretical predictions of Horst et al (31, 32). These authors calculate the coexistence curves under shear by adding to the quiescent Gibbs free energy of mixing, a term that accounts for the elastic stored energy. Under certain conditions of temperature and shear rate, the authors predict the existence of closed immiscibility gaps as well as conventional LCST (lower critical solution temperature) curves. The immiscibility "islands" are situated at temperatures lower than the LCST curve but their position changes with temperature and shear rate and in some instances they join the LCST curve. These predictions of the existence under shear of closed immiscibility islands at temperatures lower than the LCST curves, are in agreement with our observations, at least qualitatively.

According to this theoretical treatment, if a measurement were carried out at a constant shear rate and increasing the temperature across one of the predicted immiscibility islands, one would start with a clear and homogeneous sample which would become cloudy as the temperature is raised to that inside the island. As the temperature is subsequently increased the sample would "leave" the island and therefore would remix and become clear again. If the temperature were increased again, the sample would cross the second region of immiscibility and phase separate again. These are the phenomena that we have observed in our scattering measurements. The measurements in which shear is applied from low temperatures will detect the immiscibility island while those in which shear is started at higher temperatures could "miss" the island altogether. Whether these two regions of immiscibility can be detected in a single experiment would depend on how separate in temperature the two regions are (i.e., on thermodynamics) but also on kinetics. Since the measurements are carried out by continually increasing the temperature, if the extent of phase separation in the first region is large, the system might not have enough time to remix after leaving the "island" and therefore, whether the two regimes are observed in a single experiment or not could depend on the heating rate used.

Other authors have investigated the effect of shear on the miscibility behavior of the blend PS/PVME and they have reported either shear-induced miscibility (15, 17-20, 22, 23, 25, 27, 30, 31) or shear-induced immiscibility (20, 24, 26, 28, 30, 31). To our knowledge nobody has reported both effects on the same system and this could be a result of the experimental procedure followed. Most authors carry out experiments by placing a sample inside the two-phase region of the phase diagram, they then increase the shear rate and observe the change in the blend miscibility. By following this procedure it would be difficult to detect immiscibility islands since the size and position of these vary strongly with the shear rate. In our measurements we use a constant shear rate and we vary the temperature. We use the same procedure as conventional studies on miscibility systems and we always start from stable, homogeneous conditions, i.e. blends in the one-phase region. Following this procedure it is easier to detect the immiscibility islands, although it is still difficult to measure the lower and upper temperature limits of these islands in a single experiment.

Two-dimensional Light Scattering Measurements.

Two-dimensional light scattering measurements were carried out on films of Blend 3b. The quiescent cloud point temperature was 131°C as determined by one-dimensional light scattering (see Table I). Typical results are shown in Figure 3 which correspond to the scattering patterns obtained at three different temperatures and at the same shear rate $\dot{\gamma} = 3$ s^{-1}. These measurements were taken by equilibrating the samples either at 126°C (Fig. 3a), at 132°C (Fig. 3b) or 136°C (Fig. 3c) and then applying a shear rate of 3 s^{-1}. These measurements were therefore carried out at a constant temperature and changing the shear rate although the same patterns were obtained by doing measurements at a constant shear rate and increasing the temperature. The patterns in Figure 3a were taken at a temperature of 126°C. One-dimensional light scattering measurements indicate that at this temperature the system is in the one phase region both with and without shear (for a shear rate $\dot{\gamma} = 3$ s^{-1}). Before shear is started there is no scattering (only a small amount near the beam-stop) but as soon as shear is applied to the sample a clear streak pattern appears. The streak is elongated in the direction perpendicular to the applied flow which indicates the presence in the sample of anisotropic scattering entities, with the larger dimension oriented in the direction of the applied flow.

Figure 3b corresponds to the scattering patterns obtained at 132°C. At this temperature one-dimensional light scattering measurements indicate that the effect of shear is to induce miscibility and therefore the system is two-phase with no shear but one-phase under shear (for a shear rate $\dot{\gamma} = 3$ s^{-1}). When no shear is applied to the sample, an isotropic scattering pattern is observed which indicates the presence of isotropic scattering entities in the sample. As shear is applied the small angle signal disappears gradually and a streak-like pattern similar to that in Figure 3a appears.

Figure 3c corresponds to the scattering patterns obtained at 136°C. One-dimensional light scattering measurements indicate that at this temperature the system is two-phase with and without shear (for a shear rate $\dot{\gamma} = 3$ s^{-1}). When no shear is applied, an isotropic scattering pattern is observed, similarly to the case of Figure 3b although in this case the scattering is more intense because the compositions of the coexisting phases are further apart and therefore the scattering contrast is higher. As shear is applied, the small angle scattering decreases only slightly and superimposed on this there is a streak pattern similar to that of figures 3a and 3b.

The behavior of the small angle scattering is consistent with our one-dimensional light scattering results. At 126°C there is no small angle scattering

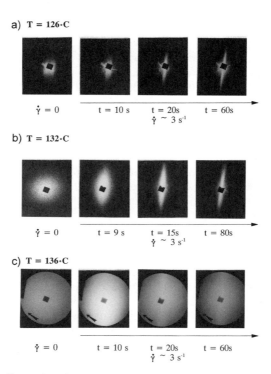

Figure 3.- Two-dimensional light scattering patterns obtained for a sample of Blend 3b for the conditions of temperature and shear rate indicated in the figure. The times indicated correspond to the time elapsed after the start of the shear. The direction of the applied flow is horizontal in the figure. (Reproduced with permission from ref. 35. Copyright 1995, Huethig & Wepf)

which is consistent with the system being one-phase either with or without shear (for a shear rate of 3 s^{-1}). At 132°C and under no shear the system is phase separated (quiescent cloud point = 131°C) which gives rise to the small angle scattering signal shown in the first picture in Figure 3b. As a shear rate of 3 s^{-1} is applied, this small angle scattering signal disappears indicating that with shear the sample remixes and becomes homogeneous. This is consistent with our one-dimensional light scattering results which indicate that the quiescent blend is two-phase and the effect of a shear rate of 3 s^{-1} is to induce miscibility. At 136°C there is a strong small angle scattering signal both with and without shear (for a shear rate of 3 s^{-1}). This is consistent with our one-dimensional light scattering observations which indicate that the quiescent system is two-phase and a shear rate of 3 s^{-1} has no effect on the miscibility behavior. The appearance of the elongated streak pattern is an unexpected phenomenon which will be explored further in the next sections. Our one-dimensional light scattering measurements detected the small angle scattering signals but not the streaks. This was because the streaks are very narrow, extend over a relatively small angular range and are oriented slightly off the direction perpendicular to the applied shear flow. Since the photodiode array is formed by discrete diodes and these are oriented in the direction perpendicular to the applied flow, it is easy to miss the signal from the streak. In addition, since the streaks extend over a relatively small angular range, they would only be detected by a few diodes near the beam stop.

In-situ Video Microscopy Measurements

Video microscopy measurements were carried out (*33, 34*) on samples similar to those in the previous section. Measurements were carried out at a constant temperature and shear rate as for the two-dimensional light scattering experiments. However, measurements carried out at a constant shear rate and ramping the temperature gave similar results. In the quiescent state, samples at temperatures below 131°C were clear and showed no structure. At higher temperatures, a spinodally decomposed phase separated structure was observed with size of the order of 1 μm. The structure was not resolved clearly because of the relatively poor resolution of the camera but it was easily identified as phase separated. As phase separation proceeded, the structure did not seem to change visibly but the samples became first cloudy and then white and the micrographs appeared considerably darker than those in the one-phase region or during the first stages of phase separation.

Under shear, for temperatures lower than approximately 126°C the micrographs were clear and showed no structure (at least within the resolution of the instrument). At approximately 126°C, structures similar to those in Figure 4a were observed. They resemble waves elongated just slightly off (a few degrees) the direction of the applied shear flow and their magnitude is approximately 100 μm x 6 μm. When observed dynamically, they resemble waves similar to a water surface seen from a height (eg. an airplane), they show crests and dips and seem to move not only along the flow but also up and down in the plane perpendicular to the film. The waves appear very quickly, approximately 5 to 10 seconds after the application of the shear. When the samples were viewed at this temperature under the naked eye they appeared perfectly clear and homogeneous.

At 132°C the samples in the quiescent state showed a spinodally decomposed phase separated structure. As shear was applied, the micron-sized structure disappeared, the micrographs became lighter and waves similar to those at 126°C appeared. With shear, the samples appeared clear and homogeneous to the naked eye.

At 136°C the samples in the quiescent state showed a spinodally decomposed phase separated structure. The colour of the micrographs was quite dark since at this temperature phase separation takes place very quickly. As shear was applied, the micron-sized structure stayed, the micrographs remained dark and waves similar to those at 126°C appeared superimposed on the smaller phase separated structure, as shown in Figure 4b. The samples appeared very cloudy to the naked eye.

At shear rates of the order of 3 s^{-1}, the waves were oriented slightly off the direction of the applied shear but at higher shear rates (10 s^{-1} to 16 s^{-1}) they rotated slightly and were parallel to the flow direction. The same phenomenon was observed for the streaks in the two-dimensional light scattering patterns. For the lower shear rates the streaks were slightly off the direction perpendicular to the applied shear flow. If the shear rate was increased, the streaks rotated so that they were perpendicular to the flow direction. If the shear rate was decreased again, the streaks rotated back to their original direction.

It was impossible to carry out video microscopy and scattering experiments simultaneously, however it is highly likely that the streaks observed by two-dimensional light scattering arise from the waves observed by video microscopy. The size, orientation and conditions under which the scattering streaks appear are consistent with the video microscopy observations.

We argue that, in our experiments, the waves and corresponding streaks are not due to phase separation but to some other effect such as fluid instabilities in the films, and there are two main reasons for this.

1.- As opposed to other authors (*40 - 42*), we have carried out measurements not only by applying shear to phase separated samples but also by shearing samples as they are heated from the one-phase region. An example of this type of experiments is the one carried out at 126°C. At this temperature without shear the sample is one-phase and homogeneous, yet within 5 to 10 seconds of applying shear we observe elongated structures of approximately 100μm x 6μm. If these structures were phase separated domains, the diffusion coefficient necessary would have to be approximately 5 to 7 orders of magnitude higher than that observed for the samples in the quiescent state.

2.- The waves, specially when seen dynamically do not look like phase separation. As mentioned earlier they look and move like real fluid waves with dips and crests. To our knowledge no other author has observed these phenomena dynamically but on quenched samples and therefore we cannot compare the different observations.

Supporting evidence for the idea that the waves are different from phase separation is provided by DSC measurements on quenched samples.

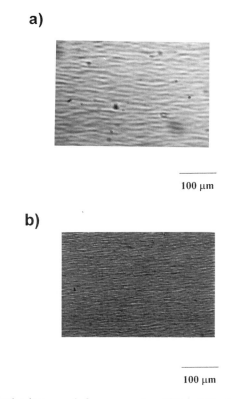

Figure 4.- a) Optical micrograph for a sample of Blend 3b sheared at 126° C and at a shear rate $\dot{\gamma} = 3$ s^{-1}. b) Optical micrograph for a sample of Blend 3b sheared at 136° C and at a shear rate $\dot{\gamma} = 3$ s^{-1}. The direction of the applied flow is horizontal in both figures. (Reproduced with permission from ref. 34. Copyright 1994, M.L. Fernandez).

DSC measurements on quenched samples

Differential scanning calorimetry measurements were carried out (*33*) on samples similar to the ones used in the previous section. A sample was sheared at 1.5 s^{-1} at 128°C and showed waves identical to those at 126°C in the previous section (see Figure 4a). The sample appeared clear to the naked eye and under the video microscope (except for the waves in the latter). After the video microscopy measurements, the sample was quickly released from the cell and quenched in iced water. The sample was then observed under a conventional optical phase contrast microscope to make sure that the wavelike structure had been preserved by the quenching procedure. The glass plates were then removed, material was taken (from the part of the sample which had been observed by microscopy) and encapsulated in a DSC pan. The thermograph obtained is shown in Figure 5a and it clearly shows a single glass transition at -15.4°C which could indicate that the system is single-phase.

A similar procedure was followed for a sample sheared at 2 s^{-1} at 134°C. In this case waves were observed superimposed on a smaller scale phase separation (see Figure 4b). The DSC thermograph obtained for this sample is shown by the bottom curve in Figure 5b and it clearly shows two glass transitions, one at -17.4°C and a second one at 44.4°C, which indicate that the system is two-phase.

From our measurements therefore it appears that the observed elongated structures are independent of phase separation. These structures appear to be similar to those reported by other authors (*40 - 42*) and one of the important questions is whether we are seeing the same effect or whether we are seeing two different effects which happen to look similar. In our case for the reasons given, we believe that these structures are not phase separated domains and they could instead arise because of fluid instabilities.

In a previous publication (*33*) we have explored this point in more detail and we have considered two fluid instabilities which could explain the observed phenomena. One is the Taylor-Saffman instability (*36*) which has been modified by Fields and Ashby (*37, 38*) for the case of a fluid between two walls, one of which is being peeled off. High viscosities of the fluid or high peeling speeds tend to enhance the instability while a high surface tension will tend to suppress it. Calculations carried out for our system give the right order of magnitude for the size of the waves. A second instability that we have considered is the Kelvin-Helmholtz (*36*) which appears when two fluids are superposed and moving at different velocities. In this case, a wave-like structure can be set up which resembles our observations. Calculations carried out for our system are also of the right order of magnitude for the waves to arise because of this instability. In both cases the instability could extend to the overall thickness of the film or be a surface effect. We believe that preferential surface segregation (*39*) of the PVME component of the blend could play an important role and indeed, the instability could be set off at the interface between a PVME-enriched surface layer and the rest of the film. These two fluid instabilities are just two examples of phenomena that could explain our observations but perhaps there are other instabilities or combinations of instabilities that could be considered (*40 - 42*).

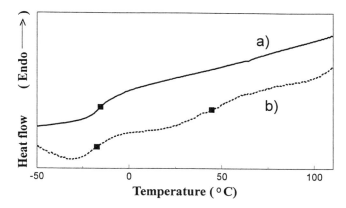

Figure 5.- a) DSC thermograph obtained for a sample of Blend 3b whose structure was similar to that of Figure 4a. The conditions of temperature and shear rate are indicated in the text. (Adapted from ref. 33) b) DSC thermograph obtained for a sample of Blend 3b whose structure was similar to that of Figure 4b. The conditions of temperature and shear rate are indicated in the text. (Adapted from ref. 33).

It is unlikely that effects such as thermal gradients in the samples or misalignment of the parallel plates are responsible for the effects observed because in this case we would also have seen the wave-like structures at temperatures below 126°C, or for the case of single polymers. However, the wavelike structure was only observed in a limited range of temperatures and they were not detected for pure PVME for the wide shear rate and temperature range investigated. The thickness of the samples used is relatively small and it is possible that our observations are restricted to the surface. We have not carried out measurements on thicker samples because of the great difficulty in producing thick films by solvent casting which are flat and solvent-free. Clearly, the effect of sample thickness has to be investigated in order to ascertain whether the waves observed are surface or bulk effects. In any case, the phenomena observed are of crucial importance not only for the study of the effect of shear on miscibility but also for general studies of polymer blend processing or adhesion.

As already mentioned, other authors (40 - 42) have reported the existence of elongated structures and of scattering streaks in sheared polymer samples which are very similar to the ones that we have observed. These authors identify the elongated structures with phase separated domains which are elongated by the shear and this is a very reasonable and plausible explanation for their observations. In general, authors start by placing the blend at a temperature at which the system is phase separated and then apply shear while we have observed the "waves" even at temperatures at which the blend is in the one-phase region. The elongated structures that other authors observe (40 - 42) are smaller than the ones reported here. In addition, there are reports of two-dimensional light scattering measurements in which the evolution of the spinodal ring or of concentration fluctuations is followed as shear is applied while in our observations, the scattering streak seems to be superimposed on the small angle scattering signal. It is possible that the phenomenon that we are observing has a different origin to that reported by other authors (*40 - 42*). It is difficult to make comparisons between the data collected by different groups because authors have looked at different systems, different molecular weights and molecular weight distributions and have followed different experimental protocols. The problem must be resolved and while it presents many complications, it is a new and worthwhile field and it is certainly very exciting.

Acknowledgments: The authors would like to thank EPSRC for the financial support received.

Literature Cited

1. Olabisi, O.; Robeson, L.M. and Shaw, M.T. *Polymer-polymer miscibility*, Academic Press, London 1978.
2. Utracki, L.A. *Polymer alloys and blends*, Henser Publications, New York, 1989.
3. Flory, P. *Principles of polymer chemistry*, Cornell Univ., Ithaca, New York, 1971, Chapter XI.
4. Huggins, M.L. *J. Am. Chem. Soc.* **1942**, *64*, 1712.
5. Binder, K. *Adv. in Polym. Sci.* **1994**, *112*, 181.
6. Bates, F.S.; Muthukumar, M.; Wignall, G.D. and Fetters, L.J. *J. Chem. Phys.* **1988**, *89*, 535.
7. Okada, M. and Han, C.C. *J. Chem. Phys.* **1986**, *85*, 5317.
8. Cahn, J.W. *J. Chem. Phys.* **1965**, *42*, 93.
9. Binder, K. *J. Chem. Phys.* **1983**, *79*, 6387.
10. Hashimoto, T.; Kumaki, J. and Kawai, H. *Macromolecules* **1983**, *16*, 641.
11. Snyder, H.L.; Meakin, P. and Reich, S. *Macromolecules* **1983**, *16*, 757.
12. Fernandez, M.L.; Higgins, J.S. and Tomlins, P.E. *Polymer* **1989**, *30*, 3.
13. Guo, W. and Higgins, J.S. *Polymer* **1991**, *31*, 699.
14. Rangel-Nafaile, C.; Metzner, A. and Wissburn, K. *Macromolecules* **1984**, *17*, 1187.
15. Hindawi, I.A. PhD Thesis, Imperial College, London, **1991**.
16. Kramer-Lucas, H.; Schenck, H. and Wolf, B.A. *Makromol. Chem.* **1988**, *189*, 1613 and 1627.
17. Mazich, K.A. and Carr, S.H. *J. Appl. Phys.* **1983**, *54*, 5511.
18. Rector, L.P.; Mazick, K.A. and Carr, S.H. *J. Macromol. Sci.-Phys. B*, **1988**, *27*, 421.
19. Katsaros, J.D.; Malone, M.F. and Winter, H.H. *Polym. Bull.* **1986**, *16*, 83.
20. Katsaros, J.D.; Malone, M.F. and Winter, H.H. *Polym. Eng. and Sci.* **1989**, *29*, 1434.
21. Lyngaae-Jorgensen, J. and Sondegaard, K. *Polym. Eng. and Sci.* **1987**, *27*, 344.
22. Lyngaae-Jorgensen, J. and Sondegaard, K. *Polym. Eng. and Sci.* **1987**, *27*, 351.

23. Cheikh-Larbi. F.B.; Malone, M.F. ; Winter, H.H.; Halary, J.L.; Leviet, M.H. and Monnerie, L. *Macromolecules* **1988**, *21*, 3532.
24. Hindawi, I.A.; J.S. Higgins; Galambos, A.F. and Weiss, R.A. *Macromolecules,* **1990**, *23*, 670.
25. Nakatani, A.I.; Kim, H.; Takahashi, Y.; Matsushita, Y.; Takano, A.; Bauer, B.J. and Han, C.C. *J. Chem. Phys.* **1990**, *93*, 795.
26. Mani. S.; Malone, M.F. and Winter, H.H. *Macromolecules*, **1992**, *25*, 5671.
27. Mani. S.; Malone, M.F. , Winter, H.H., Halary, J.L. and Monnerie, L. *Macromolecules* **1991**, *24*, 5451.
28. Hindawi, I.A.; J.S. Higgins and Weiss, R.A. *Polymer* **1992**, *33*, 2522.
29. Larson, R.G. *Rheol. Acta*, **1992**, *31*, 497.
30. Fernandez, M.L.; Higgins, J.S. and Richardson, S.M. *Trans. IChemE*, **1993**, *71A*, 239.
31. Fernandez, M.L.; Higgins, J.S.; Horst, R. and Wolf, B.A. *Polymer* **1995**, *36*, 149.
32. Horst, R. and Wolf, B.A. *Macromolecules*, **1993**, *26*, 5676.
33. M.L. Fernandez, J.S. Higgins and S.M. Richardson *Polymer* (in press)
34. M.L. Fernandez, J.S. Higgins and S.M. Richardson 1994 *IChemE Research Event*, London, January 1994.
35. M.L. Fernandez and J.S. Higgins *Die Makrom. Chemie.* (in press).
36. Chandrasekar, S. *Hydrodynamic and Hydromagnetic Instability*, Clarendon Press, Oxford, 1961.
37. Fields, R.J. and Ashby, M.F. *Phil. Mag.* **1976**, *33*, 33.
38. O'Kane, W.J. and Young, R.J. *J. Adhesion,* **1993**, *41*, 203.
39. Bhatia, Q.S.; Pan, D.H. and Koberstein, J.T. *Macromolecules* **1988**, *21*, 2166.
40. Kammer, H.W.; Kummerloewe, C.; Kressler, J. and Melior, J.P. *Polymer* **1991**, *32*, 1488.
41. Chen, Z.J.; Shaw, M.T. and Weiss, R.A. *Proc. of the ACS PMSE Div.*, **1994**, *71*, 123.
42. Hashimoto, T.; Matsuzaka, K.; Kume, T. and Moses, E. IUPAC, p 111.

RECEIVED February 9, 1995

Chapter 9

Effect of Shear Deformation on Spinodal Decomposition

Tatsuo Izumitani[1] and Takeji Hashimoto

Division of Polymer Chemistry, Graduate School of Engineering, Kyoto University, Kyoto 606-01, Japan

Time-resolved light scattering experiments have been performed on the self-assembling process of a binary mixture of poly(styrene-*ran*-butadiene)(SBR) and polybutadiene(PB) subjected to a shear deformation. Two types of experiments were performed. In both cases the samples were prepared by a solution-cast method and were first subjected to homogenization. In the first experiment the homogenized film was subjected to isothermal demixing at 75 °C for 60 min. A shear strain, γ_s, of 0 to 3 was applied to this sample in order to investigate the effects of the deformation on the spinodally decomposed structures. The peak scattering vectors parallel ($q_{m\parallel}$) and perpendicular to the shear direction ($q_{m\perp}$) were found to change with γ_s according to affine deformation. In the second experiment, the homogenized film was subjected to isothermal demixing at 75 °C for 20 min then to a shear deformation of $\gamma_s = 2.5$. The time evolution of the scattering profile of the sheared sample was observed at 75 °C and in order to investigate a memory effect of the shear deformation on further demixing process.

Studies of phase-separation kinetics are one of the intriguing research topics from both industrial and academic points of view. Over the past decade, a number of experimental studies have been made on the coarsening kinetics of phase-separated structures of polymer mixtures (*1-3*). The results have shown that the coarsening process of polymer mixtures with a critical composition can be classified into the following three stages: (i) early stage, (ii) intermediate stage, and (iii) late stage.

In the early stage of spinodal decomposition (SD), the time evolution of concentration fluctuations is well described by the linearized theory proposed by Cahn for small molecule systems (*4-6*); the concentration fluctuations grow

[1]Current address: Daicel Chemical Ind., 1239 Shinzaike, Aboshi-ku, Himeji, Hyogo 671-12, Japan

exponentially with the time t after onset of SD. In the intermediate stage of SD, the concentration fluctuations nonlinearly evolve and both the wavelength and amplitude of the dominant mode of concentration fluctuations grow with the time. In the late stage of SD, the amplitude of the concentration fluctuations reaches an equilibrium value, however the size of the phase-separated structure keeps growing with the time (7,8).

Recently, Jinnai et al. (9) reported the memory effect of the initial thermal concentration fluctuations of a mixture of deuterated polybutadiene and protonated polybutadiene on the early stage of SD by means of a small-angle neutron scattering (SANS) method. They concluded that the thermal concentration fluctuations present in the system before quenching affect both the early stage SD and the early phase of the intermediate stage SD, giving rise to a memory effect on the time evolution of the characteristic wave number, $q_m(t)$, and the maximum scattered intensity, $I_m(t)$.

The present paper reports some preliminary results concerning how initial concentration fluctuations, imposed to the system by applying a shear deformation, affect the SD process.

Experimental Method

Samples. Polybutadiene (PB) and poly(styrene-*ran*-butadiene) (SBR) were polymerized by living anionic polymerization and have a narrow molecular weight distribution. SBR contains 80 wt% butadiene and has a number average molecular weight (M_n) of 1.00×10^5 and $M_w/M_n = 1.18$ (M_w being the weight average molecular weight). PB has $M_n = 1.60 \times 10^5$ and $M_w/M_n = 1.16$. Fractions of cis, trans and vinyl linkages of the butadiene part as measured by infrared spectroscopy are 0.16, 0.23 and 0.61 for SBR and 0.19, 0.35 and 0.46 for PB, respectively.

Preparation of mixture. A binary mixture of SBR/PB = 70/30 (v/v) was dissolved into a dilute solution of toluene (7 wt% polymer solution). The 0.15 mm thick film specimens were obtained by evaporating the solvent at a natural rate at about 25 °C in a petri dish. The film specimens thus obtained were further dried under vacuum for more than one week at room temperature. The film specimens prepared were homogenized by applying "Baker's transformation" at room temperature, i.e., by repeated foldings and compressions as reported earlier (10). The homogenized specimens were quickly sandwiched between two glass plates with a 0.2 mm thick spacer and were subjected to time-resolved light scattering experiments.

Method of Shear Deformation. The shear deformation was performed on the samples which were previously subjected to homogenization and SD. The following two experiments were performed: (i) The homogenized film was first subjected to the isothermal demixing for 60 min at 75 °C, followed by rapid cooling to 25 °C, and then a shear deformation at various strain levels at 25 °C by means of a device constructed in our laboratory. Here the effects of the shear deformation on the domain structures developed by SD can be investigated. (ii) The homogenized film was subjected to isothermal SD for 20 min at 75 °C and then cooled to 25 °C using a copper block. The annealed films thus obtained were manually sheared at 25 °C with a strain, $\gamma_s = 2.5$, where γ_s is defined by $\gamma_s = x/d_0$ with x and d_0 being the shear displacement and the sample thickness, respectively. The isothermal demixing process was investigated at 75 °C on the sheared film by time-resolved light scattering. In this experiment, the shear involves shear flow and no recovery of the sample dimension took place after the removal of the external shear force.

Method of Observations. The change of the demixed structure with shear strain and time was investigated by recording the elastic light scattering patterns. A He-Ne laser of 15 mW was used as an incident beam source, and the scattered light was detected either photographically or photometrically with the incident beam along the y-axis, perpendicular to the film surface as shown in Figure 1. In the first experiment (i), the distribution of the scattering intensity before and after shear deformation was photographed using photographic films of size 85 × 65 mm². The peak scattering angle parallel to the shear direction $\theta_{m\parallel}$ (x-direction in Figure 1), or perpendicular to the shear direction $\theta_{m\perp}$ (z-direction in Figure 1) was analyzed by means of an image analyzer (Figure 2). In the second experiment (ii), the shear-deformed sample was quickly set into a sample holder and a temperature jump (T-jump) was achieved by rapid insertion of this holder into a metal block preheated to the demixing temperature of 75 °C. The isothermal demixing process was observed at real time and *in situ* by means of the time-resolved light scattering technique described elsewhere (*11*). The scattering was observed parallel to the shear direction $I_\parallel (\theta)$ and perpendicular to the shear direction $I_\perp (\theta)$, where θ is the scattering angle in the sample.

Experimental Results and Discussion

Effects of Shear Deformation on Phase-Separated Domains. Figure 3(a) shows the light scattering pattern of the undeformed film which was first homogenized and then annealed at 75 °C for 60 min ($\gamma_s = 0$) for demixing. The phase-separated samples were deformed by shear strains of $\gamma_s =$ (b) 1.3, (c) 2.5 and (d) 3.0 at 25 °C and the light scattering patterns were photographed immediately after the shear deformation. The scattering maximum arising from a periodic concentration fluctuation developed by SD shifts toward smaller scattering vectors, **q**, (defined as $q_{m\parallel}$) for the direction parallel to the shear displacement (x-direction in Figure 1). On the other hand, the peak scattering vector (defined as $q_{m\perp}$) is almost independent of γ_s for the direction perpendicular to the deformation (z-direction in Figure 1).

The characteristic parameters of the early stage of the SD were obtained in our earlier studies (*11*): mutual diffusivity $D_{app} = 60$ nm²s⁻¹, characteristic wave number (the wave number of the dominant mode of concentration fluctuations) $q_m (0) = 9.3 \times 10^{-3}$ nm⁻¹, and characteristic time t_c (defined by $t_c \equiv (D_{app} q_m(0)^2)^{-1}) = 190$ s at 75 °C. The as-cast film used in this study was first homogenized and then phase-separated at 75 °C for 60 min, corresponding to the reduced time $\tau = 19$ where τ is defined by $\tau \equiv t/t_c$. The crossover time (t_{cr}) from the intermediate stage of SD to the late stage of SD for this system is $t_{cr} = 30$ min ($\tau = 9.5$). Therefore the shear deformation was imposed at a late stage.

The angle mark in Figure 3 corresponds to a scattering angle of 10° in air. We analyzed the change of the concentration fluctuations with shear deformation from the change of the scattering patterns. The phase separation via SD generates a periodic domain structure which is directionally independent. Thus the scattering pattern shows a circular ring as shown in Figure 3(a), with intensity maximum at a particular **q** defined as q_{m0}. Upon shear deformation along the x-axis, the circular ring is transformed into an elliptical ring, with long axis oriented parallel to the z-axis as shown in Figures 3(b) to 3(d). The wave numbers $q_{m\parallel}$ and $q_{m\perp}$ at the intensity

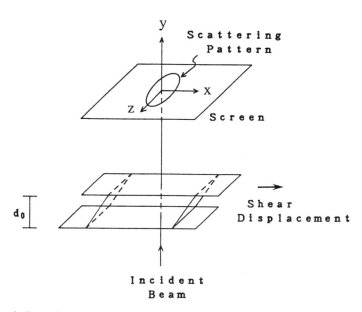

Figure 1. Experimental set-up and coordinate system Oxyz used in this work.

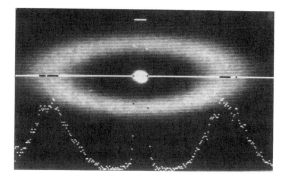

Figure 2. Example of the analysis of a light scattering pattern of a sheared sample with an image analyzer. The profile shown below the pattern represents the intensity profile along the equatorial direction of the scattering pattern. The scattered intensity was divided into 256 digital levels and plotted in a linear scale.

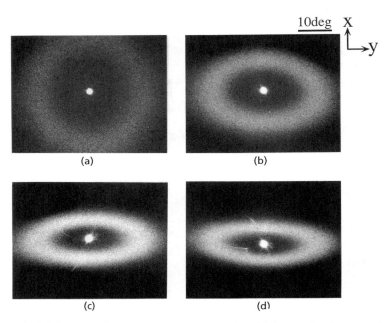

Figure 3. Light scattering patterns for the sample which was first homogenized and then phase-separated by annealing at 75 °C for 60 min (a), and its change after shear deformation with shear strains of $\gamma_s = 1.3$ (b), 2.5 (c) and 3.0 (d) at room temperature. The angle mark corresponds to a scattering angle θ equal to 10° in air.

maximum scattered parallel and perpendicular to the shear are related to the periodicities $\Lambda_{m\|}$ and $\Lambda_{m\perp}$ of the phase-separated domains parallel and perpendicular to the shear deformation, respectively: $q_{m\|} = 2\pi/\Lambda_{m\|}$ and $q_{m\perp} = 2\pi/\Lambda_{m\perp}$.

If the shear deformation induces a simple *affine deformation* of the domain structure, the strain dependence of the scattering vector should be given by (see Appendix)

$$q_{m\|}(\gamma_s)/q_{m0} = (1 + \gamma_s^2)^{-1/2} \tag{2}$$

$$q_{m\perp}(\gamma_s)/q_{m0} = 1 \tag{3}$$

Figure 4 shows $q_{m\|}(\gamma_s)/q_{m0}$ (circles) and $q_{m\perp}(\gamma_s)/q_{m0}$ (triangles) plotted as a function $(1 + \gamma_s^2)^{-1/2}$ on a linear scale. The results indicate that the changes of the scattering vectors $q_{m\|}(\gamma_s)$ and $q_{m\perp}(\gamma_s)$ almost obey the affine deformation.

The sample shear deformation was imposed in a few minutes. This time scale is much shorter than that of the structural evolution via SD (thus no significant coarsening of the phase-separating structure occurs during the deformation) but is longer than that of stress relaxation and hence than that of molecular orientation relaxation (thus the shear deformation involves shear flow).

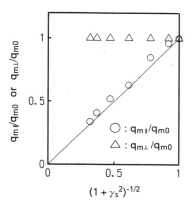

Figure 4. Change of the peak scattering vectors parallel ($q_{m\|}$ / q_{m0}) or perpendicular to the shear direction ($q_{m\perp}$ / q_{m0}) relative to that of the undeformed specimen with $(1 + \gamma_s^2)^{-1/2}$. The films were first homogenized and then demixed for 60 min at 75 °C and were subjected to the shear deformation with various shear strains γ_s at 25 °C.

Spinodal Decomposition after Shear Deformation

Time Change of Scattering Pattern. Figure 5(a) shows the light scattering pattern for the film that was first homogenized and then demixed at 75 °C for 60 min. Part b to d in Figure 5 show the light scattering patterns of the samples which were further annealed for 30 min (b), 120 min (c) and 300 min (d) at 75 °C respectively. The peak scattering maximum shifts toward smaller scattering angles, indicating a coarsening of the domains via SD.

Figures 5(e) to (h) show the time change of the light scattering patterns for the samples deformed by shear. Figure 5(e) shows the light scattering pattern obtained immediately after shear deformation with a strain of $\gamma_s = 2.4$ on a specimen which was first homogenized and then annealed for 60 min at 75 °C. The scattering pattern obtained immediately before the deformation corresponds to pattern 5(a). Figure 5(f) to 5(h) show the scattering patterns of the further annealed specimens at 75 °C for 30 min (f), 120 min (g) and 300 min (h) after the shear deformation, respectively. Thus the patterns (b) to (d) and (f) to (h) correspond to the time evolution of the scattering patterns without and with shear deformation, respectively. The results indicate that the scattering maximum perpendicular to the shear direction $\theta_{m\perp}(t; \gamma_s)$ shifts toward a smaller scattering angle approximately at the same rate as that of the undeformed specimen. On the other hand, the scattering maximum parallel to the shear direction $\theta_{m\|}(t; \gamma_s)$ hardly changes until $\theta_{m\perp}(t; \gamma_s)$ becomes nearly equal to $\theta_{m\|}(t; \gamma_s)$ (Figure e to g).

Time Evolution of Scattering Profile. Figure 6 shows the time-evolution of the scattering profiles at 75 °C after homogenization. Figure 6(b) indicates the early-to-intermediate stage of SD and (a) shows the intermediate-to-late stage of SD. This phenomenon which reflects the coarsening of the phase-separating domains was discussed earlier (*11-13*).

Figure 7 shows the time evolution of the scattering profile parallel to shear deformation. The film was first homogenized, then phase separated by annealing at

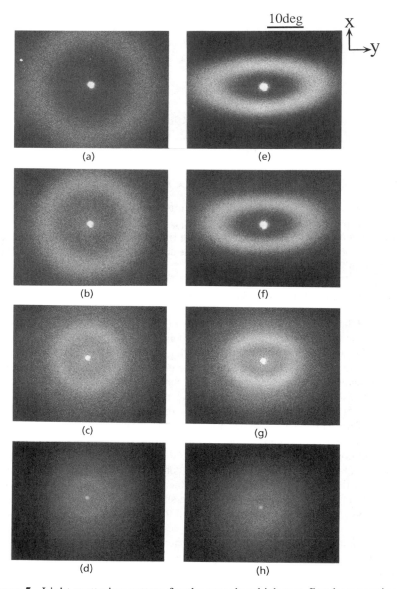

Figure 5. Light scattering pattern for the sample which was first homogenized and then annealed at 75 °C for 60 min (a), and for the samples that were further annealed for 30 min (b), 120 min (c) and 300 min (d) at 75 °C. Light scattering pattern immediately after the shear deformation ($\gamma_s = 2.4$) of the sample first homogenized and then annealed at 75 °C for 60 min (e), and for the deformed samples which were further annealed for 30 min (f), 120 min (g) and 300 min (h) at 75 °C.

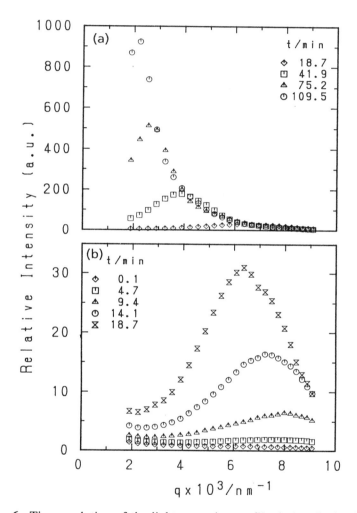

Figure 6. Time evolution of the light scattering profile during the isothermal demixing at 75 °C of the homogenized film without shear deformation.

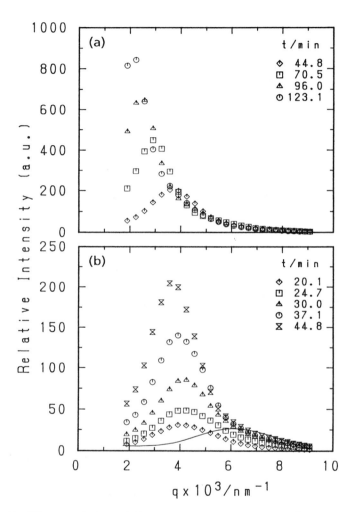

Figure 7. Time evolution of light scattering profile at 75 °C of the shear-deformed specimen parallel to the shear direction ($\gamma_s = 2.5$). The specimen was first homogenized, then annealed for 20 min at 75 °C for SD (time zero is set right after the onset of SD at 75 °C), followed by quenching to room temperature where a shear strain of $\gamma_s = 2.5$ was imposed to the specimen, and finally quickly transferred to the temperature enclosure controlled at 75 °C for further phase separation. The line in part (b) corresponds to the scattering profile of the film immediately before the shear deformation.

75 °C for 20 min; this demixing time corresponds to $\tau = 6.3$ (intermediate stage of SD). The specimen was rapidly cooled to 25 °C, a shear strain of $\gamma_s = 2.5$ was manually imposed at 25°C, and finally a T-jump was achieved by quick insertion of the deformed film into a metal block preheated to 75 °C for observation of the further progress of the phase separation. Figure 7(b) corresponds to the evolution of the scattering profile shortly after shear deformation (20.1 to 44.8 min). The time indicated in Figure 7 refers to the time after the onset of phase separation at 75 °C of the homogenized film, and the shear deformation was applied to the specimen 20 minutes after the onset of the phase separation. Although the peak scattering intensity increases with time, the scattering vector $q_{m\parallel}$ hardly changes with time t during t = 20.1 (immediately after the shear deformation) to 37.1 min. The profile shown by a line in 7(b) corresponds to that immediately before shear deformation. The shear deformation shifts the scattering vector of the intensity maximum toward smaller q from q_{m0} to $q_{m\parallel}$.

Figure 8 shows the time evolution of the scattering profile perpendicular to the shear deformation. Figure 8(b) corresponds to the time evolution of the profile from t = 20.1 (immediately after the shear deformation) to 44.0 min, and Figure 8(a) to that of the profiles at a later stage. The profile shown by a line in Figure 8(b) corresponds to that obtained at t = 20 min immediately before the shear deformation. This indicates that the scattering profiles immediately before and after shear deformation were almost the same, and, as time elapses, the peak continuously shifts toward smaller q similar to the undeformed case (Figure 6).

Figure 9(a) shows the time change of the scattering vector for the undeformed case (open square, q_m ($\gamma_s = 0$)) and those for the deformed specimen parallel to (open triangle, $q_{m\parallel}(\gamma_s = 2.5)$) and perpendicular to the shear direction (open circle, $q_{m\perp}$ ($\gamma_s = 2.5$)) in a double logarithmic scale. The scattering vector parallel to the shear direction decreases toward smaller q (i.e., from q_m ($\gamma_s = 0$) to $q_{m\parallel}(\gamma_s = 2.5)$) upon applying shear deformation at the time indicated by an arrow marked "Deformation" in the figure. $q_{m\parallel}$ stays nearly constant until a certain time, t_a, at which $q_{m\parallel} \cong q_{m\perp}$. At $t > t_a$, $q_{m\parallel}$, $q_{m\perp}$ and q_{m0} become equal and they decrease with t. Thus the memory effect of the shear deformation disappears at $t > t_a$. Afterwards the coarsening depends purely on the temperature. On the other hand, the change of the scattering vector perpendicular to the shear deformation shifts toward smaller q similar to the undeformed case. Naturally, the shear deformation does not affect the structure in the neutral direction.

Figure 9(b) shows the time changes of the maximum scattering intensity, in a double logarithmic scale, for undeformed specimen (open square, I_m ($\gamma_s = 0$)), and those for the deformed specimen, parallel to (open triangle, $I_{m\parallel}(\gamma_s = 2.5)$) and perpendicular to the shear direction (open circle, $I_{m\perp}$ ($\gamma_s = 2.5$)). It should be noted that the maximum scattering intensities are not affected by the shear deformation.

Scaled Structure Factor. The scattered intensity I(**q**) can be generally expressed by

$$I(\mathbf{q}) = I(q_x, q_y, q_z)$$
$$\sim \langle \eta^2 \rangle (q_{mx} q_{my} q_{mz})^{-1} S(q_x/q_{mx}, q_y/q_{my}, q_z/q_{mz}) \qquad (4)$$

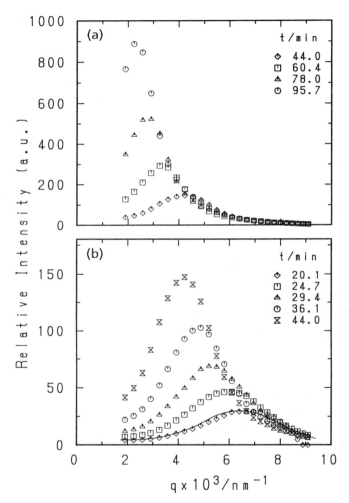

Figure 8. Time evolution of light scattering profiles at 75 °C of the shear-deformed film ($\gamma_s = 2.5$) perpendicular to the shear direction. Solid line in part b corresponds to the scattering profile before shear deformation. The thermal history and the deformation procedure are same as those in Figure 7.

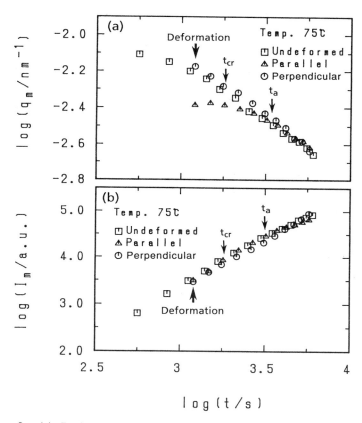

Figure 9. (a) Peak scattering vector q_m plotted against time in double logarithmic scale for the undeformed case (open square), parallel to the shear direction (open triangle) and perpendicular to the shear direction (open circle). (b) Peak scattering intensity I_m against time in double logarithmic scale for the undeformed case (open square), parallel to the shear direction (open triangle) and perpendicular to the shear direction (open circle). The times t_{cr} and t_a are the crossover time between the intermediate and late stage and the time beyond which $q_{m\parallel}$ becomes identical to $q_{m\perp}$ respectively.

where q_x, q_y and q_z are the components of the scattering vector **q** parallel to x, y, and z axes, respectively, and q_{mx}, q_{my} and q_{mz} are the components of $\mathbf{q_m}$ parallel to the three coordinate axes, respectively.

The scaled structure factor is defined by

$$F(X, Y, Z) \equiv I(\mathbf{q})q_{mx}q_{my}q_{mz}$$
$$\sim \langle \eta^2 \rangle S(X, Y, Z) \tag{5}$$

where

$$X \equiv q_x/q_{mx}, \quad Y \equiv q_y/q_{my}, \quad Z \equiv q_z/q_{mz} \tag{6}$$

and $\langle \eta^2 \rangle$ and S are the mean-squared fluctuations of refractive index and the scaling function, respectively. S characterizes a shape of the deformed phase-separated structure. The scaled structure factor parallel to the shear direction $F_\parallel(X)$ is described by,

$$\begin{aligned} F_\parallel(X) &\equiv F(X, Y = 0, Z = 0) \\ &= I(q_x, 0, 0)q_{mx}q_{my}q_{mz} \\ &\equiv I_\parallel(q)q_{mx}q_{my}q_{mz} \\ &\sim \langle \eta^2 \rangle S(X, 0, 0) \end{aligned} \tag{7}$$

Here we define $I(q_x, 0, 0)$ as $I_\parallel(q)$ for simplicity.

The scaled structure factor perpendicular to the shear direction $F_\perp(Z)$ is described by,

$$\begin{aligned} F_\perp(Z) &\equiv F(X = 0, Y = 0, Z) \\ &= I(0, 0, q_z)q_{mx}q_{my}q_{mz} \\ &\equiv I_\perp(q)q_{mx}q_{my}q_{mz} \\ &\sim \langle \eta^2 \rangle S(0, 0, Z) \end{aligned} \tag{8}$$

Again we define $I(0, 0, q_z)$ as $I_\perp(q)$ for simplicity. If the change of $\mathbf{q_m}$ with the shear obeys affine deformation, we find (see Appendix)

$$q_{my} = q_{mz} = q_{m0} \tag{9}$$

$$q_{mx} = q_{m0}(1 + \gamma_s^2)^{-1/2} \tag{10}$$

and

$$q_{mx}q_{my}q_{mz} = q_{m0}^3(1 + \gamma_s^2)^{-1/2} \tag{11}$$

Then the scaled structure factor is given by,

$$F_\|(X) = I_\|(q)q_{mx}q_{mz}^2$$
$$= I_\|(q)q_{m0}^3(1+\gamma_s^2)^{-1/2} \qquad (12)$$

and

$$F_\perp(Z) = I_\perp(q)q_{mx}q_{mz}^2$$
$$= I_\perp(q)q_{m0}^3(1+\gamma_s^2)^{-1/2} \qquad (13)$$

Figure 10 shows the scaled structure factor F plotted against the reduced scattering vector $\tilde{Q}$ where $F \equiv I(q)q_m^3$ and $\tilde{Q} \equiv q/q_m$ for the undeformed specimen (a), $F \equiv F_\|(X)$ and $\tilde{Q} \equiv q/q_m$ for the structure factor parallel to the shear displacement (b), and $F \equiv F_\perp(Z)$ and $\tilde{Q} \equiv Z$ for the structure factor perpendicular to the shear direction (c). The scaled structure factor for the undeformed case was well explored and described for this mixture (*12, 13*). In the late stage covered in Figure 10, the scaled structure factor near $\tilde{Q} = 1$ becomes universal with the time, viz., the global structure is scaled with a single length parameter $\Lambda_m(t) \equiv 2\pi/q_m(t)$ which grows with time. Thus the dynamical scaling hypothesis (*14*) is applicable for this system too. However in the early phase of the late stage covered in this experiment, the local structure still changes with time, e.g., the interface thickness decreases with the time. Therefore the structure cannot be scaled with a single length parameter. As a result the scaled structure factor at $\tilde{Q} > 1$ becomes non universal and increases with the time. It was found that at much later time of the late stage (late stage II) (*15*) the interface thickness reaches an equilibrium value and its contribution to the scaled structure factor becomes negligible (i.e., the interface thickness relative to the domain size becomes negligible) so that the whole structure factor at $\tilde{Q} \leq 10$ becomes universal with time and can be scaled with a single length parameter.

The scaled structure factor perpendicular to the shear direction (curve c) is essentially identical to that of the undeformed case (curve a) over the whole range of $\tilde{Q}$ covered here. Therefore the shear appears not to affect the self-assembly along the neutral axis. The scaled structure factor parallel to the shear direction is also approximately identical to that of the undeformed case or that perpendicular to the shear direction, despite the fact that the characteristic wave number $q_{m\|}$ is much smaller than $q_{m\perp} = q_{m0}$ or the characteristic wavelength $\Lambda_{m\|} = 2\pi/q_{m\|}$ is much larger than $\Lambda_{m\perp} = \Lambda_{m0}$. Thus the shear deformation appears to change only the characteristic wave numbers $q_{m\|}$ and $q_{m\perp}$. A closer observation of the structure factor parallel to the shear direction shows the following features: (i) The structure factor near $\tilde{Q} = 1$ (i.e. $\tilde{Q} \leq 1.5$) is essentially identical to that of the undeformed specimen (curve a) or to that perpendicular to the shear direction; thus the global structure is really scaled with the characteristic wave numbers $q_{m\|}$ and $q_{m\perp}$. (ii) The structure factor at $\tilde{Q} \geq 1.5$ is slightly smaller in intensity than in the undeformed case, which implies that the interface thickness relative to the domain size parallel to the shear direction is slightly greater than in the undeformed case. Thus the shear deformation tends to affect the interface structure parallel to the shear displacement.

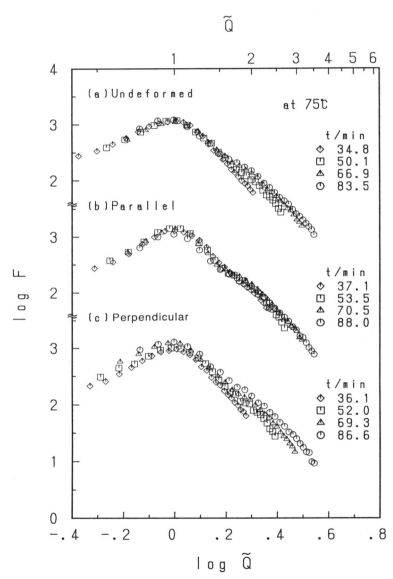

Figure 10. Scaled structure factor F plotted against reduced scattering vector $\tilde{Q}$ in a double logarithmic scale for the undeformed case (a), parallel to the shear direction (b), and perpendicular to the shear direction (c).

Appendix

Here we attempt to predict effects of the shear deformation on the domain structure developed by SD on the basis of affine deformaion. The concentration fluctuations $\eta_0(\mathbf{r})$ before the shear deformation are transformed into $\eta(\mathbf{r}')$ after the affine shear deformation where $\mathbf{r}$ and $\mathbf{r}'$ are the displacement vectors before and after the deformation. The scattered intensity $I_0(\mathbf{q})$ before the deformation is given by

$$I_0(\mathbf{q}) \sim \int \tilde{\eta}_0^2(\mathbf{r}) \exp[i(\mathbf{q}\cdot\mathbf{r})] d\mathbf{r} \tag{A1}$$

where $\tilde{\eta}_0^2(\mathbf{r})$ is autocorrelation function of $\eta_0(\mathbf{r})$. The scattered intensity $I(\mathbf{q})$ after the deformation is given by

$$I(\mathbf{q}) \sim \int \tilde{\eta}^2(\mathbf{r}') \exp[i(\mathbf{q}\cdot\mathbf{r}')] d\mathbf{r}' \tag{A2}$$

Noting that

$$\mathbf{r}' = \mathbf{r} + \Delta\mathbf{r} \tag{A3}$$

with $\Delta\mathbf{r} = (\gamma_s y, 0, 0)$ for the shear displacement along x direction as shown in Figure 1, equation (A2) is rewritten by

$$I(\mathbf{q}) \sim \int \tilde{\eta}_0^2(\mathbf{r}) \exp\{i[\mathbf{q}\cdot(\mathbf{r}+\Delta\mathbf{r})]\} d\mathbf{r} \tag{A3}$$

Since

$$\mathbf{q}\cdot(\mathbf{r}+\Delta\mathbf{r}) = q_x(x+\gamma_s y) + q_y y + q_z z$$
$$= (\mathbf{q} + \Delta\mathbf{q})\cdot\mathbf{r} \tag{A4}$$

with

$$\Delta\mathbf{q} \equiv (0, \gamma_s q_x, 0) \tag{A5}$$

we obtain

$$I(\mathbf{q}) = I_0(\mathbf{q} + \Delta\mathbf{q}) \tag{A6}$$

from equations (A1) and (A3) to (A5). Thus if the scattering maximum occurs at $\mathbf{q}_0$ satisfying

$$q_x^2 + q_y^2 + q_z^2 = q_{m0}^2 \tag{A7}$$

before the shear deformation, it occurs at $\mathbf{q} = (q_x, q_y, q_z)$ satisfying

$$q_x^2 + (\gamma_s q_x + q_y)^2 + q_z^2 = q_{m0}^2 \tag{A8}$$

after the affine deformation.

For the light scattering observation in the plane of $q_y = 0$, we have

$$\frac{q_x^2}{q_{m0}^2/(1+\gamma_s^2)} + \frac{q_z^2}{q_{m0}^2} = 1 \tag{A9}$$

Thus we have the elliptical spinodal ring with

$$q_{m\parallel} = q_{m0}/(1+\gamma_s^2)^{1/2} \tag{A10a}$$

$$q_{m\perp} = q_{m0} \tag{A10b}$$

which are identical to equations (2) and (3), respectively. For the light scattering observation in the plane of $q_x = 0$, we have

$$q_y^2 + q_z^2 = q_{m0}^2 \tag{A11}$$

In the undeformed sample the scattering maximum exists on the surface of a sphere of radius q_{m0}. Equations (A7) and (A8) imply that the shear deformation transforms this sphere into an ellipsoid with its three principal axes along x', y' and z' as shown in Figure 11 in which z and z' axis concide. The two Cartesian coordinates are obtained by the rotation around z or z' axis through the angle α,

$$2\alpha = \tan^{-1}(-2/\gamma_s) \tag{A12}$$

The ellipsoid given by equation (A8) is rewritten by

$$\frac{q'_x{}^2}{q_{m0}^2/\left[\frac{(\gamma^2+2)-\gamma(\gamma^2+4)^{1/2}}{2}\right]} + \frac{q'_y{}^2}{q_{m0}^2/\left[\frac{(\gamma^2+2)+\gamma(\gamma^2+4)^{1/2}}{2}\right]} + \frac{q'_z{}^2}{q_{m0}^2} = 1 \tag{A13}$$

with the principal coordinate Ox'y'z' where q'_x, q'_y and q'_z are the components of **q** along x', y' and z' axes, respectively. Naturally the volume of the ellipsoid is identical to that of the sphere. Thus an ideal experiment to investigate the effect of the shear deformation on the phase-separated domains should be carried out as follows. We send the incident beam along the neutral axis, i.e., z-axis, and observe the scattering intensity distributions as a function of q_x' and q_y' along the principal axes of the ellipsoid, x' and y' axes. Since this type of experimental conditions is extremely difficult to set up, we observed and analyzed the scattering intensity distributions as a function of q_x and q_y, which is useful for qualitative investigation.

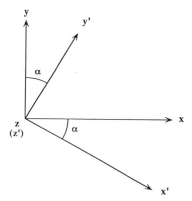

Figure 11. Cartesian coordinate system Ox'y'z' appropriate for the specimens subjected to the shear displacement along x-axis. The z and z' axes are identical and the neutral direction for the shear deformation. The angle α is given by equation A12.

Acknowledgments.

The authors express their thanks to Japan Synthetic Rubber Co. Ltd. for providing the samples. This work was partially supported by a Grant-in-Aid for Scientific Research from the Ministry of Science and Culture Japan (05453149) and by scientific grants from Japan Synthetic Rubber Co. and Yokohama Rubber Co., Japan.

Literature Cited

1. Hashimoto, T. In "*Current Topics in Polymer Science. Vol II*"; Ottembrite, R. M.; Utracki, L. A., Inoue, T., Eds.; Hanser:Munich, New York, 1986; pp 199-242.
2. Hashimoto, T. *Phase Transition* **1988**, *12*, 47.
3. Hashimoto, T. In "*Materials Science and Technology. Vol 12 Structure and Properties of Polymers*", Cahn, R. W.; Haasen, P.; Kramer, E. J. Eds., Thomas, E. L. Vol. Ed., VCH: Weinheim, 1993, Chapter 6.
4. Cahn, J. W.; Hilliard, J. E. *J. Chem. Phys.* **1958**, *28*, 258.
5. Cahn, J. W. *J. Chem. Phys.* **1965**, *42*, 93.
6. de Gennes, P. G. *J. Chem. Phys.* **1980**, *72*, 4756.
7. Hashimoto, T.; Itakura, M.; Hasegawa, H. *J. Chem. Phys.* **1986**, *85*, 6118.
8. Hashimoto, T.; Itakura, M.; Shimidzu, N. *J. Chem. Phys.* **1986**, *85*, 6773.
9. Jinnai, H.; Hasegawa, H.; Hashimoto, T. *J. Chem. Phys.* **1993**, *99*, 4845, 8145.
10. Hashimoto, T.; Izumitani, T.; Takenaka, M. *Macromolecules* **1989**, *22*, 2293.
11. Izumitani, T.; Hashimoto, T. *J. Chem. Phys.* **1985**, *83*, 3694.
12. Izumitani, T.; Takenaka, M.; Hashimoto, T. *J. Chem. Phys.* **1990**, *92*, 3213.
13. Takenaka, M.; Izumitani, T.; Hashimoto, T. *J. Chem. Phys.* **1990**, *92*, 4566.
14. Binder, K.; Stauffer, D. *Phys. Rev. Lett.* **1974**, *33*, 1006.
15. Takenaka, M.; Hashimoto, T. *J. Chem. Phys.* **1992**, *96*, 6177.

RECEIVED January 25, 1995

Chapter 10

Domain Structures and Viscoelastic Properties of Immiscible Polymer Blends Under Shear Flow

Yoshiaki Takahashi and Ichiro Noda

Department of Applied Chemistry, Nagoya University, Chikusa-ku, Nagoya 464-01, Japan

The relationship between domain structures and viscoelastic properties of a 1:1 blend of two immiscible polymers under steady shear flow and after step increase of shear rate were studied. The sample was a mixture of polydimethylsiloxane and polyisoprene which have almost the same viscosity. At steady states, domains were elipsoidal and their diameters were rather uniform and inversely proportional to shear rate. The shear stress and first normal stress difference were almost proportional to shear rate. After a step increase of shear rate, domains in the initial steady state were extremely elongated, which then ruptured to shorter ones and gradually changed to the final steady-state structure. An undershoot of shear stress and an overshoot of normal stress followed by a gradual change to steady-state values were observed during these processes. These results can be qualitatively explained by the change in the distribution of unit normal vectors of the interface.

In immiscible polymer blends, there exist various domain structures of component polymers, which deform, rupture and aggregate under flow fields, so that their viscoelastic properties become very complicated.

Many experimental and theoretical studies on viscoelastic properties of immiscible polymer blends have been carried out with reference to suspension and emulsion systems, which were first theoretically studied by Einstein (1) and Taylor (2), respectively. Most of the studies have been recently summarized by Utracki (3). Despite many studies, the relationship between domain structures and viscoelastic properties of immiscible polymer blends are not elucidated especially when the compositions of the blends are close to 1:1, that is, concentrated blends. Further experimental and theoretical studies on concentrated blends are needed. Recently, two such theories have been proposed independently.

Onuki (4) presented a theory on the viscoelastic properties of phase-separating fluids near the critical point taking into account of interfacial tension. In the weak shear case, he predicted that enhancements of the shear stress, $\Delta\sigma_{12}$, and first normal stress difference, N_1, are proportional to shear rate, $\dot{\gamma}$, and the excess viscosity, $\Delta\eta(=\Delta\sigma_{12}/\dot{\gamma})$, and normal viscosity $(=N_1/\dot{\gamma})$ become of order $\eta\phi$ for ellipsoidal domains, where ϕ is volume fraction of domains.

$$\Delta\eta/\eta\phi = 1 \sim 2 \qquad (1)$$

$$(N_1/\dot{\gamma})/\eta\phi = 3 \sim 6 \qquad (2)$$

Here the viscosities, η, of the existing two phases are assumed to be the same for near-critical fluids. It is to be noted that N_1 of phase-separating fluids is much larger than for a one-phase system. Onuki also noted that these ideas should be applicable to other systems which have domain structures such as polymer blends even far from critical point, as long as η of the two phases are the same.

Doi and Ohta (5, 6) proposed a constitutive equation of textured materials by considering the time evolution of the area and orientation of the interface and the interfacial tension opposing the deformation of 1:1 mixtures of immiscible Newtonian fluids with the same viscosity. They predicted interesting scaling relations, of which two typical relations are described below.

For a steady shear flow, they predicted that both shear stress, σ_{12}, and first normal stress difference, N_1, are proportional to the shear rate, $\dot{\gamma}$.

$$\sigma_{12} \propto \dot{\gamma} \qquad (3)$$

$$N_1 \propto |\dot{\gamma}| \qquad (4)$$

These relations are the same as those predicted by Onuki (equations 1 and 2).

Another scaling relation was presented for a step change of shear rate, where the shear rate is changed from an initial shear rate, $\dot{\gamma}_i$, to a final shear rate, $\dot{\gamma}_f$, at time t=0. In this case, they predicted that plots of the transient stresses, $\sigma(t;\dot{\gamma}_i;\dot{\gamma}_f)$, divided by $\sigma(\dot{\gamma}_i)$ against the strain, $\dot{\gamma}_f t$, yield a universal curve independent of $\dot{\gamma}_i$, as long as the shear rate ratio, $\dot{\gamma}_f/\dot{\gamma}_i$, is maintained constant, i.e.,

$$\sigma(t;\dot{\gamma}_i,\dot{\gamma}_f)/\sigma(\dot{\gamma}_i)=f(\dot{\gamma}_f t,\dot{\gamma}_f/\dot{\gamma}_i) \qquad (5)$$

We have studied the viscoelastic properties of binary immiscible Newtonian fluid mixtures (7) and immiscible polymer blends (8), made of components with almost the same viscosity, over a wide range of composition under steady shear flow and the transient state after a step change of shear rate. It was confirmed that N_1 is observed even for the Newtonian fluid mixtures and the above scaling relations are applicable for the excess shear and normal stresses at steady and transient states for both systems.

In the theory of Doi and Ohta (5, 6), moreover, it is pointed out from dimensional analysis that the characteristic length of domains under steady shear flow, R, is given by

$$R \propto \Gamma/\eta\dot{\gamma} = \Gamma/\sigma_{12} \qquad (6)$$

even for concentrated blends, where Γ is the interfacial tension. The same relation is also derived by Onuki (4) for the radius of an ellipsoidal domain in the weak-shear case. However, this relation has not been examined by experiments for concentrated polymer blends under steady shear flow.

In transient experiments, an undershoot of shear stress and an overshoot of normal stress were observed after a step increase of shear rate. The existence of undershoot and overshoot cannot be explained by the constitutive equation proposed by Doi and Ohta (5, 6), which agrees with the experiments in terms of scaling relation.

Moreover, it was observed that the exponents in shear-rate dependencies of σ_{12} and N_1 become slightly lower than unity in the high shear rate region for polymer blends. The deviations from the linear proportionality occur at different shear rates for σ_{12} and N_1 and also for different compositions of the samples. The shear rates for the deviations are always lower for σ_{12} than for N_1 (8).

In order to examine equation 6 and to know the reason why the undershoot and overshoot exist after the step increase of shear rate and why the deviation occurs in steady-state measurements, it is essential to observe directly the structure changes under flow and discuss them in comparison with the rheological properties. In this work, therefore, we studied the domain structures of a 1:1 blend of two immiscible polymers having almost the same viscosity, under steady shear flow and after a step increase of the shear rate to clarify the relationship between their structures and viscoelastic properties.

Experimental

The two component polymer samples used were a polydimethylsiloxane (PDMS) KF96H-60000 from Shinetsu Chemical Co. and a polyisoprene (PI) LIR-20000 from Japan Synthetic Rubber Co., Ltd.. As shown in Figure 1, these materials have linear viscoelastic properties in the experimental range of shear rates; σ_{12} is proportional to shear rate and N_1 is proportional to the square of shear rate. It should be noted that N_1 was measured at relatively high shear rates only. The viscosities and densities of the PDMS and PI samples at 25 °C are tabulated in Table I. The interfacial tension between these polymers determined by the pendant drop method is 1.7 dyn/cm.

Table I. Characteristics of Component Polymers

	$M_W \times 10^4$	η (Pa·s) at 25°C	ρ (g/cm^3) at 25°C
KF96H-60000	5	60	0.976
LIR-20000	2.5	50	0.912

The blend sample was prepared by stirring weighed amounts of polymers in a vessel with a spatula for a few minutes. The composition of the blend was 1:1 by weight. Bubbles in the sample were removed by keeping the sample *in vacuo* for about 10 minutes.

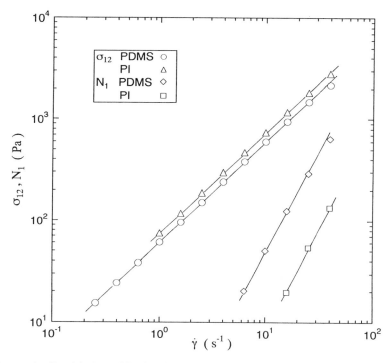

Figure 1. Double logarithmic plots of shear stress σ_{12} and first normal stress difference N_1 against shear rate $\dot{\gamma}$ for component polymers.

Viscoelastic measurements were carried out with a Rheometrics Ltd. RMS800 mechanical spectrometer using 5 cm diameter cone-plate geometry with 0.04 radian cone-angle at room temperatures (22 to 24 °C).

Reflective image of domain structures of the blend sample were observed with a Hyper microscope VH-6110 equipped with a 100 w halogen lamp as a light source from Keyence Corporation. Six different magnifications for the camera probe, which is connected to the main body with optical cables can be used in this system. The highest magnification is 1000 times. The camera probe was attached to a Weissenberg rheogoniometer R17, equipped with a modified drive unit (9) which enables a step change of shear rate. The cone-plate used was a 10 cm diameter quartz plate and a stainless steel cone with 0.017 radian cone-angle. Since the temperature controlled bath and torque transducer were removed from the R17 to get enough room to attach the camera probe, the observations were carried out at room temperatures (22 to 24 °C) which were almost the same as for the viscoelastic measurements. The camera probe was set perpendicularly to the quartz plate, so that the observed structures were projections onto the flow plane. The observed structures were continuously recorded on a VCR and analyzed by using the pictures printed with a video printer at certain experimental conditions. The times needed to observe the structures at the steady states were determined on a basis of the viscoelastic measurements. The highest shear rate in the viscoelastic measurements (10 s^{-1} for the blend) was limited by edge fracture of the sample, while that in the structure observation (2 s^{-1} for the blend) was limited by both domain size and its flow rate.

Results

Figures 2 and 3 show the change of domain structures observed when a steady shear flow was started after a long rest time and when steady shear flow was stopped, respectively. Before the start of flow, domains were spherical and the distribution of diameters was broad. After the onset of steady shear flow, the domains were elongated and aggregated to form long rod-like structures. The rod-like domains then ruptured to shorter ones and gradually approached the steady state (within the order of hundred seconds for typical shear rates). An undershoot of shear stress and an overshoot of normal stress were observed in the corresponding viscoelastic measurement, as reported previously (7). After cessation of the steady shear flow, the ellipsoidal domains quickly relaxed (within ten seconds) to spherical ones and also the stresses relaxed almost instantaneously, in contrast to the approach to steady state. The diameters of the spherical domains were rather uniform and the larger the shear rate before cessation, the smaller the diameters are. At rest state, the spherical domains aggregated to become macrodomains at an extremely low rate.

All these observations are qualitatively the same as for immiscible Newtonian fluid mixtures (7). As pointed out in a previous paper (7), the effect of flow history is quite strong when the second shear rate is smaller than the initial shear rate and the rest time between the two flows is not long enough to erase the memory of the initial flow. In such a case, the undershoot of shear stress and the overshoot of normal stress were not observed. To have the reproducibility when we go back to the lower shear rate, we must wait at least one night. When the second shear rate is higher than the initial shear rate, however, the memory is erased in a rather short time. Therefore, most of measurements at steady shear were carried out in the increasing order of shear rate. Moreover, transient data in a step change of shear rate were well reproducible.

Figure 4 shows double logarithmic plots of shear stress, σ_{12}, and first normal stress difference, N_1, against shear rate, $\dot{\gamma}$, at steady states for the blend. It is clear that both quantities of the blend are enhanced compared to those of the component polymers shown in Figure 1. The enhancement of N_1 is larger than that of σ_{12} and both are almost proportional to shear rate, in accordance with the theoretical predictions (equations 1 to 4), though the deviation of σ_{12} from equation 3 is observed at higher shear rate. All the features in Figure 4 are similar to those observed for PDMS/polybutadiene(PB) blends (8).

Figure 5 shows plots of the rescaled transient stresses, $\sigma(t;\dot{\gamma}_i,\dot{\gamma}_f)/\sigma(\dot{\gamma}_i)$, against strain, $\dot{\gamma}_f t$, at a constant shear rate ratio ($\dot{\gamma}_f/\dot{\gamma}_i = 2$). Here, the measured σ_{12} and N_1 were used instead of the excess σ_{12} and N_1 values, but these rescaled stresses correspond to the rescaled excess stresses when the viscosity is Newtonian and the N_1 values of the components are small compared to the measured N_1 (8). In Figure 5, the rescaled transient shear and normal stresses nearly reduce to each single line, if the shear rate ratio is maintained constant. An undershoot of shear stress and an overshoot of normal stress were observed after the step increase of shear rate. All these features are similar to those observed in previous studies (7, 8).

Figure 6 shows a typical example of the successive change of domain structures from one steady state (0.111 s^{-1}) to another (0.203 s^{-1}). Figures 6a and 6d show

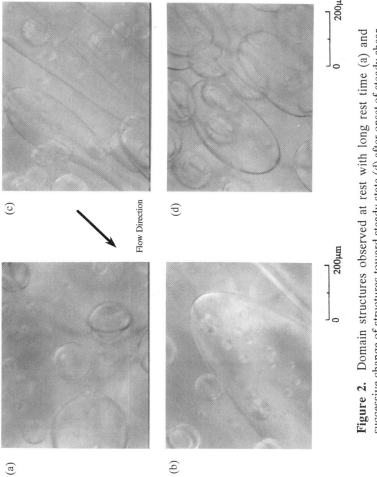

Figure 2. Domain structures observed at rest with long rest time (a) and successive change of structures toward steady state (d) after onset of steady shear flow ($\dot{\gamma}=0.111$ s^{-1}). (b) and (c) show transient structures 30 and 230 s after the start of flow, respectively. (c) corresponds to the condition when the peak of normal stress (or the valley of shear stress) was observed.

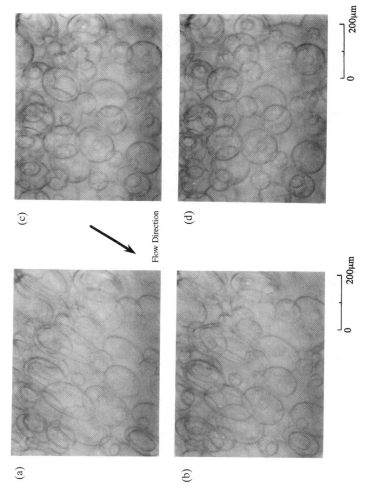

Figure 3. Domain structures observed after cessation of steady shear flow at time = 0. (a) steady state $\dot{\gamma}$ = 0.203 s^{-1} (b) t=0 s (c) t=5 s (d) t=20 s.

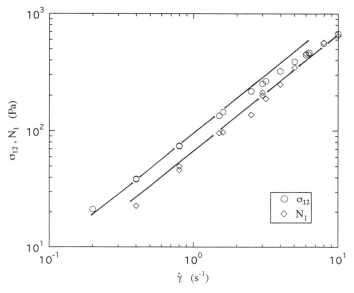

Figure 4. Double logarithmic plots of shear stress σ_{12} and first normal stress difference N_1 against shear rate $\dot{\gamma}$ for the 1:1 blend of PDMS/PI.

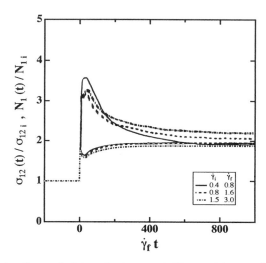

Figure 5. Plots of rescaled transient normal (lines with overshoot) and shear (lines with undershoot) stresses after step change of shear rate. The initial and final shear rates for different data lines are denoted in the Figure. The shear rate ratio is 2.

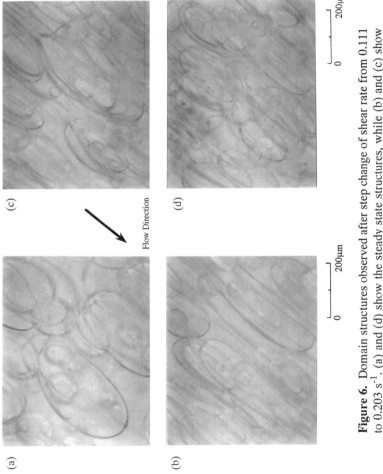

Figure 6. Domain structures observed after step change of shear rate from 0.111 to 0.203 s^{-1}. (a) and (d) show the steady state structures, while (b) and (c) show transient structures 70 and 140 s after step change of shear rate, respectively. (b) and (c) correspond to conditions slightly before and immediately after the peak of normal stress (or the valley of shear stress).

domain structures at steady states with different shear rates, while Figures 6b and 6c show transient domain structures from the initial to final steady state. At the steady states, there exist both large and rather small domains, elongated in flow direction. At first sight, the large domains appear to have rather uniform diameters. To characterize the features of these domain structures, the shorter diameter, a, and the longer diameter, b, perpendicular and parallel to the flow direction, respectively, were measured on the printed video pictures.

Figure 7 shows histograms of a, b, and aspect ratio b/a at 0.203 and 1.11 s^{-1}. All the histograms show their maxima and the a and b values at the maximum decrease with increasing shear rate, while the maximum of the aspect ratios is almost independent of shear rate. As shown in the histograms of a and b, the distribution of domains, n, are not symmetric and widen at the side of the smaller sizes. This implies that there are the maximum cut-off values for these diameters under steady shear flow. The value of the maximum aspect ratio (about 1.7) may correspond to the critical value for rupture of the ellipsoidal domain (10). The existence of a few thick and extremely elongated domains suggests that the domains continually aggregate and rupture under flow. These features are common at the all shear rates.

Figures 6b and 6c correspond to the states slightly before and immediately after the peak of the normal stress (or the valley of the shear stress) in Figure 5. It is observed that the large domains in the initial steady state are extremely elongated when shear rate is changed. Then the elongated rod-like domains starts to rupture to shorter ones and gradually change to the structure at the final steady state. This feature is the same as that observed after the onset of steady shear flow after a long rest, shown in Figure 2. Therefore, the existence of an overshoot or undershoot in the transient stresses may correspond to the formation of long rods and their rupture.

Discussion

As shown in the histograms at steady state (Figure 7), there exist the domains which are larger or smaller than the main domains. The larger domains have larger interfacial areas, but their numbers are relatively small. The number of the smaller domains are not small, but they have smaller interfacial areas. Therefore, we may assume that the contributions of these domains to the viscoelastic properties are relatively small and the domains having the maximum numbers in these histograms are mainly responsible for the viscoelastic properties.

As already mentioned, the observed structures are projections onto the flow plane, so that a is the actual size, but b is slightly different from the true values depending on the tilt angle of the domains, which is unknown, but may be small. Since b/a is almost constant at steady state, the characteristic value of b is closely related to the characteristic value of a. Therefore, we regard the average value of a around the maximum as the characteristic length of the ellipsoidal domains under steady shear flow and hereafter we refer to it as the drop size.

Figure 8a shows double-logarithmic plots of the drop size against shear rate. The data of 1:1 mixture of immiscible Newtonian fluids (silicone oil/resin oil) having almost the same viscosities (7) are also shown for comparison. The drop sizes in two samples are different, but both are almost inversely proportional to shear rate.

To examine the difference in the drop sizes, the data in Figure 8a are replotted in double logarithmic form against Γ/σ_{12} in Figure 8b. This figure indicates that the

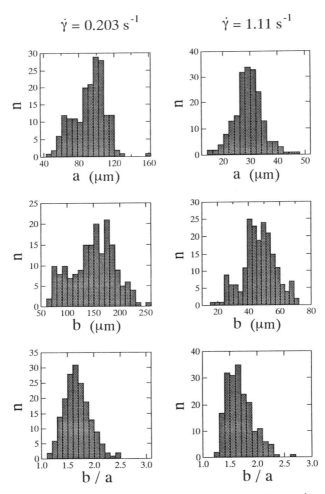

Figure 7. Histograms of diameters a and b and aspect ratio b/a at $\dot{\gamma} = 0.203$ and $1.11\ s^{-1}$.

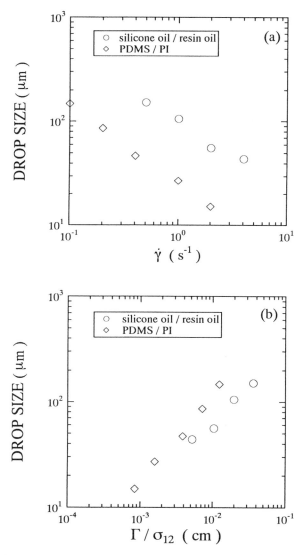

Figure 8. (a) Double logarithmic plots of a against γ for PDMS/PI and silicone oil/resin oil, and (b) double logarithmic plots of a against Γ/σ_{12} for the data in Figure 8a.

difference in drop size of two different systems are approximately explained by the difference of interfacial tension.

The excess stress tensor due to the interface is determined by the distribution of unit normal vector **n** of the interface (4-6), and the excess shear stress and excess first normal stress difference are given by

$$\Delta\sigma_{12} \propto <n_1 n_2> \tag{7}$$
$$\Delta N_1 \propto <n_2^2 - n_1^2> \tag{8}$$

where < > is the average on the interface and 1 is the direction parallel to the flow. As shown in Figure 6 (and also in Figure 2) in the transient experiments, domains are first largely elongated and then ruptured after the step increase of shear rate. In this process, **n** becomes nearly perpendicular to the flow direction when the domains become long rods so that $<n_1 n_2>$ become small, while $<n_2^2 - n_1^2>$ remains large. Consequently, the excess shear stress due to interfacial tension becomes very small, while the excess first normal stress difference becomes large. When the long rods are ruptured and approach the steady state ellipsoidal domains, $<n_1 n_2>$ and $<n_2^2 - n_1^2>$ become close to each other. This may be the reason why the undershoot of σ_{12} and the overshoot of N_1 exist after a step increase of shear rate.

We cannot have a definite conclusion on the deviation from the scaling law for η at high shear rates since we could not observe the structures in that shear rate region. However, we may speculate that the deviation is caused by the change of domain structures from ellipsoidal ones observed at low shear rate in this work to rod-like or string ones as reported by Hashimoto et. al. (11) for polymer solutions at high shear rates. Onuki pointed out that $<n_x n_y>$ becomes small, while $<n_y^2 - n_x^2>$ remains large for such string domains (12).

Literature Cited

1. Einstein, A. *Ann. Phys.* **1905**, *17*, 549.
2. Taylor, G. I. *Proc. Roy. Soc.* **1932**, *A138*, 41.
3. L. A. Utracki, *Polymer Alloys and Blends;* Hanser Publishers: Munich, 1989.
4. Onuki, A. *Phys. Rev. A* **1987**, *35*, 5149
5. Doi, M.; Ohta, T. *J. Chem. Phys.* **1991**, *95*, 1242.
6. Doi, M. In *Rheology of Textured Materials;* Garrido, L., Ed.;Complex Fluids, Lecture Notes in Physics; Springer: NY, 1993, pp 221.
7. Takahashi, Y.; Kurashima, N.; Noda, I.; Doi, M. *J. Rheol.* **1994**, *38*, 699.
8. Takahashi, Y.; Kitade, S.; Kurashima, N.; Noda, I. *Polym. J. (Tokyo)* **1994**, *26*, 1206.
9. Takahashi, Y.; Isono, Y.; Noda, I.; Nagasawa, M. *Macromolecules* **1986**, *19*, 1217.
10. Taylor, G. I. *Proc. Roy. Soc.* **1934**, *A146*, 501.
11. Takebe, T.; Fujioka, K.; Sawaoka, R.; Hashimoto, T. *J. Chem. Phys.* **1990**, *93*, 5271.
12. Onuki, A. *Europhys. Lett.* **1994**, *28*, 175.

RECEIVED January 25, 1995

Chapter 11

Effects of Flow on the Structure and Phase Behavior of Polymer Blends

Z. J. Chen, N. G. Remediakis, M. T. Shaw, and R. A. Weiss

Department of Chemical Engineering and Polymer Science Program, University of Connecticut, Storrs, CT 06269–3136

The effect of flow on the structure and phase behavior of polystyrene/poly(vinyl methyl ether) (PS/PVME) and polybutadiene/polyisoprene (PB/PI) blends was studied with a rheo-optical instrument that consisted of a two-dimensional light-scattering (SALS) system coupled with a parallel-plate rheometer. When shear was applied to the blends following spinodal decomposition, the spinodal-type structure was stretched in the flow direction, and a spinodal ring in the light scattering pattern was correspondingly shrunk in the flow direction. The distorted spinodal ring was shown to be the origin of the H-shaped scattering patterns previously reported for sheared blends when the angular range was insufficient to observe the entire ring. The characteristic dimensions of the fluid structure parallel and normal to the flow-direction calculated from the scattering pattern were consistent with the morphology observed by optical microscopy.

The morphology of a polymer blend influences its mechanical properties and is an important factor in determining the suitability of the blend for commercial applications (*1*). Although the influence of flow fields, such as encountered in common polymer processing operations, on the orientation of polymer molecules and the morphological structure of blends is well-known, only recently have researchers begun to consider the effect of flow on phase behavior (*2* and references therein). Shear flow can perturb the location of the phase boundary of a partially miscible blend, and both shear-induced phase mixing (*3-6*) and demixing (*6, 7*) have been reported.

A variety of techniques have been used to study the phase behavior of polymer blends, though turbidity measurements are historically the most common. For a quiescent blend, turbidity measurements provide an unambiguous determination of the cloud point, as long as the difference in the refractive indices of the two components is sufficient to scatter visible light. When the blend is undergoing flow, however, turbidity is unable to separate thermodynamic effects, such as phase changes, and hydrodynamic effects, such as the deformation, orientation or breakup of a dispersed phase. For this reason, alternative techniques that can provide additional information on the structure of the fluid are needed to understand better the relationship between flow and the phase behavior of polymer blends.

Small-angle light scattering (SALS) is a well-established method for investigating the shape and size of particles (*11*) and the phase separation kinetics of polymer mixtures (*12*). Thermodynamic and non-thermodynamic effects of flow can be resolved from time-dependent changes of the scattering patterns (*13-16*). Several laboratories have recently constructed rheo-scattering instruments for acquiring two-dimensional (2-D) SALS patterns of the polymer blends undergoing simple shear flow (*17-20*).

A common observation in rheo-SALS studies of phase-separated polymer mixtures is a bright-streak pattern oriented along the meridian normal to the flow-direction (*20, 21,* Chen, Z. J.; Wu, R-J.; Shaw. M. T.; Weiss, R. A.; Fernandez, M.; Higgins, J. S. *Polym. Eng. Sci.*, in press). In several studies the streak is reported to disappear at a critical shear rate, and that observation is taken as evidence for the phenomenon of flow-induced mixing (*20, 21,* Chen, Z. J.; Wu, R-J.; Shaw. M. T.; Weiss, R. A.; Fernandez, M.; Higgins, J. S. *Polym. Eng. Sci.*, in press).

The most common explanation of the bright-streak scattering pattern is that of highly elongated domains oriented in the flow-direction. The development of these oriented domain has not been characterized, and the mechanism by which the streak disappears is not clear. For example, does the disappearance of the streak scattering pattern represent a true thermodynamic transition associated with phase mixing? Other unanswered questions regarding the bright streak, especially in those instances when the streak scattering pattern does not disappear, is whether the domains are stretched infinitely long in the flow direction or do they eventually break-up due to capillary instabilities.

To better understand the details of structural changes and disappearance of domains (mixing), we have used rheo-SALS techniques to study polymer blends under simple shear flow (*22,* Chen, Z. J.; Wu, R-J.; Shaw. M. T.; Weiss, R. A.; Fernandez, M.; Higgins, J. S. *Polym. Eng. Sci.*, in press). A curious result of our initial work was the observation of an H-shaped scattering pattern (or dark-streak pattern) that developed when a blend was sheared above the critical point (LCST system), and the subsequent transformation of the H-shaped pattern into a bright-streak pattern as shear was continued.

This paper considers the origin of the H-shaped and streak scattering patterns for sheared polymer blends. The relationship between the blend morphology and the light scattering patterns was determined by using rheo-SALS measurements, independent dark-field microscopy, and static light scattering measurements on specimens that were quenched into glasses after shearing in the melt.

Experimental Details

Materials. Blends of polystyrene with poly(vinyl methyl ether) (PS/PVME) and polybutadiene with polyisoprene (PB/PI) were studied. Both systems exhibit lower critical solution temperature (LCST) behavior. The refractive-index difference is large for PS/PVME, but very small for PB/PI. Thus, PB/PI exhibits little multiple scattering, but the greater contrast between PS and PVME facilitates microscopy investigation of that blend. PS/PVME is a well-studied model blend system, while PB/PI is an elastomer system with important industrial applications.

PS/PVME Blends. The characteristics of the component polymers and the critical conditions of the blend are given in Table I.

Table I. Characteristics of the Components and the Blend of PS/PVME.

Component	$M_n \times 10^{-3}$	M_w/M_n
PS	135	1.13
PVME	99	2.13

Critical Composition: 75 wt% PVME, Critical Temperature: 123 °C

A 25/75 (w/w) blend of PS/PVME was prepared by dissolving both polymers in toluene and casting the solution onto a poly(tetrafluoroethylene) (PTFE) surface. The films were dried in a vacuum oven at 50°C for about two weeks. Film samples, 0.05 mm thick, were used for both microscopy and light scattering studies.

PB/PI System. The characteristics of the component polymers are given in Table II.

Table II. Characteristics of the Components of the PB/PI Blend.

Component	T_g(°C)	% 1,2-PB	% 3,4-PI	$M_n \times 10^{-3}$	M_w/M_n
PB	-85	24	-	164	1.02
PI	-63	-	8	125	1.38

A 40/60 (w/w) PB/PI blend was prepared by dissolving the polymers in toluene and casting the solution onto PTFE. The spinodal temperature of this blend system was determined to be less than 61 °C since a spinodal structure developed after 7 hours at that temperature.

Rheo-SALS Studies. A rheo-SALS instrument was designed to study the two-dimensional small-angle light scattering (SALS) of polymer mixtures undergoing shear flow. Specific details of the original design and operation of the instrument have been described elsewhere (Wu, R.; Shaw, M. T.; Weiss, R. A. *Rev. Sci. Instrum.*, submitted), but a few of the key features are provided here. The light source is a polarized 10-mW He-Ne laser (Uniphase model 1105P) with a wavelength of 632.8 nm. The beam passes through two pinholes of diameters of 0.5 mm and 0.75 mm, respectively, to eliminate stray light, and a series of neutral-density filters (optical density range from 0.1 to 4) to attenuate the light intensity. A prism mirror directs the beam normal to the rheometer's parallel plate fixture, which consists of a glass top plate and a steel bottom plate with a small quartz window. The scattered light from the sample is imaged on a screen made from a uniform translucent film, and the scattering pattern is focused by a macrofocus lens (CANON 100-mm micro lens FD) onto a two-dimensional CCD array detector (Princeton Instruments model TE/CCD-576T/UV). The signal from the detector is taken by a controller interface (model ST-130) and saved directly to a microcomputer. CSMA data acquisition software from Princeton Instruments is used to acquire the scattering image. The rheometer used is a Rheometrics Mechanical Spectrometer (RMS) model KMS-71C. Torque, normal force, and temperature measurements are recorded by a CYRDAS 802/IBM PC data acquisition system.

Microscopy and Light Scattering Studies after Shear Quench. To compare the real-space structures of polymer blends with light-scattering images (Fourier transform space), optical micrographs and light-scattering patterns were obtained for blend samples quenched from different experimental conditions. To prepare samples for this study, a small amount of blend was transferred to a microscope slide, and the sample was covered with another glass slide. It was then pressed at room temperature to form a thin film. The sample between the two glass slides was heated in a microscope hot stage, and shear was applied. After shearing, the sample was quenched to room temperature by removing it from the hot stage. This sample was used for separate microscopy and light scattering measurements.

A Nikon optical microscope was used to obtain micrographs of the PS/PVME blend, while an Olympus phase-contrast microscope was used for the PB/PI blend. The light scattering patterns were obtained using a He-Ne laser and a Polaroid camera. The scattered light from a sample was imaged on a white screen. The sample to detector distance was 965 mm, which provided a range of scattering vector $q \approx 0$ -8.4 μm^{-1} where

$$q = 4\pi \sin\theta/\lambda \qquad (1)$$

λ is the wavelength of the light (0.6328 μm) and θ is one-half the scattering angle.

Results and Discussions

Microscopy and Light Scattering Studies after Shear-Quench Experiment.

PS/PVME blends. The blend sample was first heated on the microscope hotstage without shear to about 3 °C above the cloud point. A bicontinuous, two-phase structure typical of spinodal decomposition was observed, and this grew more distinct with increasing annealing time. At about 30 minutes after phase separation began, the sample was quenched quickly to room temperature for microscopy and light scattering studies. The structure relaxation was sufficiently slow that the morphology of the melt was retained in the quenched sample. Figure 1a shows the morphology of the blend sample. The characteristic size of the morphology measured from the micrograph is about 4 μm. The corresponding light scattering pattern of this blend is shown in Figure 1d, and the "spinodal ring" is at $q = 1.5$ μm^{-1}. This corresponds to a characteristic wavelength of the concentration fluctuations in real space,

$$d = 2\pi/q \qquad (2)$$

of 4.2 μm, which agrees well with the size scale measured from the micrograph.

The sample was returned to the microscope hot stage and shear was applied by displacing the upper glass slide. The sample was again quenched to room temperature. Figure 1b is an optical micrograph of the sheared blend and Figure 1e is the corresponding light scattering pattern. The displacement of the upper glass slide was measured by microscopy and the shear strain was $\gamma = 4$ (0.2 mm displacement).

The blend sample was returned to the hot-stage and an additional displacement of 1.3 mm was applied to the upper glass slide, providing a total strain of 30. The bicontinuous phases were elongated into a fibrillar structure as shown in the Figure 1c. The light scattering pattern for this sample is shown in Figure 1f.

The light scattering patterns in Figures 1e and 1f show a deformed ring that appears to be a product of the spinodal ring seen in Figure 1d. In fact, this "squashed" ring explains the dark and bright streak patterns that were previously observed in a rheo-optical study of this blend (22) and it may explain the origin of similar light scattering patterns of sheared blends that have been reported by other research groups (6, 7, 9, 10). It was shown (22) that an H-shaped light scattering pattern developed upon shearing this same PS/PVME blend; the struts of the H were aligned normal to the flow-direction. As the shear strain increased, the H-pattern progressed to a dark-streak and finally a bright-streak pattern. The light scattering data in Figure 1 now show that the observation of "streaks" in our original rheo-SALS studies was a consequence of insufficient angular range. That is, at low shear strains, the q-range accessed by the instrument was too small to see the tips of the squashed "spinodal" ring, while at higher strain, the ring was squashed so much that the inner region was not resolved. As a result, we observed what appeared to be first a dark-streak and then a bright-streak pattern. The question remains whether the dark- and bright-streak patterns reported by the other groups have the same origin.

The micrographs in Figure 1 are similar to those reported by Fernandez et al. (Fernandez, M. L.; Higgins, J. S.; Richardson, S. M. *I. Chem. Eng. Research Event*, 1994, in press), though our explanation differs. They attributed the structure to a surface instability; the bulk of the sheared blend was reportedly homogeneous and one-phase. In contrast, the structures shown in Figure 1 were characteristic of both the surface and off-surface morphologies of our sheared specimens, which was determined by varying the focus of the microscope. It is also clear from Figures 1a-c that the sheared morphology was related to the initial spinodal decomposition, i.e., the bicontinuous structure in Figure 1a.

Figure 1b shows that although the bicontinuous structure was stretched in the flow

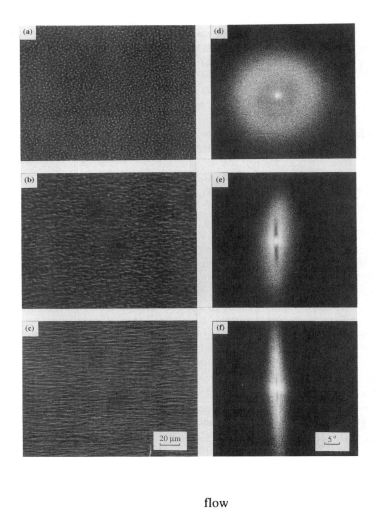

flow
→

Figure 1. The phase structures of the PS/PVME blend seen under the microscope (a-c), along with the corresponding light scattering images (d-f). a, d after phase separation for about 30 min., no shear; b, e after a shear strain of about 4; c, f after a shear strain of about 30.

direction by the shear flow, the size of the structure in the vorticity direction was virtually unchanged. The larger size in the flow-direction corresponds to the increased scattering at lower q for light scattered in the flow-direction and this produces what we describe as a "squashed" spinodal scattering ring, Figure 1e. For $\gamma = 30$, the two-phase morphology was further aligned parallel to the flow direction, Figure 1c, and the light scattering maximum in the flow direction moved to even lower q, Figure 1f. The light scattering pattern in Figure 1f looks at first like a bright streak, but upon closer examination, the interior of the squashed ring can be seen.

The scale of the concentration fluctuations in the sheared blend can be estimated from the scattering parallel and normal to the flow-direction, which corresponds to the flow and vorticity directions, respectively, in real-space. The characteristic size of the concentration fluctuations calculated from the scattering maxima parallel and normal to the flow direction in Figures 1d, 1e and 1f are given in Table III. These values compare well with the corresponding dimensions measured from the micrographs in Figure 4, which are also listed in Table III.

Table III. The Characteristic Size (μm) of the Concentration Periodicity Calculated from the Scattering Maxima and Measured from Micrographs in Figure 1.

	$\gamma = 0$		$\gamma = 4$		$\gamma = 30$	
	parallel to flow	normal to flow	parallel to flow	normal to flow	parallel to flow	normal to flow
scattering maxima	4.2	4.2	18.9	3.7	43	3.7
micrographs	4	4	12 - 21	3.8	31 - 38	3.8

A curious result of this study is that we observe a periodicity in the flow direction even when the morphology is highly elongated. The length of this periodicity may depend on the viscosity ratio of the phases, the interfacial energy, and the applied shear rate, and future research will consider the effects of these variables.

PB/PI Blends. The blend was formed into a film, 0.075 mm thick, by pressing it between two glass microscope slides at room temperature. Because of the small refractive-index difference between PI and PB, a phase-contrast microscope was used to observe the morphology of the blend. The magnification used was 400×. After being heated on the microscope hot-stage for about 180 min. at 81°C, which was well within the spinodal decomposition region, the blend exhibited bicontinuous morphology typical of spinodal decomposition, Figure 2a, with a characteristic size of ca. 4 µm. The corresponding light scattering pattern is shown in Figure 2b.

Figure 2c shows the micrograph and Figure 2d shows the corresponding light scattering pattern of the blend for a shear strain of $\gamma = 6.6$. As with the PS/PVME blend described above, shear deformation stretched the bicontinuous structure in the

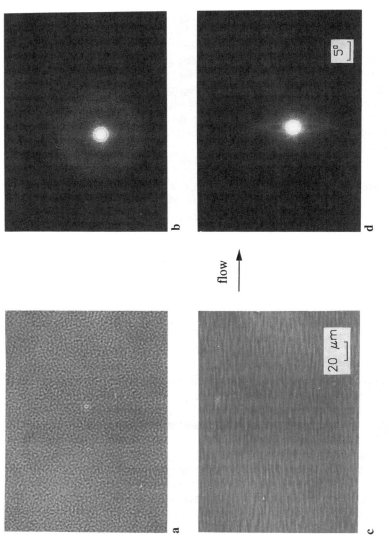

Figure 2. The phase structures of a near-critical polybutadiene/polyisoprene blend. Microscopy (a and c), and the corresponding light scattering images (b and d). a, b after phase separation for about 180 min. at 81 °C, no shear; c, d after a shear strain of about 6.6.

flow direction. The characteristic length in the flow direction measured from the micrograph increased to about 30-35 μm, which agrees with the size of 29 μm calculated from the position of the maximum in the scattering intensity, $q = 0.217$ μm^{-1}. The squashed spinodal ring scattering pattern in the Figure 2d is similar to that of Figure 1e.

Rheo-SALS Studies. The same PB/PI blend was also studied with the rheo-SALS instrument, which allowed real-time measurements of the light scattering patterns, and therefore the blend structure, during simple shear flow.

Two different experiments are reported here. Both had the same initial condition and final state. The initial condition was room temperature and quiescent state. The final state was a temperature of 85 °C and a shear rate of 0.03 s^{-1}. The paths of heating and shearing were different: In the first experiment, the shear was applied about 60 minutes after the sample was heated to 85 °C, while in the second experiment, the shear was applied 90 minutes before the sample was heated to the same temperature.

Path 1: Shear Jump at Constant Temperature into the Two-Phase Region. A 1.3-mm-thick film was first heated to 85 °C for about 1 hour without shear. A spinodal scattering ring developed and slowly shifted to smaller scattering angle while annealing, which indicates that the spinodal decomposition was in the intermediate stage. The position of the scattering maximum was at $q = 3.28$ μm^{-1}, as shown in Figure 3a, which corresponds to a characteristic size of 1.92 μm. The black spot in the center of the scattering pattern is due to the beam stop and the bright halo around the beam stop is due to the unscattered incident beam and parasitic scattering.

After allowing the phase separation to proceed for 1 hour, steady shear was applied to the sample. The shear rate at the optical path was 0.03 s^{-1}. Figure 3 shows time-resolved light scattering patterns of the flowing polymer blend. After 18 s, the spinodal ring was slightly deformed, Figure 3b; and, as the shearing continued, the ring deformed even more, Figures 3c-e. Figure 3i shows the history of the shear stress calculated at the position of the optical path, and the times at which Figures 3a-e were obtained are noted on the curve. The shear stress became steady after about 500 s, which corresponded to $\gamma = 15$.

With longer shearing, the deformed spinodal ring finally became a bright streak, as shown in Figure 3f, and the dark region could not be resolved due to the limitation of the angular resolution of the rheo-SALS instrument. However, this bright streak was not the steady-state result. At about 2024 s, which corresponded to $\gamma = 61$, the bright streak became more diffuse, Figure 3g, and eventually at t = 2624 s ($\gamma = 79$) the squashed ring was again observed. The reappearance of the squashed ring was followed by a second time period in which the separation of the top and bottom of the ring decreased with increasing shear strain. In this regime, the "streak" pattern was not quickly reestablished and after 6000 s ($\gamma = 180$), the interior of the ring was still resolved.

The characteristic wavelength of the concentration fluctuations in the flow and the vorticity directions may be calculated from equation 2, where $q = q_{max}$ is the wave vector where the maximum in scattered light intensity occurs. For the flow direction, q_{max} was often obscured by the beam stop and the incident beam transmission, e.g., in

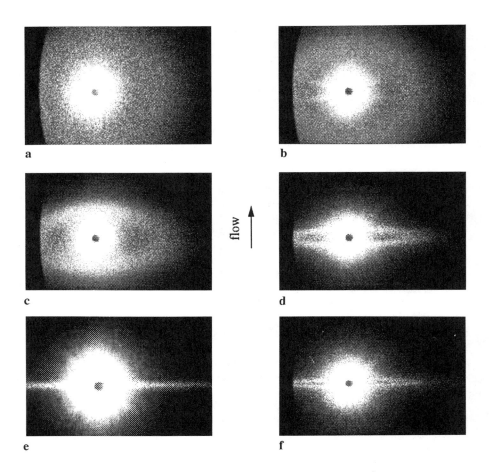

Figure 3. Time-resolved light scattering patterns of PB/PI at 85 °C and a shear rate of 0.03 s^{-1}. (a) 0 s (b) 18 s (c) 50 s (d) 200 s (e) 524 s (f) 1184 s (g) 2024 s (h) 2624 s (i) shear stress of the sample. Times when images a-e were taken are indicated on the stress-time graph.

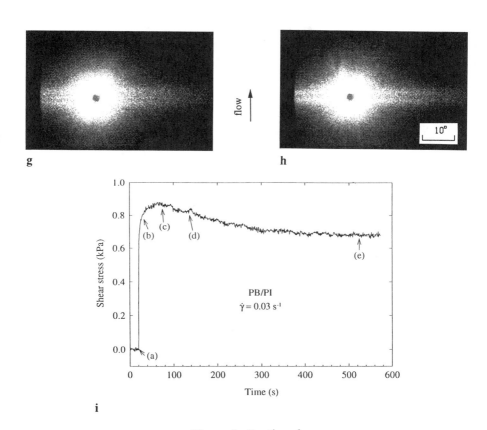

Figure 3. *Continued*

Figures 3d, 3e, 3h. In those instances, q_{max} in the flow direction was estimated as one-half the separation of the visible components of the ring on either side of the beam stop.

The relative change in d, defined as

$$\Delta \equiv (d - d_o) / d_o \tag{3}$$

where d_o was the characteristic wavelength of the concentration fluctuations before shear was applied, is compared with the shear strain in Figure 4, where both open circles and filled circles are calculated from the scattering patterns. The open circles track the behavior of the initial spinodal ring, while the filled circles correspond to times after the reappearance of the squashed ring. The solid line has a slope of 1, while the dashed line is the regression result of the filled circles, which has slope of 0.25. There are no data between the open circles and filled circles because the dark region was not resolved in the scattering patterns for those times, e.g., in Figure 3f.

The characteristic length increased by an order of magnitude at relatively low strains, e.g., for $\gamma = 19$, $d = 39$ μm compared with $d = 1.91$ μm for $\gamma = 0$, and the relationship between Δ and γ was linear up to a shear strain of $\gamma = 20$, as shown by the open circles of Figure 4. This result indicates that at least at the beginning of the shear flow, the morphology deformed affinely. For $20 < \gamma < 60$, the characteristic length could not be determined from the light-scattering pattern, since only a bright streak was observed, as shown in Figure 3f. At $\gamma \approx 60$, the bright streak defused and the subsequent re-establishment of the squashed ring suggests that breakup of the elongated morphology occurred. This was followed by a restretching of the morphology and the width of the interior of the ring again decreased with increasing shear strain. The characteristic length of the structure in the flow direction is plotted as filled circles in Figure 4.

In Figure 4, the slope of the line through the filled circles is 0.25 instead of 1 as it was for the first region of stretching of the morphology. This indicates that in that case the stretching was not affine. One explanation for this is that a rotating macro structure exists.

When the shear rate was increased to 0.7 s^{-1}, the ring was no longer observed (Figure 5), which means that the characteristic length became much larger than could be seen by our instrument. In this case, we cannot determine whether the concentration fluctuations had a finite wavelength or were infinite. The former would be consistent with a rotating macro structure, while the latter would be more consistent with a uniform velocity field.

Path 2: Temperature Jump under Steady-State Shearing. The same blend of PB/PI was also studied under steady-state shear following a temperature jump. A 1.3 mm thick sample was loaded into the parallel-plate fixture and a shear rate of 0.03 s^{-1} was applied at room temperature; no phase separation was detected. Figure 6a shows the scattering pattern after shear was applied at room temperature. Again the black spot in the center of the scattering pattern is due to the beam stop and the bright halo around the beam stop is due to the tail of the unscattered incident beam and parasitic scattering.

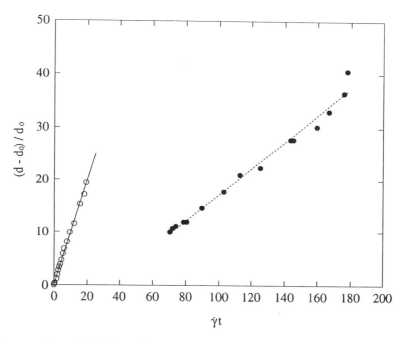

Figure 4. Variation with shear strain of the relative change of the characteristic length in the flow direction, $(d - d_o)/d_o$ as measured from the light scattering images.

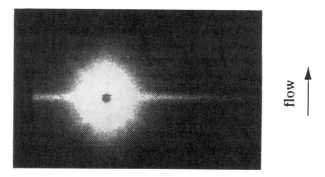

Figure 5. Light scattering pattern of PB/PI at 85 °C and a shear rate of 0.7 s^{-1}.

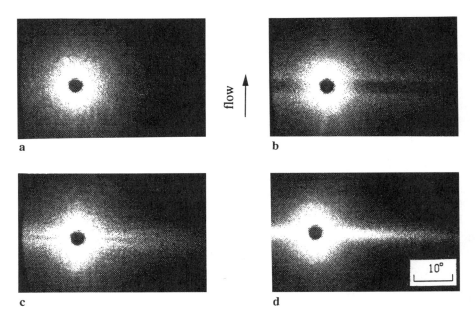

Figure 6. Time-resolved light scattering patterns of PB/PI at a shear rate of 0.03 s^{-1}. (a) 84 min. of shearing, at room temperature; (b) 120 min. of shearing and 30 min. of heating at 85 °C; (c) 164 min. of shearing and 74 min. of heating at 85 °C; (d) 410 min. of shearing and 380 min. of heating at 85 °C.

Figure 6b shows the scattering pattern that developed 30 min. after the temperature was jumped to 85 °C. The pattern was H-shaped. Here, phase separation occurred simultaneously with the deformation of the two-phase structure. The spacing of the two wings of the H-pattern decreased with time, as shown in the Figure 6c.

Comparing the Figure 6b with Figure 3a, it is noticed that the characteristic length in the flow direction for Figure 6b is much larger than that in Figure 3a, although the heating time for Figure 6b was shorter.

Apparently shear flow not only elongates the phase-separated structure, but also may cause anisotropic phase separation.

After longer shearing, the spacing of two wings decreased further and, finally, the dark region between the two wings could not be resolved due to the angular limitations of our SALS detector. Figure 6d shows the bright streak after the wings merged. This pattern persisted for more than 180 min. and the intensity of the bright streak did not change significantly. This suggests that the concentration periodicity in the flow direction did not increase infinitely in length as the bright streak did not keep shrinking. One hypothesis concerning this phenomenon is that there exists a rotating macro structure. Clearly more experiments should be done to comprehend this phenomenon.

Acknowledgment

This work was supported by the Polymer Compatibilization Research Consortium at the University of Connecticut and the State of Connecticut Critical Technologies Grant. We thank Dr. A.F. Halasa of the Goodyear Tire and Rubber Company for kindly supplying the polybutadiene and polyisoprene samples. The PB/PI blend was originally prepared by Dr. Albert Narh. ZJC thanks the Rubber Division of the American Chemical Society for a John D. Ferry Fellowship.

Literature Cited

1. Utracki, L. A. *Polymer Alloys and Blends*; Hanser Publ.: NY, 1989.
2. Larson, R. G. *Rheol. Acta* **1992**, *31*, 497-520.
3. Mazich, K. A.; Carr, S. H. *J. Appl. Phys.* **1983**, *54*, 5511-5514.
4. Langaae-Jorgenson, J.; Sondergaard, K. *Polym. Eng. Sci.* **1987**, *27*, 344-350.
5. Mani, S.; Malone, M. F.; Winter, H. H.; Halary, J. L.; Monnerie, L. *Macromolecules* **1991**, *24*, 5451-5458.
6. Hindawi, I. A.; Higgins, J. S.; Weiss, R. A. *Polym.* **1992**, *33*, 2522-2529.
7. Mani, S.; Malone, M. F.; Winter, H. H. *Macromolecules* **1992**, *25*, 5671-5676.
8. Katsaros, J. D.; Malone, M. F.; Winter, H. H. *Polym. Eng. Sci.* **1989**, *29*, 1434-1445.
9. Kammer, H. W.; Kummerloewe, C.; Kressler, J.; Melior, J. P. *Polym.* **1991**, *32*, 1488-1492.
10. Wu, R.; Shaw, M. T.; Weiss, R. A. *J. Rheol.* **1992**, *36*, 1605-1623.
11. Bohren, C. F.; Huffman, D. R. *Absorption and Scattering of Light by Small Particles*; John-Wiley & Sons: N.Y, 1983.
12. Snyder, H. L.; Meakin, P.; Reich, S. *Macromolecules* **1983**, *16*, 757-762.

13. Wu, X-L.; Pine, D. J.; Dixon, P. K. *Phys. Rev. Lett.* **1991**, *66*, 2408-2411.
14. Beysens, D.; Gbadamassi, M.; Boyer, L. *Phys. Rev. Lett.* **1979**, *43*, 1253-1256.
15. Takebe, T.; Sawaoka, R.; Hashimoto, T. *J. Chem. Phys.* **1989**, *91*, 4369-4379.
16. Takebe, T.; Hashimoto, T. *Polym. Commun.* **1988**, *29*, 261-263.
17. News, J. M.; Weiss, R. A.; Shaw, M. T. *Proc. Ann. Meeting North America Thermal Analysis Soc.*, 1990.
18. Nakatani, A. I; Waldow, D. A.; Han, C. C. *Rev. Sci. Instrum.* **1992**, *63*, 3590-3598.
19. Matsuzaka, K; Hiroshi, J.; Hashimoto, T. *Polym. Mat. Sci. Eng., Am. Chem. Soc.* **1994**, *71*, 121-122.
20. Werner, D. E.; Fuller, G. G.; Frank, C. W., in *Theoretical and Applied Rheology, (Proc. XIth Int. Congr. on Rheol.,)*; Moldenaers, P., Keunings, R., Eds.; Elsevier Science Publishers B. V.: 1992, p 402.
21. Takebe, T.; Fujioka, K.; Sawaoka, R; Hashimoto, T. *J. Chem. Phys.* **1990**, *93*, 5271-5280.
22. Chen, Z. J.; Shaw, M. T.; Weiss, R. A. *Polym. Prepr., Am. Chem. Soc., Div. Polym. Chem.* **1993**, *34*(2), 838-839.

RECEIVED February 22, 1995

Chapter 12

Influence of Interface Modification on Coalescence in Polymer Blends

K. Søndergaard and J. Lyngaae-Jørgensen

The Danish Polymer Center, Department of Chemical Engineering, Technical University of Denmark, Building 229, DK-2800 Lyngby, Denmark

> Interfacial modifiers (compatibilizers), such as block and graft copolymers have been found to yield a pronounced influence on the morphology development in blends. In the reviewed work the influence of A-B copolymers on coalescence (agglomeration) in A:B blends have been revealed by electron microscopy and rheo-optical measurements. Divergent criteria and experimental evidence appear in literature on the copolymer molecular weight if the requirement is the copolymer to effectively modify the interfacial properties. Branch lengths, composition and volume fraction of copolymer do influence the "steric stabilization" effect. Graft copolymers have been generally found to be somewhat more effective in inhibiting coalescence and, thus minimizing the particle size and its distribution. Rheo-optical measurements on a blend containing a block copolymer (BC) revealed that the stabilization effect of the BC is not unconditional during the flow: coalescence is prevented for a time which decreases with increasing shear rate due to a removal of the BC away from the interface. The origin of the observed behavior is outlined based on various mechanisms.

Multiphase polymer systems constituting polymer blends and alloys have been subject of increasing academic and industrial interest since the early 1970's (*1*). Blending of components with complementary properties has increasingly constituted a more cost effective alternative to synthesizing new polymers. Since most systems are immiscible the properties are to a large extent affected by their morphology. The multiphase nature is, however, directly desirable in many applications. The morphology resulting from a blend process depends on the thermodynamic, rheological and interfacial properties of the constituent components, the composition, and the blending conditions. In this respect the interface modification (compatibilization) plays a key role on the microrheology and, thus the resulting phase size and mechanically on the interfacial adhesion.

In the literature the modifiers are often referred to as compatibilizers, interfacial agents or surfactants. The term compatibilization in this context does not mean that an immiscible blend is rendered miscible, although the addition of a modifier may expand the miscible region in the phase diagram. In the frame work of this presentation it merely implies a local activity of the modifier at the interface in the strong segregation region of the phase diagram.

From the point of view of morphology, the blends can roughly be divided in two classes: discrete and cocontinuous systems. When analyzing the effects of various parameters on the morphology, a clear distinction between these classes should be observed in avoiding misinterpretation of data. In this presentation only discrete systems of particles will be discussed, i.e., systems with volume fraction of the dispersed phase below the percolation threshold for spheres (theoretically ~ 15 vol%) (2,3).

The dispersion of fluid particles in various flow fields has been the subject for numerous investigations. In spite of these efforts the mechanism of droplet instability and break-up in viscoelastic systems has not yet been fully clarified. The influence of increasing particle concentration is that coalescence processes have to be taken into account. The effect of flow-induced coalescence of the dispersed phase has not always been considered in the past.

Most theoretical as well as experimental work have been based on Newtonian drops in Newtonian media. Few contributions deal with viscoelastic drops in Newtonian or viscoelastic media. In most studies only the behavior of non-interacting droplets has been considered while in practice the dispersed phase concentrations are too high to exclude interactions between droplets. Exact studies dealing with single drop studies have been conducted, but only few involved doublet interactions, while multiparticle interactions and coalescence have been treated semiempirically.

The problem of development of domain size distribution in selected flows has been generally formulated by Valentas et al. (4) in the so-called "general population balance equation" (5). Silberberg and Kuhn (6), Tokita (7), Elmendorp and van der Vegt (8) and Fortelny and Kovar (9), Lyngaae-Jørgensen et al. (10), and Huneault et al. (11) formulated simplified models taking both break-up and coalescence processes into account.

For polymer blends and alloys, solutions of the general population balance equations have not been published. A number of publications deal with the influence of volume fraction of the dispersed phase on polymer blend rheology, drop size and stability, viz., van Oene (12), Utracki (13), Favis and Willis (14), Chen et al. (15), and Sundararaj and Macosko (16).

The topics considered have been summarized in a number monographs and reviews; in the last decade, notably: Han (17), Elmendorp and van der Vegt (18), Utracki (19), Wu (20), Plochocki (21), Utracki (1), Meijer and Janssen (22), Utracki and Shi (23), and Søndergaard and Lyngaae-Jørgensen (24).

Background

In molten simple binary immiscible blends the dispersed phase has a tendency to coalescence which leads to macro domains, broad size distribution, and ultimately inferior mechanical properties. It is experimentally established and commonly recognized that the dispersed particles in dispersive multiphase polymer systems, their size and distribution are the results of a competitive process between break-up and coalescence. At given flow conditions an invariant morphology can result that represents a balance existing between droplet break-up and flow-induced coalescence. The larger particles are deformed and broken up by the stresses of the flow field, while concurrently droplets collide and fuse to temporarily form large domains.

The major mechanism of collision and coalescence of particles are considered to be flow (Figure 1) but the effect may also take place in quiescent systems, the proposed origins being Brownian motion, dynamics of concentration fluctuation, sedimentation, temperature gradients, etc. Some of these effects may also be present during flow and they may be either enhanced or suppressed by it (*25*). However, considering the relative high viscosity and large particles present in polymer blends, Brownian motion scarcely play a major role in coalescence since the particle diffusion is inversely related to particle diameter and matrix viscosity. A mechanism likely to occur in polymer blends is the so-called Ostwald ripening: the large particles grow at the expense of small ones by diffusion of molecules from the smaller particles with high interfacial energy to the larger ones. This process is enhanced by flow and can be described by a scaling law (*27*). The influence of other mechanisms on the blend coalescence is discussed later in a reviewed study.

Interface Modification. Compatibilization of simple binary blends is commonly recognized to have a pronounced and favorable influence on interfacial properties. In order to stabilize the morphology and thus prevent gross aggregation, compatibilizers (e.g. block or graft copolymers) may be added or formed in-situ to obtain a decreased coalescence rate and an enhanced interfacial adhesion. The effect of the various methods of interface modification on the fluid properties can be considered as qualitatively equivalent although some quantitative differences may exist as will be outlined in what follows.

The purpose of incorporating compatibilizers is generally to improve the physical properties that can be attributed to the following major effects: (i) to reduce the interfacial tension between phases in order to achieve an enhanced deformability of large domains and thus upon break-up gain a finer particle size dispersion during blend mixing, (ii) to stabilize the interface, thus prevent the particles against coalescence and gross aggregation during blending, and (iii) to obtain enhanced adhesion between phases in the solid state.

Compatibilizers consisting of block or graft copolymers, has been found to enhance the interfacial interaction in simple binary blends as revealed in earlier pioneering work (*27-31*). These copolymers most often contain segments that are chemically identical or similar to the main components of the blend so as to obtain some segment miscibility in each bulk phase. The molecular weight of block copolymers are frequently chosen equal to or higher than that of the blend components, if the desired purpose is for the copolymer to be located at the interface between the phases (*32-39*). However, divergent criteria and experimental evidence appear in the literature. At favorable conditions even small quantity of the copolymer can yield a pronounced influence of the interfacial properties (*30-32*).

While diblock copolymers are expected to be more effective with regard to interface adhesion in comparison to graft copolymers (because of fewer conformational restraints for the penetration of each block segment into the respective bulk polymer phases) graft copolymers can be synthesized by a simpler route, e.g. by in situ reaction during mixing or processing. Efforts within this latter area have been progressing considerably. Recent reviews on this topic have been given by Xanthos (*40*).

Inhibited coalescence in compatibilized blends is caused by steric stabilization of the interface. The force associated with the particle motions is not sufficient to squeeze and drain the interphase layer of the copolymer when the particles collide. This is also related to the yield stress usually associated with block copolymers which causes a so-called immobilization of the interphase.

The basic factors that affect the interfacial properties has been qualitatively described in terms of scaling arguments as outlined by Leibler (*41*). Mean-field theories of polymer interfaces has been developed by Hong and Noolandi (*42-45*) and Shull and Kramer (*46*). A mean-field approximation was also proposed by Löwenhaupt and Hellmann (*47*). The theory of Leibler is valid for nearly miscible systems whereas the mean-field theories apply to highly immiscible ones. Nolandi et al. also calculated the critical concentration of block copolymer for micelle formation in the bulk of a homopolymer (*43*). A later work deals with the case in which the blocks differ from the homopolymers (*45*). The amount of a symmetrical copolymer residing at the interface was expressed as $\phi_{c,0} \sim \phi_c \exp(X_c \chi \phi_p/2)$ presumed that $(X_c \chi \phi_p/2) \leq 1$, where ϕ_c is the bulk copolymer volume fraction, X_c is the degree of polymerization of the copolymer, ϕ_p is the volume fraction of homopolymer A or B (assumed equal), and χ is the Flory-Huggins interaction parameter between A and B segments. For a concentrated polymer-polymer system with a symmetrical block copolymer an analytical expression for the interfacial tension reduction was given by

$$\Delta v = d \, \phi_c \left[\frac{1}{2}\chi + \frac{1}{X_c} - \frac{1}{X_c}\exp(X_c \chi/2) \right] \quad (1)$$

where d is the width at half height of the copolymer concentration profile, reduced by the Kuhn statistical length (d may be left out as an adjustable parameter) and the remaining parameters defined as previously.

A subject of prime interest is experimentally to determine whether the copolymer actually resides at the interface and its effectiveness. Direct study of polymer interfaces are difficult. The techniques involved are various forms of microscopy (UV fluorescence, phase contrast, elemental electron loss spectroscopy, etc.) cited in (*48*); diverse scattering techniques that can be used to get information on the interface of bulk samples of blends and more conveniently on block copolymers (*61*); IR spectroscopy (*62*) and x-ray microanalyses (*61*) that provide more detailed information on the composition profile; techniques based on scattering of ion beams that can be used to investigate the interface of bulk samples (64-66); and methods involving light spectroscopic ellipsometry (67-71), x-ray (72) or neutron (73) reflection that allows short range details of layered interfaces to be obtained.

Shull et al. (46) directly measured by forward recoil spectrometry the equilibrium segregation of a deuterated diblock copolymer to interfaces between the immiscible homopolymer phases. Predictions from their mean-field theory were quantitatively accurate for values of the copolymer chemical potential below a certain limiting value related to the formation of block copolymer micelles.

Reflection techniques have been used in several studies. The main purposes have been to measure the interfacial thickness between immiscible polymers (69,70), interface segregation, concentration profiles (46,75,76), segment distribution (77) in blends with added copolymers or to measure the interfacial thickness in reactive blends (71).

Some recent attempts to measure the effect of the modifier are mechanical (15,55) and rheological (56,57) testing or measurements of the interfacial tension (15, 51-54,58-60).

The influence of addition of a block copolymer (with block segments identical to the homopolymers) to blends of PS and PMMA on rheology was investigated by Valenza et al. (56). In a delineation of viscosity versus shear rate it was shown that compatibilization shifted the flow curve to a higher level. The prediction of the zero shear viscosity composition relationship by a relation of Choi and Schowalter (74) was in good agreement with the experimental data for the compatibilized blends.

Decreases in the interfacial tension (and, thus the interfacial energy) in the presence of copolymers have been observed experimentally in several studies (48-54). All observed an initially pronounced reduction of the interfacial tension at small concentration of the copolymer followed by a leveling off, which was indicative of saturation of the interface by the modifier. An example is shown in Figure 2. Experimentally the critical micelle concentration (CMC) can be determined from the plot of ν versus concentration as the intersection of a straight line drawn at low concentration and the leveling off line. Fleisher et al. (54) additionally investigated the molecular weight and end group type on interfacial tension.

To obtain more information on the topic of interface characterization and properties the reader may consult two recent monographs in the area edited by Feast et al. (78) and a monograph edited by Sanchez (79).

Theories and Past Studies on Coalescence

Unmodified Systems. Several theories have been proposed to describe interactions and coalescence of particles. A comprehensive treatment of the subject is given in a review by Chesters (80). The theories of coalescence are either based on equilibrium thermodynamics or hydrodynamics: the former are considering quiescent systems in which coalescence results from minimization of the total free energy by reduction of the interface area whereas the latter are considering flowing systems. For the case of unmodified systems many experimental studies have shown that the observed particle size is larger than predicted by the theories (7-9,14,21,81,82).

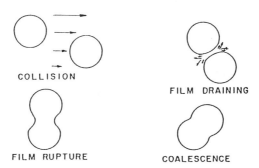

Figure 1. Sketch of a shear induced coalescence event of dispersed Newtonian droplets.

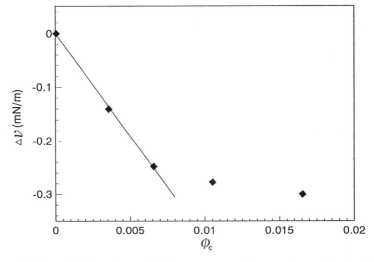

Figure 2. Interfacial tension reduction versus copolymer volume fraction for the system polystyrene/1,2 polybutadiene/P(S-b-B). Solid line is theoretical data below CMC (*44*). (Data from ref. 51).

Consistent reports confirm that the droplet size increases considerably with increasing concentration of the dispersed phase. Elmendorp and van der Vegt (8) observed that only for very low concentrations (<0.5 vol%) the domain size approached that predicted by Taylor's theory (83). Consequently coalescence of particles has to be taken into account.

A set of complex population balance equations was developed by Valentas (4); these are commonly used to predict the domain size distribution in emulsions. Tokita used a simplified version of these equations to predict the dependence of domain size on concentration in rubber blends (7). Both the approaches by Valentas and Tokita failed to predict the particle size quantitatively because of the key problem: the probability that collisions result in a coalescence are not directly accessible.

Fortelny et al. (81) developed a relation for equilibrium droplet size in steady state shearing flow following a procedure similar to that of Tokita (7). A more pronounced coalescence was observed than that predicted.

Utracki carried out an extensive study on shear coagulation of latexes under steady state shear flow, including a wide range of variables (73). An expression relating the coagulation time to these variables was derived. However, in order to estimate the coagulation time a constant must be determined by running several controlled experiments. The theory has recently been extended to polymer-polymer systems and implemented in a model of morphology evolution during blending (11). The diameter rate change of spherical drops due to coalescence was expressed as:

$$dD/dt = C \ D^{-1} \phi^{8/3} \dot\gamma \qquad (2)$$

where C is the previously mentioned constant, ϕ is the volume fraction of the dispersed phase, and $\dot\gamma$ is the shear rate.

A model based on three differential equations describing the dispersion processes during steady state and transient shear flow of viscoelastic liquid drops in a viscoelastic matrix was proposed by Lyngaae-Jørgensen et al. (10,24). These equations also included interactions between nonspherical drops. The rate equations for the dispersion process were based on several assumptions outlined in (24). Adjustable model parameters were estimated by fitting the model to transient data of average aspect ratio of prolate spheroids obtained from rheo-optical light scattering measurements. The number rate of domains coalescing was defined as $dN_c/dt \equiv (\phi/\bar{v})/\theta_c$ where ϕ is volume fraction of drops, $\bar{v}$ the average drop volume, and θ_c the average life time of drops.

The coalescence rate of drops using an expression for the average life time of drops by Silberberg and Kuhn (6) was expressed as:

$$dN_c/dt = 6\pi^{-2} \ D_n^{-3} \ \phi^2 \ \dot\gamma \ P \qquad (3)$$

where D_n is the number average radius of spherical drops, $\dot\gamma$ is the shear rate, and P is the probability that the collision will lead to coalescence. At constant volume fraction, the probability of coalescence, P, was assumed to be proportional to the time of interaction between two drops:

$$P \propto p/(\dot{\gamma}\tau_D) \tag{4}$$

where p is the drop aspect ratio, τ is a characteristic time for the squeeze out process and interdiffusion. Thus $P \to 1$ for high values of $p/(\tau_D \dot{\gamma})$, $\to$ zero at low values and variable for $p \sim \dot{\gamma}\tau_D$. For $\eta_2 > \eta_1$ the value τ was expressed as (85):

$$\tau \sim \frac{\eta_o M_c H \rho}{c^2 R T} \tag{5}$$

where η_o is the zero shear viscosity, M_c a critical molecular weight, $H \equiv \overline{M}_w/\overline{M}_n$, ρ is density, c is polymer concentration ($\sim \rho$ for melts), T is temperature and R is the gas constant; all parameters for the continuous phase (neat blends).

Roland and Böhm (82) investigated shear induced coalescence in a two phase rubber blend by small angle neutron scattering. The scattering invariant (not dependent on size and shape of particles) was used to estimate the extend of coalescence. It was found that the coalescence effect proceeds at a very rapid rate during mixing in a two-roll mill. The rheological properties of both phases and the flow field used in blending greatly influenced the process. It was concluded that the coalescence rate depends on the same conditions that favor droplet break-up.

Effect of Compatibilization. It is to be expected that the coalescence probability strongly depends upon the degree of mobility of the interface. As previously mentioned, immobilization of the interface can be obtained by addition of compatibilizers to the blend. Elmendorp and van der Vegt investigated theoretically and experimentally the influence of shear induced coalescence on the morphology in polymer blends (8,9). In the theoretical part a procedure was proposed to estimate the coalescence probability. It was estimated that this probability is only significant if the interface between polymers exhibits a high degree of mobility.

For immobile interfaces Elmendorp obtained the following equation for the coalescence time (8):

$$t_c = \frac{3}{4} \frac{\kappa}{\dot{\gamma}} \left(\frac{F R}{2\pi v} \right) \left(\frac{1}{h_{cr}^2} - \frac{1}{h_o^2} \right) \tag{6a}$$

For partly mobile interfaces (the most likely in interface modified blends) the following equation was obtained by Chester (80):

$$t_c = \frac{1}{4} \frac{\kappa \lambda}{\dot{\gamma}} \left(\frac{F R}{2\pi v} \right)^{1/2} \left(\frac{1}{h_{cr}} - \frac{1}{h_o} \right) \tag{6b}$$

with $\kappa = \dfrac{\eta_c R \dot{\gamma}}{v}$, $h_{cr} = \left(\dfrac{r_a^2 A}{4\pi^3 v}\right)$ and $h_o \approx R/2$

where κ is the capillarity number, R is the droplet radius, $\dot{\gamma}$ is the shear rate, λ is the viscocity ratio, η_c the viscosity of the continuous phase, v the interfacial tension, F is a force (a hydrodynamic force for the case of a flow field), h_{cr} is the critical distance, r_a is the radius of contact area, and A the Hamaker constant. A rough estimate on h_{cr} for polymer systems is 50 nm (*8*).

It was shown that the coalescence probability decreases considerably with increasing particle size and capillarity number (*8*). The Equations by Elmendorp and Chesters were, however, derived for an isolated pair of droplets and, consequently, they do not take the effect of concentration into account.

A first approximation of the probability of coalescence may be expressed by $P = \exp(-t_c/t_i)$, where t_c is given by equation 6 and t_i is the time of interaction which may be taken as proportional to the inverse shear rate (*86*). An alternative approach is to use equation 4 with all parameters in equation 5 for the interfacial layer.

Emulsification curves in accordance with the concept of Djakovic et al. (*87*) have recently also been determined for blends (*88-93*). The emulsification curve essentially follow the evolution of the dispersed phase size with interface modifier concentration as shown in Figure 3.

The critical conditions for obtaining a quasiequilibrium particle size were determined and a critical interface area was estimated according to the following expression (*93*):

$$A_{cr} = 9\dfrac{M_A + M_B}{N_A \bar{R}_{ee}^2} \dfrac{3\phi_c}{R} \qquad (7)$$

where M_A and M_B are the molecular weights of A and B blocks, respectively, ϕ_c is the block copolymer volume fraction, N_A is Avogrados number, R_{ee} is the average size of the end to end vector of the block chain extending into the matrix, and R is the radius of the droplet.

Considering the qualitative similarity between plots of, respectively, v and ΔD versus copolymer concentration a direct proportionality between interfacial tension reduction and particle size reduction may be argued: $\Delta D \cong C\Delta v$, where the expression for Δv is given by equation 1. It should be noted that the relation is only valid below CMC.

Procedures have been developed to produce graft copolymers by free radical chain transfer to methylmethacrylate monomers to yield an interfacial modifier for binary blends of PS and PMMA (*63*). A decrease in particle size of the dispersed phase and a significant improvement of mechanical properties confirmed the effect of the graft copolymer. Recently Macosko et al. (*99*) compared premade versus various reactively formed block copolymers in blends with basic components of PS and PMMA blended in a parallel plate mixer. They found the latter copolymers to be most effective in reducing the particle size (Figure 4). In the reactive case the blend was stabilized more quickly and a more narrow size distribution was obtained. This may be ascribed to an effect of the graft copolymer already at the creation of the interface between the basic components.

The influence of electron beam irradiation of the dispersed phase on the microrheology of immiscible blends was investigated by van Gisbergen and Meijer (95) and by Valenza et al. (96). The coalescence of particles were retarded to a large extent but not as much as expected.

Investigations of the coalescence and deformability have been carried out on a commercial research blend of poly(styrene-co-acrylonitrile) (SAN), EPDM rubber, and a graft copolymer of those (97). A rheo-optical technique was used for the measurements. The most conspicuous result was that a minor shear-induced agglomeration took place. This was particularly surprising considering that the rubber phase was slightly crosslinked and the graft copolymer was present at the interface. Similar results have also been obtained in comprehensive investigations on ABS polymer systems (98). The results showed that the efficiency of the graft copolymer is, beside the graft level (GL), dependent on the copolymer molecular weight (M_c). At low GL an increase of M_c caused a pronounced decrease of coalescence, while at higher GL the coalescence actually increased above a threshold level of M_c.

Using a model system of PS and PMMA with a fixed composition ratio 10/90, the influence of adding a block copolymer with components chemically equivalent to the homopolymers was examined by the authors in simple shear flow (88). In this system also deformation and break-up of domains were present that were accounted for using a simulation model as outlined in (88). Adjustable model parameters were estimated by fitting the model to transient data of average aspect ratio of prolate spheroids obtained from rheo-optical measurements for the purpose of retrieving specific parameters. From the simulation it was concluded that addition of a block copolymer lowers the interfacial tension as well as the probability of coalescence due to collision. The block copolymer addition also seemed to stabilize the fibrils and to lower the tendency to percolation of the minor phase.

A Microrheological Study of Coalescence. Experiments referred so far can be divided into investigations at quiescent conditions (e.g. annealing under influence of gravitional force), in complex morphologies or complex flow fields, whereas less attention has been devoted to specific flow fields.

In the latest work of the authors of this chapter light scattering measurements were used in order to quantitatively examine the effects of immobilization of the interface on the particle size (99). In the reviewed coalescence studies, attention was focused on the influence of a premade block copolymers on the interfacial properties of a model blend. The presence of the block copolymer at the interface was not measured directly, but the effect of the copolymer was indirectly revealed by light scattering measurements during annealing and shearing. The investigations were performed with a laser rheo-optical unit in connection with a Rheometrics Mechanical Spectrometer.

The model system chosen was a binary blend of polystyrene (PS) and

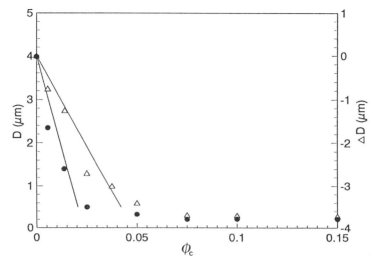

Figure 3. Domain diameter, D, and increment, ΔD, versus copolymer volume fraction for the system polystyrene/polymethylmethacrylate/P(S-b-MMA). M_w of copolymer (kg/mol): Δ 200; • 750. Solid lines are theoretical data according to ΔD = C Δv. (Data from ref. 90)

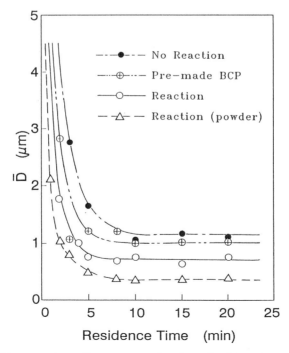

Figure 4. Disperse phase diameter as a function of mixing time measured by small-angle light scattering (94).

polymethylmethacrylate (PMMA) in absence and in the presence of a block copolymer. Blends of this components are known to display pronounced two phase morphologies. By virtue of some similarity of the solubility parameters, one might assume that blends of these components posses some miscibility region. A rather sharp interface was, however, observed by neutron reflection measurements (69) at ambient temperature whereas a slightly increasing interphase thickness with temperature was determined by ellipsometric analysis (71). Some material characteristics are given in Table I.

Table I. Selected Material Characteristics

	PS	PMMA
Density (kg/m^3)	1.04	1.19
Refractive index	1.49	1.59
Solubility parameter (kJ$^{1/2}$m$^{-3/2}$)	590	610

Commercial grades of polystyrene (PS), ($\overline{M}_w$ = 225 kg/mol; $\overline{M}_w/\overline{M}_n$ = 3) and polymethylmethacrylate (PMMA), ($\overline{M}_w$ = 87 kg/mol; $\overline{M}_w/\overline{M}_n$ = 2) were used as test materials. In this presentation the bulk polymer composition is confined to PS/PMMA (wt/wt%): 90/10. Beside the pure blend, a second sample was prepared by addition of 2 wt% block copolymer with components chemically equivalent to the homopolymers constituting the blend with block lengths smaller than both homopolymers. Samples with higher copolymer concentration were found not to give any significant change of properties. The block copolymer had a styrene content of 75% with 5% of the total mass as neat PS, ($\overline{M}_w$ = 91 kg/mol; $\overline{M}_w/\overline{M}_n$ = 1.8). Thus, in the blend concerned the longest chain of block copolymer penetrated the PS matrix phase. The asymmetrical block copolymer was chosen so as to obtain a more sensitivity measure of the influence of flow on the interfacial stability.

The samples were prepared by melt blending components in a mixer for 20 min and 10 rpm at 180 °C to obtain a dispersion of PMMA drops in a matrix. The minor component possessed the highest viscosity (zero shear viscosity ratio ~ 8) so as to exclude the effect of droplet deformation and break-up in the coalescence measurements. Rheological data for the blends was published in (56).

A rheo-optical set-up for simultaneous rheological and optical measurements was used in the experimental work; a more detailed description can be found in reference 24. The unit was built into a Rheometrics Mechanical Spectrometer. It allows rapid detection of angular distribution of scattered light from multi-phase polymer melts during shear flow while the sample is irradiated by a laser beam. In addition to the transmittance, the light detector is measuring the light in two orthogonal directions simultaneously with one direction in that of the flow. The measurements were conducted in a closed cone and plate shear cell to obtain a shear flow field at higher shear rates than usually accessible. The sample was placed between a glass cone and a fixed plate. The accessible shear rates varied from 0.01 to approximately 1000 s^{-1}. Dimensions of the glass geometry was: diameter 28 mm and angle of cone 0.01 rad.

The data acquisition system and analysis software made it possible to construct the light scattering pattern (isointensity contours and the scattering profiles) based upon two-dimensional measurements so as to provide a basis for visual as well as quantitative evaluation of the results.

The approximation applied to estimate the particle size was that of Debye-Bueche (*100*). An isotropic system with random two-phase structure and sharp interfaces was assumed. Here, a random structure refers to a system in which particles of random size are randomly distributed in a continuous phase. The intensity of the total scattered light from this structure can be expressed in a simplified form by the equation:

$$I(q) = C \langle \eta^2 \rangle a_c^3 S(qa_c) \tag{8}$$

where C is a constant, q is the magnitude of the scattering vector, $\langle \eta^2 \rangle$ is the mean square fluctuation of scattering contrast, a_c is a correlation distance which can be related to the particle size, and S is a scaling function defined by $S(x) = (1 + x^2)^{-2}$. For a structure consisting of randomly distributed spherical domains of radius R the following relation apply

$$a_c = (4/3)(1 - \phi)\phi R \tag{9}$$

Under the assumptions specified above this equation correlates quantitatively the domain size with the correlation distance. The light scattering data yielded and underestimation of particle size due to multiple scattering and due to a decrease in contrast between phases during flow not fully accounted for by the model. Accordingly, the data was corrected for these effects by approximative procedures. A detailed description of a stringent procedure for correction of multiple scattering is given in (*24*) chapter 3.

Prior to analyzing the extend of shear induced coalescence, investigations were carried out to elucidate any concomitant effects. The effects subject to closer examination were:
1. Ostwald ripening (large particles grow in expense of small ones).
2. Stokes flow (gravitational effects).
3. Viscous flow (particle motions caused by irregular shaped interfaces) also referred as interfacial driven coalescence.
4. Marangoni flow (temperature gradient induced particle motions)

Initially recordings were performed during annealing on neat blend platelet without any pressure on the samples for up to 3 hours. As can be ascertained from the lower curve in Figure 5 no significant domain growth could be detected. This excluded domain growth and coalescence caused by any of the processes 1-2 listed above. These observations was also confirmed by electron micrographs of samples. By applying a small shear strain to the platelet in order to induce irregular shaped collision doublets and then leaving the sample completely relaxed for one hour, did not result in any effect due to a self-acting interfacial tension driven coalescence (item 3). This conclusion is supported by the constant level of the integrated intensity after cessation of shear as shown in Figure 6. To elucidate the effect of the remaining process (item 4) measurements were carried out with temperature gradients up to 60 °C/mm across the sample thickness. Size distribution analysis of micrographs from the original and the gradient annealed samples only indicated a slight increase of particle size in the latter one.

The results of shearing measurements at low shear rates shown in Figure 5 indicate that in the neat blends domain growth starts immediately after onset of shearing. The increase of domain size presumably levels off at some upper limit which was assumed to only increase slowly with shear rate. The main effect of an increased shear rate is expected to be an increased rate of domain growth up to a maximum where the contact time is reduced below a critical limit and the contact area becomes too large for the coalescence to occur.

The results on the blend with block copolymer indicated a more complex behavior. Figure 7 shows recordings of time resolved traces of integrated light scattering signals for various shear rates. The signals were initially constant for periods of time decreasing with increasing shear rate followed by a marked increase. This change was ascribed to an onset of coalescence. The interpretation of the light scattering signals was further supported by micrographs of quenched samples (see Figure 8a,b,c). From Figure 8c it is noted that micelles or mesophases have been formed in the dispersed phase (PMMA). The shear and time dependent onset of coalescence were tentatively attributed to one or more of the following effects:

a. Changes in the block copolymer molecular weight due to cleavage in the interface or shear-induced enhanced miscibility between components that thermodynamically favored diffusion of copolymer molecules away from the interface (into the PMMA domains).
b. Block copolymer chains gradually pulled out by frictional forces acting during the flow on the chains extending into the matrix phase in accordance with a pull out mechanism suggested by Henderson and Williams (84). In this case the copolymer is expected to be distributed in the matrix.
c. Entrapment of the copolymer at the interface of colliding particles and formation of micelles in the domains.

In order to effectively discriminate between the above mechanisms one method is to selectively label the block copolymer and use a technique which can sense that.

The stability data are summarized in Figure 9. Beyond the stability limit the system essentially behaves as a neat blend. The inclusions observed in the domains after onset of coalescence is in case (b) that some of the matrix phase has been entrapped between colliding particles.

The abrupt changes observed in the signals (Figure 7) at inception and at cessation of shear were ascribed to changes in the optical properties (birefringence) due to orientation and relaxation, respectively, of primarily the continuous phase.

The rheo-optical measurements clearly monitored the effect of the block copolymer. The results from the shearing measurements showed that domain growth

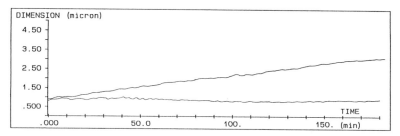

Figure 5. Particle diameter versus time for PS/PMMA-90/10 at 210 °C. Lower curve: annealing; upper curve: $\dot{\gamma} = 0.03$ s^{-1} (Adapted from ref. 99).

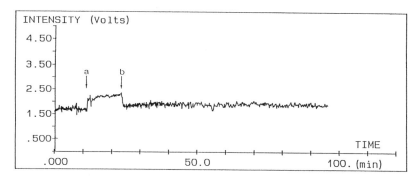

Figure 6. Test of interfacial driven coalescence initially induced by shear. The curve reflects the integrated light scattering intensity. a. inception of shear; b. cessation of shear and subsequent annealing at 210 °C. (Adapted from ref. 99).

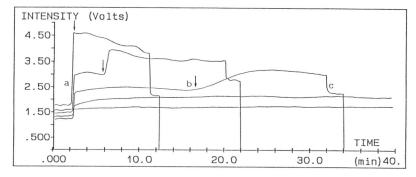

Figure 7. Traces of integrated light intensity signals versus time at 210°C with shear rates related to curves (from bottom): 8.5 s^{-1}; 34 s^{-1}; 85 s^{-1}; 340 s^{-1}; 850 s^{-1}. Arrows indicate onset of coalescence. (Reproduced with permission from ref. 99. Copyright 1995 Butterworth-Heinemann Ltd.).

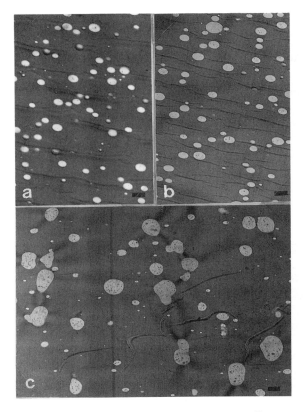

Figure 8. Micrographs of sample in Figure 7 corresponding to various stages on the curve recorded at 85 s^{-1}: a. before inception of shear; b. before onset of coalescence; c. after cessation of shear. x 4000; bar: 1 μm. (Adapted from ref. 99).

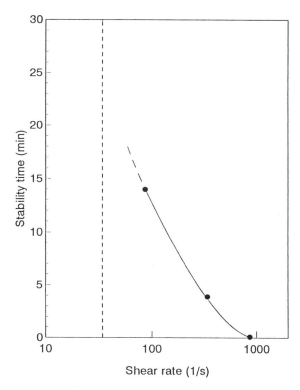

Figure 9. Stability time of diblock copolymer at the interface of a PS/PMMA-90/10 blend versus shear rate at 210 °C. The vertical dashed line indicates the limit below which the copolymer resides at the interface for at least 2h. (Adapted from ref. 99).

is immediately initiated in the blend without copolymer while initially hardly any growth occurred in the blends with block copolymer. In the latter case the onset of coalescence depended on time and shear rate. The measurements showed that the primary and dominant source of coalescence is that due to flow in highly immiscible high-molecular-weight systems in the strong segregation region.

The work indicated that the interfacial stability is not unconditional during the flow. Criteria for optimal concentration and block chain lengths for the block copolymer to reside in thermodynamic equilibrium on the interface at static conditions should also take the influence of flow
into account.

Summary

General conclusions drawn from the observations reviewed in this chapter are summarized as follows:

- the probability of collision increases with particle size (the projected cross sectional area of the particles in the flow direction) (8,9), whereas the probability of coalescence levels off (8).
- smaller particles coalesce more readily owing to easier removal of the intervening film (86).
- there is a nearly linear relation between mixing time and extend of coalescence (8).
- several observations confirm that the particle size increases with concentration.
- a viscosity increase of the continuous phase causes a reduced coalescence (70).
- an increase of the dispersed phase molecular weight and thus viscosity inhibits coalescence (70).
- increasing the shear rate up to a critical level results in an increase of coalescence. At higher shear rates the coalescence efficiency is expected to be reduced, presumably due to a reduced contact time and an enlarged contact area.
- upon addition of a block copolymer the coalescence is strongly inhibited, the onset of coalescence being dependent on time and shear rate (99).

Acknowledgments

The authors work reviewed was supported by the Danish Council for Scientific and Industrial Research and the Danish Council of Technology.

Literature Cited

1. Utracki, L. A. *Polymer Alloys and Blends, Thermodynamics and Rheology*, Hanser Publishers: Munich, 1989.
2. Lyngaae-Jørgensen, J.; Utracki, L. A. *Makromol. Chem., Macromol. Symp.* **1991**, *48/49*, 189.
3. Utracki, L. A. *J. Rheol.* **1991**, *35*, 1615.
4. Valentas, K. J.; Bilous, Q.; Amundson, N. R. *Ind. Eng. Chem. Fundam.* **1966**, *5*, 271.
5. Coulaloglou, C. A.; Taularides, L. L. *Chem. Eng. Sci.* **1977**, *32*, 1289.
6. Silberberg, A.; Kuhn, W. *J. Polym. Sci.* **1954**, *13*, 21.
7. Tokita, N. *Chem. Technol.* **1979**, *50*, 292.
8. Elmendorp, J. J.; van der Vegt, A.K. *Polym. Eng. Sci.* **1986**, *26*, 1332.
9. Fortelny, I.; Kovár, J. *Eur. Polym. J.* **1989**, *25*, 317.
10. Lyngaae-Jørgensen J.; Søndergaard, K.; Utracki, L.A.; Valenza, A. *Polym. Networks Blends* **1993**, *3*, 167.
11. Huneault, M. A.; Shi, Z. H.; Utracki, L. A. *Polym. Eng. Sci.* **1995**, *35*, in press.
12. Van Oene, H. In *Polymer Blends*; Paul, D. R.; Newman, S., Eds.; Academic Press: New York, 1978.
13. Utracki, L. A. *Polym. Eng. Sci.* **1983**, *23*, 602.
14. Favis, B. D.; Willis, J. M. *J. Polym. Sci. Part B. Polym. Phys.* **1990**, *28*, 2259.

15. Chen, C. C.; Fontan, E.; Min K.; White, J. L. *Polym. Eng. Sci.* **1988**, *28*, 69; Chen, C. C.; White, J. L. *Polym. Eng. Sci.* **1993**, *33*, 923.
16. Sundararaj, U.; Macosko, C. W. Submitted to *Macromolecules*.
17. Han, C. D. *Multiphase Flow in Polymer Processing*, Academic Press: New York, 1981.
18. Elmendorp, J. J.; van der Vegt, A. K. In *Two-Phase Polymer Systems*; Utracki, L. A., Ed.; Prog. Polym. Process. Vol. 2, Hanser Publishers: Munich, 1991, Vol. 3; p 165.
19. Utracki, L. A. In *Current Topics in Polymer Science*; Ottenbrite, R. M; Utracki, L. A.; Inoue, S., Eds.; Hanser Publishers: Munich, 1987.
20. Wu, S. *Polym. Eng. Sci.* **1987**, *27*, 335.
21. Plochocski, A. P.; Dagli, S. S.; Andrews, R. D. *Polym. Eng. Sci.*, **1990**, *30*, 741; Plochocski, A. P. *Report on IUPAC Working party 4.2.1.*, 1993.
22. Meijer, H. E. H.; Janssen, J. M. H. In *Mixing and Compounding - Theory and Practical Progress*; Manas-Zloczower I.; Tadmor, Z., Eds.; Progr. Polym. Process.; Hanser Publishers: Munich, 1994, Vol. 4.
23. Utracki, L. A.; Shi, Z. H. *Polym. Eng. Sci.* 1992, *32*, 1824; Shi, Z. H.; Utracki, L. A.; *ibid*, 1834.
24. *Rheo-Physics of Multiphase Polymer Systems: Characterization by Rheo-Optical Techniques* Søndergaard, K.; Lyngaae-Jørgensen, J., Eds.; Technomic Publishing Co.: Lancaster PA, 1995.
25. Stasiak, W.; Cohen, C. *J. Phys. Chem.* **1993**, *98*, 6510.
26. Ratke, L.; Thieringer, W. K. *Acta Metall.* **1985**, *33*, 1793
27. Paul, D. R. In *Polymer Blends*; Paul, D. R.; Newman, S., Eds.; Academic Press: New York, 1978, Vol. 2, p. 35.
28. Gaylord, N. G. *Adv. Chem. Ser.* **1975**, *142*, 76.
29. Kraus, G. In *Polymer Blends*; Paul, D. R.; Newman, S., Eds.; Academic Press: New York, 1978, Vol. 2, p. 243.
30. Heikens, D.; Hoen, N.; Barentsen, W.; Piet, P.; Ladan, H. *J. Polym. Sci. Polym. Symp.* **1978**, *62*, 309.
31. Rudin, A. *J. Rev. Macromol. Chem.* **1980**, *C19*, 267.
32. Fayt, R.; Jérôme, R.; Teyssie, P. In *Multiphase Polymers: Blends and Ionomers*; Utracki L. A.; Weiss, R. A., Eds.; ACS Symp. Ser. 395, ACS: Washington D.C., 1989, p. 38.
33. Riess, G.; Kohler, J.; Tournut, C.; Brandert, A. *Makromol. Chem.* **1967**, *101*, 58.
34. Riess, G.; Solivet, Y. *Adv. Chem. Ser.* **1975**, *142*, 243.
35. Inoue, T.; Soen, T.; Hashimoto, T.; Kawai, H. *Macromolecules* **1970**, *3*, 87; and in *Block Polymers*; Aggarwal, S. L., Ed.; Plenum: New York, 1970, p. 53.
36. Meier, D. *ACS Polym. Prepr.* **1977**, *18*, 340.
37. Jiang, M.; Huang, X.; Yu, T. *Polymer* **1983**, *24*, 1259.
38. Jiang, M.; Cao, X.; Yu, T. *Polymer* **1986**, *27*, 1923.
39. Xie, H., Liu, Y., Jiang, M. and Yu, T. *Polymer* **1986**, *27*, 1928.
40. Xanthos, M. *Polym. Eng. Sci.* **1988**, *28*, 1392; *ibid* **1991**, *31*, 929.
41. Leibler, L. *Makromol. Chem., Makromol. Symp.* **1988**, *16*, 1.
42. Hong, K. M.; Noolandi, J. *Macromolecules* **1981**, *14*, 727.

43. Noolandi, J.; Hong, K. M. *Macromolecules* **1982**, *15*, 482.
44. Noolandi, J.; Hong, K. M. *Macromolecules* **1984**, *17*, 1531.
45. Vilgis, T. A.; Noolandi, J. *J. Makromol. Chem., Macromol. Symp.* **1988**, *16*, 225.
46. Shull, K. R.; Kramer, E. J. *Macromolecules* **1990**, *23*, 4769.
47. Lövenhaupt, B.; Hellmann, G. P. *Colloid Polym. Sci.* **1990**, *268*, 885.
48. Wu, S. *Polymer Interfaces and Adhesion*; Marcel Dekker: New York, 1982.
49. Patterson, H. T.; Hu, K. H.; Grindstaff, T. H. *J. Polym. Sci. Part C,* **1971**, *34*, 31.
50. Gailard, P.; Ossenbach-Sauter, M.; Riess, G. *Makromol. Chem. Rapid Commun.* **1980**, *1*, 771.
51. Anastasiadis, S. H.; Gancarz, I.; Koberstein, J. T. *Macromolecules* **1989**, *22*, 1449.
52. Elemans, P. H. M.; Janssen, J. M.; Meijer, H. E. M. *J. Rheol.* **1990**, *34*, 1311.
53. Wagner, M.; Wolf, B. A. *Polymer,* **1993**, *34*, 1460.
54. Fleisher, C. A.; Koberstein, J. T.; Krukonis, V.; Webmore, P. A. *Macromolecules* **1993**, *26*, 4172.
55. van Ballegooie, P.; Rudin, A. 1989. *Makromol. Chem.* **1989**, *190*, 3153.
56. Valenza, A.; Utracki, L. A.; Lyngaae-Jørgensen, J. *Polym. Networks Blends,* **1991**, *1*, 71.
57. Gleinser. W.; Braun, H.; Friedrich, C.; Cantow H.-J. *Polymer* **1994**, *35*, 128.
58. Graebling, D.; Muller, R. *Colloids and Surfaces* **1991**, *55*, 89.
59. Gramespacher, H.; Meissner, J. *J. Rheol.* **1992**, *36*, 1127.
60. Søndergaard, K.; Lyngaae-Jørgensen, J.; Valenza, A. *Abstracts, The Fourth Eur. Symp. on Polymer Blends*, May 24-26, Capri, 1993, pp. 231-234.
61. Hashimoto, T.; Shibayama, M.; Kawai, H. *Macromolecules* **1980**, *13*, 1327.
62. Klein, J.; Briscoe, B. J. *Proc. Roy. Soc. A, London,* **1979**, *365*, 53.
63. Goldstein, J. I.; Newbury, D. E.; Echlin, P. E.; Coy, D. C.; Fiori, C.; Lifshin, E. *Scanning Electron Microscopy and X-Ray Microanalysis*; Plenum: New York, 1981.
64. Chu, W. K.; Mayer, J. W.; Nicolet, M. A. *Back-Scattering Spectrometry*, Academic Press: New York, 1978.
65. Shull, K. R.; Kramer, E. J.; Hadziioannou G.; Tang, W. *Macromolecules* **1990**, *23*, 4780.
66. Jones, R. A. L.; Kramer, E. J.; Rafailovich, M. H.; Sokolov, J.; Schwarz, S. A. *Interfaces between Polymers, Metals and Ceramics*; DeKoven, B. M.; Gellman, A. J.; Rosenberg, R, Eds.; *Mat. Res. Soc. Symp. Proc. 153*, 132.
67. Arwin, H.; Aspues, D. E. *Thin Solid Films* **1986**, *138*, 195.
68. Azzam, R. M. A.; Bashara, N. N. *Ellipsometry and Polarized Light*; Elsevier Sci. Publ.: Amsterdam, 1987.
69. Walsh, D. J.; Sauer, B. B.; Higgins, J. S.; Fernandez, M. L. *Polym. Eng. Sci.* **1990**, *30*, 1085.
70. Kresler, J.; Higashida, N.; Inoue, T.; Heckmann, W.; Seitz, F. *Macromolecules* **1993**, *26*, 2090.
71. Yukioka, S.; Inoue, T. *Polymer,* **1994**, *35*, 1182.
72. Sinha, S. K.; Sirota, E. B.; Garoff, S.; Stanley, H. B. *Phys. Rev. B.* **1988**, *38*, 2297.

73. Penfold, J.; Ward, R. C.; William, W. G. *J. Phys. E. Sci. Instr.* **1987**, *20*, 1411.
74. Choi, S. J.; Schowalter, W. R. *Phys. Fluids* **1975**, *18*, 420.
75. Werner, S. A.; Klein, A. G. In *Neutron Scattering*; Celotta, R. et al., Eds.; Academic Press: New York; Chapter IV.
76. Russell, T. P.; Karim, A.; Mansour, A.; Felcher, G. P. *Macromolecules* **1988**, *21*, 1890.
77. Anastasiadis, S. H.; Menelle, A.; Russell, T. P.; Satija, S. K.; Felcher, G. P. *ACS Polymer Prepr.* **1990**, *31(2)*, 77.
78. *Polymer Surfaces and Interfaces*; Feast, W. J.; Monroe, S., Eds.; Wiley: Chichester, 1987; *Polymer Surfaces and Interfaces II*; Feast, W. J.; Richards, R. W.; Monroe, H. S., Eds.; Wiley: Chichester, 1993.
79. *Physics of Polymer Surfaces and Interfaces*; Sanchez, I. C., Ed.; Butterworth-Heinemann: Boston, 1992.
80. Chesters, A. K. *Trans IChemE* **1991**, *Vol 69, Part A*, 259.
81. Fortelny, I.; Michalkova, D.; Koplikova, J.; Navratilowa, E.; Kovár, J. *Angew. Makromol. Chem.* **1990**, *179*, 185.
82. Roland, C. M.; Böhm, G. G. A. *J. Polym. Sci., Polym. Phys. Ed.* **1984**, *22*, 79.
83. Taylor, G. I. *Proc. Roy. Soc. A, London* **1932**, *138*, 41; *ibid* **1934**, *146*, 501.
84. Utracki, L. A. *J. Colloid. Interf. Sci.* **1973**, *24*, 185.
85. Lyngaae-Jørgensen, J. *Adv. Rheol.* **1984**, *3*, 503.
86. Ross, S. I.; Verhoff, F. H; Curl, R. I. *Ind. Eng. Chem. Fundl.* **1977**, *16*, 371; *ibid* **1978**, *17*, 101.
87. Dokic, P.; Radivojevic, P.; Sefer, I.; Sovilj, V. *Coll. Polym. Sci.* **1987**, *265*, 993.
88. Favis, B. D. *Polymer, Polym. Comm.* **1994**, *35*, 1552.
89. Favis, B. D. *PMSE Proc. ACS* **1993**, *69*, 180.Djakovic, L.;
90. Thomas, S.; Prud'home, R. E. *Polymer* **1992**, *33*, 4260.
91. Wagner, M.; Wolf, B. A. *Polymer* **1993**, *34*, 1460.
92. Tang, T.; Huang, B. *Polymer* **1994**, *35*, 281.
93. Matos, M.; Lomellini, P.; Tremblay, A.; Lepers, J.-C.; Favis, B. D. *Lecture, The Eur. Reg. Meet. PPS, Strasbourg, Aug. 29-31*, 1994.
94. Nakayana, A.; Inoue, T.; Guigan, P.; Macosko, C. W. *ACS Polym. Prepr.* **1993**, *34(2)*, 840.
95. van Gisbergen J. G. M.; Meyer, H. E. H. *J. Rheol.*, **1991**, *35*, 63.
96. Valenza, A.; Spadaro, G.; Calderaro, E.; D. Acierno, D. *Polym Eng. Sci.* **1993**, *33*, 845; *ibid* 1336.
97. Søndergaard, K.; Lyngaae-Jørgensen, J. In *Polymer Rheology and Processing*; Collyer A. A.; Utracki, L. A., Eds.; Elsevier Appl. Sci.: London, 1990, pp. 107-155.
98. Chang, M. C. O.; Nemeth, R. L. *ACS Polymer Prepr.* **1994**, *71(2)*, 739 and ref. therein.
99. Søndergaard, K.; Lyngaae-Jørgensen, J.; Utracki, L. A. Submitted to *Polymer*, 1994.
100. Debye, P.; Anderson, H. R.; Brumberger, H. *J. Appl. Phys.*, **1957**, *28*, 679.
101. Henderson, C. P.; Williams M. C. *J. Polym. Sci. Lett.* **1979**, *17*, 257.

RECEIVED February 3, 1995

Chapter 13

Orientation Behavior of Thermoplastic Elastomers Studied by ^{2}H-NMR Spectroscopy

Alexander Dardin[1], Christine Boeffel[1], Hans-Wolfgang Spiess[1], Reimund Stadler[2], and Edward T. Samulski[3]

[1]Max Planck Institute for Polymer Research, Ackermannweg 10, 55021 Mainz, Germany
[2]Institut für Organische Chemie, Johannes-Gutenberg-Universität, Becherweg 18-20, 55099 Mainz, Germany
[3]Department of Chemistry, University of North Carolina, Chapel Hill, NC 27599

> In functionalized polybutadienes extended supramolecular junction zones are formed by self-aggregation of functional units via directed hydrogen bonds. The unusual ^{2}H-NMR lineshapes as well as the large quadrupolar splittings indicate highly oriented network chains in the stretched samples. Angular dependent measurements and inversion-recovery experiments prove the uniaxiality of the deformation and the homogeneity of the polymer backbone dynamics. A quantitative analysis of the lineshapes according to present theories is not possible but the spectra can be understood assuming a gradient of orientation along individual network chains. For the highest strain ratios of $\lambda > 10$ the ^{2}H-NMR spectra give evidence of completely stretched chain segments which are located near the supramolecular clusters.

The orientation behavior of elastomers has been studied extensively by means of a wide variety of different methods, e.g. FT-IR dichroism and strain birefringence. As a very powerful tool, ^{2}H-solid-state NMR was used to monitor dynamics and orientation on a molecular level simultaneously. Not only rigid (1) or liquid crystal (2) systems but also highly mobile materials such as elastomers (3) or even polymer melts (4) can be investigated. Mainly covalently endlinked PDMS networks were studied as models for 'ideal' networks. In order to understand the results of the NMR experiments different approaches have been developed based on short-range nematic-like orientational couplings (5), deformation behavior of single chains (6), anisotropic junction fluctuations (7) or the effect of a screened potential (8). Turning towards more complex systems, the orientation behavior of statistically crosslinked polybutadienes and thermoplastic elastomers, such as poly(styrene-b-butadiene-b-styrene) block copolymers and segmented polyurethanes has also been investigated by means of ^{2}H-NMR (9).

An unusual type of thermoreversible network was realized in flexible polydienes such as polybutadiene, which are attached by polar functional units (10).

The use of 4-(3,5-dioxo-1,2,4-triazolin-4-yl) benzoic acid (U4A, Figure 1, I) or -isophthalic acid (U35A, Figure 1, II), respectively, as functional groups results in a self assembling of these units via directed non-covalent interactions.

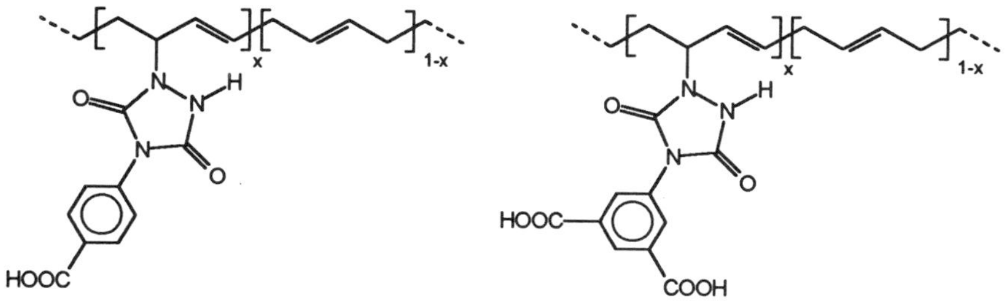

Figure 1. Functional groups used for functionalization of polybutadiene: 4-(3,5-dioxo-1,2,4-triazolin-4-yl) benzoic acid (I) and 4-(3,5-dioxo-1,2,4-triazolin-4-yl) isophthalic acid (II)

For the former, a set of hydrogen bondings leads to the formation of two-dimensional extended, ordered supramolecular structures (Figure 2), which can be considered as the effective junction zones in a polybutadiene matrix.

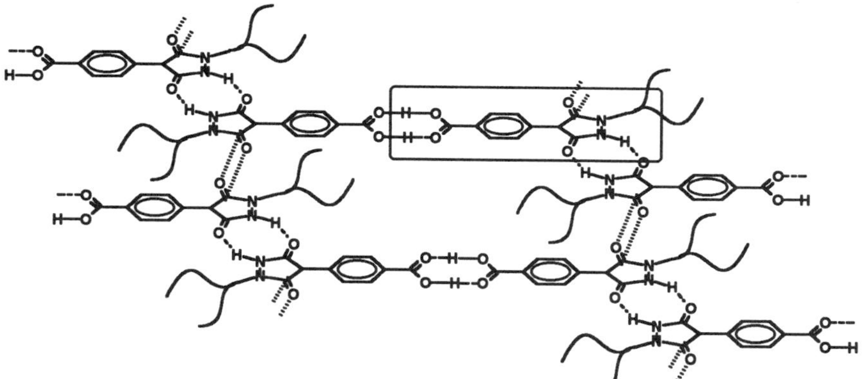

Figure 2. Supramolecular structure formed of 4-(4-carboxyphenyl)-3,5-dioxo-1,2,4-triazolidin-1-yl (urazole) units by hydrogen bondings (---) and dipole-dipole interactions (///) as taken from the X-ray analysis of trans-butene-U4A (11). A single urazole unit is framed.

Alternatively, a two dimensional structure may also be constructed by lateral aggregation of the hydrogen bond chains via weak urazole-aryl-H-hydrogen bonds (Seidel U.; Hellmann J.; Schollmeyer D.; Hilger C.; Stadler R. in preparation). In an analogous way, the U35A units are also assumed to form ordered structures via hydrogen bonds, capable of interacting with one more functional site. Although the exact structure is not solved yet, there is evidence for similar interactions as in the U4A clusters (*12*, Hellmann J.; Hilger C.; Stadler R. *Polymers for Advanced Technologies*, in press.). The aggregates formed by both urazole units undergo an order-disorder transition at temperatures of 65-80 °C (U4A-units) and 130-185 °C (U35A-units), respectively, which can be detected by differential scanning calorimerty (DSC), FT-IR spectroscopy and small angle X-ray scattering (SAXS) (*13 -16*).

In recent investigations of these systems, it was demonstrated that the mechanical properties are essentially determined by the structure of the functional units, i.e. by the organization of the supramolecular aggregates as well as by the degree of functionalization. It has also been shown that the molecular dynamics of the polybutadiene chains are influenced by the presence of the extended junction zones (*17*).

In this contribution we report on investigations of the orientation behavior of the network chains by means of different ^{2}H-solid-state-NMR methods.

Experimental

Sample Preparation. Deuterated 1,4-polybutadiene was synthesized by anionic polymerization of 1,1,4,4-d_4-butadiene in cyclohexane with sec.-Butyllithium as initiator. The deuterated butadiene was obtained by ^{1}H-^{2}H-exchange on butadienesulfone and subsequent thermal decomposition. The synthesis of the functional groups is described elsewhere (*12,13*).

The functionalization of the polybutadiene was carried out in solution and the solvent was allowed to evaporate slowly from the solution after complete reaction in petri-dishes to give completely transparent films. The films were cut to 70 mm x 5 mm stripes. Table I summarizes the samples used in the NMR investigations.

^{2}H-NMR Measurements. Solid state deuteron NMR data were recorded on a Bruker MSL-360 spectrometer operating at a frequency of 55,82 MHz. The spectra were obtained by a solid-echo pulse sequence ($90°_{\pm x}$-τ_1-$90°_{\pm y}$-τ_1-acquire-recycle delay) followed by a Fourier transformation starting at the echo maximum as described elsewhere (*18*). The r.f.-pulse length for a 90°-pulse was always 2 µs and the number of accumulations varied from 1024 to 30000 depending on the extension ratio of the sample. All experiments are performed at room temperature with no further temperature control.

For the experiments with stretched samples at different strains a new probe was designed (Figure 3) which allows a stepwise uniaxial deformation of the sample inside the r.f-coil with perpendicular orientation of the sample with respect to the external magnetic field. The body of the device is manufactured from Teflon to fit to a standard Bruker high power NMR probe. The sample film can be fixed on one end with two Vespel clamps (d). The other end is fixed on the top brass roll (f) with a screw-clamp.

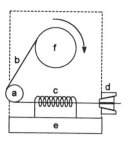

Figure 3. Schematic view of the stretching device (a,f: Brass rolls, b: Sample, c: 5 mm r.f. coil, d: Sample holder, e: Connection to NMR probe)

In these static experiments, the strain ratio of the sample was adjusted by manually turning the roll (f) clockwise. This roll can be fixed every 1/12th of a full rotation. The extension ratio of the sample was then determined by measuring the distance of two marks on the film and was checked after each experiment to assure that no slipping of the film occured during a single experiment. For details of the mechanical properties of the samples refer to references 10 and 12.

Due to the brass rolls the external magnetic field homogeneity was impaired resuling in a linewidth of about 70 Hz for a D_2O sample compared with a linewidth of about 15 Hz obtained with the standard high-power NMR setup.

Inversion-recovery experiments were performed with the standard pulse sequence in combination with a solid-echo sequence ($180°_{\pm x,y}$-t_1-$90°_{\pm x,y}$-τ_1-$90°_{\pm y,y}$-τ_1- acquire-recycle delay). The delay time, t_1, was varied from 0,5 to 50 ms and for every t_1 a spectrum was obtained after Fourier transformation.

Table I. Investigated samples

Polymer	$\overline{M}_n$ / g/mol[a]	Functional unit	Degree of modification / mole-%[b]	Sample name
dPB	100000	U4A	2	dPB-U4A-2
dPB	100000	U4A	4	dPB-U4A-4
dPB	100000	U35A	2	dPB-U35A-2
dPB	100000	U35A	4	dPB-U35A-4

[a] The letter 'd' denotes the deuterated compound. The molecular weight is determined by means of gel permeations chromatography (GPC) with polystyrene standard.
[b] The degree of modification is given in mole-% with respect to the number of double bonds of the polybutadiene.

²H-NMR Background

Due to the coupling of the deuteron quadrupolar moment (spin I=1) with the electric field gradient tensor (*efg*) along the C-²H-bond the resulting NMR frequency is directly related to its orientation in the external magnetic field B_0. Thus, the frequency can be expressed in terms of the Euler angles of the C-²H-bond in the laboratory frame

$$\omega = \omega_0 \pm \frac{\delta}{2}\left(3\cos^2\vartheta - 1 - \eta\sin^2\vartheta\cos 2\varphi\right) \tag{1}$$

where δ denotes the static quadrupolar splitting constant (164 kHz for polyethylene (*19*) and η the asymmetry parameter of the *efg* tensor which is usually zero in the absence of mobility.

While an isotropic distribution of C-²H-bond directions for static systems gives rise to the well known Pake pattern, in the presence of molecular motion the quadrupolar interaction can be averaged.

In rubbers, the correlation times of molecular motion is fast ($\tau_c=10^{-7}$s) compared to the inverse width of the spectrum, δ^{-1}, and thus the averaging is nearly complete resulting in a narrow single line on the order of 100 Hz in width at half height. In uniaxially strained elastomers, molecular dynamics becomes anisotropic with respect to the stretching direction and therefore the averaging along this direction is incomplete. As a result, a quadrupolar splitting is observed which is usually expressed in terms of the second Legendre polynomials:

$$\Delta v = \frac{3\delta}{2}\langle P_2(\cos\Phi)\rangle\langle P_2(\cos\Theta)\rangle P_2(\cos\Omega) \tag{2}$$

The first factor specifies the orientation of a single C-²H-bond with respect to chain's end-to-end vector, the second gives the orientation of the end-to-end vector with respect to the stretching direction, and the third gives orientation of the stretching direction with respect to the external magnetic field. The brackets denote time averages.

The classic theory of segmental orientation developed by Kuhn and Grün (*20*) predicts a linear dependence between the orientation of a statistical segment (Kuhn segment), the inverse of the number of segments, n^{-1}, and the strain function, $(\lambda^2-\lambda^{-1})$

$$\langle P_2(\cos\theta)\rangle_{KG} = \frac{1}{5n}\left(\lambda^2 - \frac{1}{\lambda}\right) + \ldots \tag{3}$$

assuming uncorrelated segmental orientation, affine deformation and Gaussian statistics of the network chains. Thus, it is expected to be valid only for small strains while for non-Gaussian behavior at larger strains, the higher order terms of equation 3 have to be taken into account.

Newer theories use mean-field concepts to account for the nature of the observed splittings in the ²H-NMR spectra of stretched elastomers. Sotta and

Deloche (5) report that the orientation of a single chain is not sufficient to give rise to a splitting in the ^{2}H-NMR spectra of strained elastomers. By introducing a 'nematic-like' field for orientational coupling they allowed the network chains to interact with each other. The result is a residual, incompletely averaged quadrupolar coupling which gives rise to a splitting as observed in the spectra. With the 'nematic-like' field approach the spectra of stretched PDMS-network could be explained qualitatively.

Another approach was given by Brereton (8) who used a 'screened potential' as an interaction parameter. In both models (5,8) the resulting quadrupolar splitting is given by

$$\Delta v = \frac{3 v_Q R^2}{5 n \langle R_0 \rangle^2} \left\{ \langle P_2(\cos\alpha) \rangle - C \left(\lambda^2 - \frac{1}{\lambda} \right) P_2(\cos\Omega) \right\} \quad (4)$$

where C denotes the interaction parameter for the 'nematic-like' interaction or the screened potential, respectively. v_Q is the quadrupolar coupling constant, R and R_0 the chain's end-to-end vector in the stretched and unstretched state and n the number of statistical segments.

Since these calculations account mainly for the origin of the observed quadrupolar splitting, other methods have been applied to this problem, e.g. Monte-Carlo methods were used to ascertain the actual chain conformations and to calculate the complete lineshape of strained elastomers for covalently crosslinked polybutadiene networks (21).

Experimental Results

Stretching experiments. In Figures 4 and 5, the ^{2}H-NMR spectra of dPB-U4A and dPB-U35A are shown for different extension ratios, λ. All samples exhibit single absorption lines in the unstretched state ($\lambda=1$) indicating an isotropic motional averaging on the NMR timescale. However, the linewidths at half height of about 800 to 1000 Hz, respectively, show that the motional averaging is not complete but depends on the crosslinking density, i.e. the number of functional groups along the polymer backbone. Upon uniaxial stretching, molecular motion and, therefore the time average becomes anisotropic with respect to the orientation direction. For small strains, this results only in a line broadening. At larger strains, a quadrupolar splitting can be observed. Further increase of the deformation ratio up to $\lambda=8$ leads to unusual complex lineshapes which exhibit small shoulders and broad components up to spectral widths of about 40 kHz in the U35A system.

In Figure 6 the quadrupolar splittings at the maximum of the spectra are plotted against the strain function. For the dPB-U4A samples, the splitting first increases for both crosslinking degrees with increasing strain up to $\lambda=5$ and then levels off to splittings of about 800 Hz for dPB-U4A-4% and 1400 Hz for dPB-U4A-2%, respectively. In comparison, the networks with the U35A units show even larger splittings up to 3500 Hz for the 4% functionalized sample at $\lambda=19$.

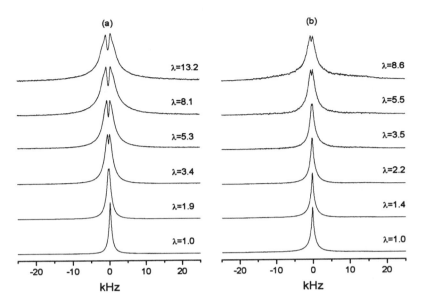

Figure 4. ^{2}H-NMR spectra of stretched (a) dPB-U4A-2% and (b) dPB-U4A-4%

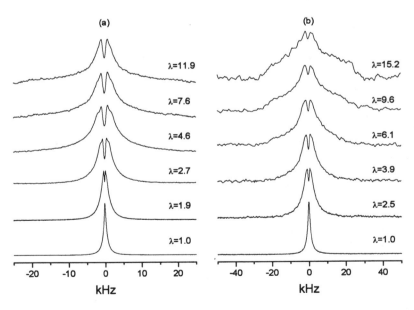

Figure 5. ^{2}H-NMR spectra of stretched (a) dPB-U35A-2% and (b) dPB-U35A-4%

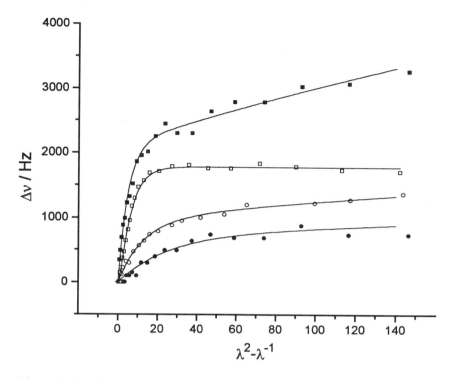

Figure 6. Quadrupolar splittings in the maxima of the spectra of dPB-U4A-2% (O) dPB-U4A-4% (●), dPB-U35A-2% (□) and dPB-U35A-4% (■)

Variation of stretching direction. According to equation 2, the splittings are supposed to depend on the angle of the stretching direction with respect to the external magnetic field. The spectra in Figure 7 show the ^{2}H-NMR spectra of dPB-U4A-4% and dPB-U35A-4% for $\lambda=4,5$ and $\lambda=4,1$, respectively. The spectra show the expected behavior. If the stretching direction is parallel to B_0, the quadrupolar splitting is about twice the splitting compared to the perpendicular setup. For both crosslinking degrees, the quadrupolar splittings disappear if the sample is oriented at the 'magic angle' (54.7°) and the lineshapes collapse to single narrow lines of 1897 and 2099 Hz in width, respectively. It has to be emphazised that even the very broad components in the spectra in the range of ±20 kHz and more vanish for an orientation at the magic angle. Nevertheless, the splittings are not strictly proportional to $(3\cos^2\Omega-1)$ as predicted by theory.

Inversion-recovery experiments. To obtain additional information about the presence of different dynamic states in the stretched elastomers, T_1-relaxation experiments were performed at various strains. Since the T_1-relaxation time itself is not the quantity of interest but the relaxation of every component in a spectrum, i.e. the relaxation at every frequency, we used the inversion-recovery technique which is

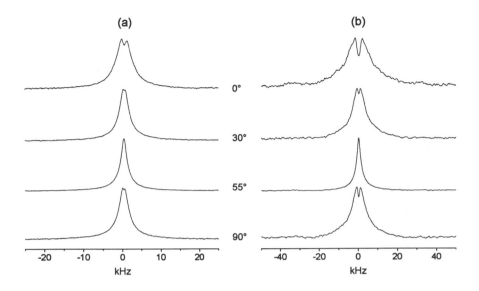

Figure 7. ^{2}H-NMR spectra of stretched (a) dPB-U4A-4% at $\lambda=4.5$ and (b) dPB-U35A-4% at $\lambda=4.1$ for various angles of the stretching direction with respect to B_0.

very sensitive for detecting differences in longitudinal relaxation rates. After the inversion of magnetization by a 180 degree pulse, spin relaxation is allowed for different durations and solid-echo spectra were recorded for various relaxation times. In the presence of inhomogeneous molecular mobility, 'faster' chain segments result in higher relaxation rates, i.e. smaller T_1-relaxation times, while 'slower' segments exhibit lower relaxation rates. Consequently, in such a case, the spectra should show positive and negative parts after a certain delay indicating chain segments with high and low molecular mobility. As an example, an inversion-recovery experiment for dPB-U35A-2% is shown in Figure 8 for different delays between 0,5 and 50 ms after the inversion. At no time can positive and negative parts in the spectra be detected simultaneously. At t=8 ms the intensity is almost zero for all frequencies, only a little intensity in the center of the spectra can be observed.

Discussion

In the isotropic state all samples show narrow single lines due to the fast molecular motion of the polybutadiene segments at temperatures 100 K above T_g. As mentioned before, the linewidths of 800 to 1000 Hz indicate that the averaging is not completely liquidlike. This can be understood by studies of molecular dynamics, which demonstrated that the mobility is heterogeneous along a network chain with completely immobile polybutadiene segments in the neighborhood of the supramolecular aggregates (*17*). Consequently, the orientational averaging is determined by the amount of functional groups per network chain. This leads to

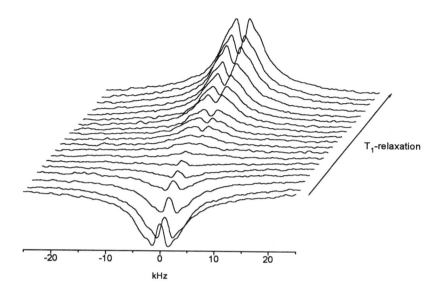

Figure 8. Inversion recovery experiments of dPB-U35A-2% at λ=3

different linewidths for different degrees of functionalization. In addition, the lineshapes can not be explained by single Lorentzian lines but only as a superposition of multiple Lorentzians. Due to this large linewidth, a splitting cannot be identified immediately after stretching the sample but only above extension ratios of about λ=2.

As can be seen in Figure 6 the quadrupolar splittings from the maxima of all spectra are neither proportional to the strain function nor to the inverse number of statistical segments in the case of dPB-U4A as predicted by the theories (equations 3 and 4). The splittings first increase indicating the increase of orientation in the beginning of the stretching experiment. In a second region (above extension ratios of about λ=5) the splitting becomes less dependent on λ. In covalently crosslinked network with a PDMS (5) or polybutadiene backbone (22) - where the connection between two network chains is permanent and 'point-like' - a linear correlation between the quadrupolar splitting and the strain function was observed. Deviations from the linear dependence between quadrupolar splitting and the strain function have also been reported (22). However, these deviations were rather small and could be attributed to a non-Gaussian network deformation behavior at larger strains.

The behavior of the investigated samples dPB-U4A and dPB-U35A can be related directly to the nature of the extended junction zones. Small strains result first in chain orientation which is accompanied by a co-orientation of the supramolecular clusters as shown by IR dichroism experiments for PB-U4A (23). At large strains, the external force is sufficient to pull single functional units out of the aggregates and stress is removed without further chain orientation. Stress-strain experiments and FT-IR spectroscopy investigations (23) confirm this interpretation, the former showing a yield behaviour at strain ratios between 3 and 5, in the latter, a decrease of the characteristic absorption bands for the supramolecular structures can be detected.

Nevertheless, up to this point it is still remarkable that for the U4A modified samples the splitting is larger for the lower degree of functionalization. This would lead to the assumption that the observed splittings in the maxima are not a direct measure of the orientation parameter $\langle P_2(\cos\vartheta)\rangle$ of the network chains. As a consequence, not only the detectable quadrupolar splittings, but also the complete lineshapes have to be taken into account in order to quantify the network orientation.

In contrast to covalently crosslinked PDMS networks, where the spectra show relatively narrow signals even in the stretched state, in the urazole-functionalized samples the linewidth increases dramatically with increasing strain resulting in the complex lineshapes shown in Figures 4 and 5. These lineshapes can not be explained by only one doublet of Lorentzian lines with a single splitting to account for the orientation parameter.

Several explanations for this phenomenon could be given. Samulski (3, 24) discussed a distribution of molecular dynamics where the orientation is equal for all network chains but the quadrupolar interaction is averaged differently depending on the different timescales of molecular motion. Similar results have been reported earlier in order to explain broad components in the ^{2}H-NMR spectra of styrene-d_8-styrene block copolymers (25) and were interpreted with different segmental mobilities of the center parts and the dangling ends of polymer chains. On the other hand, it is instructive to note that Poon and Samulski could not distinguish differences between lineshapes generated by assuming a distribution of molecular dynamics from those created with uniform mobility but a distribution of local order parameter - 'static heterogeneity' caused by different degrees of local preaveraging within a chain segment (26). Another concept given by Gronski et al. also uses homogeneous mobility and a distribution of orientation, so that higher oriented network chains, preferably shorter chains in a statistically crosslinked system, contribute to the broad parts in the spectra (9).

In order to assign the experimental results to one of the models, the angular-dependent measurements and T_1-inversion-recovery experiments can be used. As can be seen in Figure 7 the spectra of all samples collapse to narrow single lines if the stretching direction is at an angle of 54.7 ° with respect to the external magnetic field. For angles of 0 ° and 90 °, respectively, the complex lineshapes are obtained and the splitting for the spectra at 0 ° are about twice the value as for the spectra at 90 °. This behavior indicates that the orientation of network chains is uniaxial along the strain direction and no additional orientation, e.g. perpendicular to the stretching direction, takes place. A second feature is that at the magic-angle even all broad components in the spectra disappear and thus, these shoulders in the spectra can be attributed to be a consequence of the network orientation and not of reduced molecular mobility. The T_1-inversion-recovery experiments confirm this assumption. All components in the spectra relax homogeneously, i.e. chain segments which contribute to the different frequencies in the spectra exhibit equal or very similar correlation times of molecular motion. That means the concept using a distribution of chain dynamics can be ruled out and the broad components in the spectra are a result of chain orientation, exclusively. Furthermore, as already mentioned, the complex lineshapes cannot come from only a single orientation, which would give rise to a single doublet, and so a orientation distribution must be taken into consideration. Therefore, it becomes clear that the splitting in the maxima of the spectra is only a measure of this fraction of network chains which are the least oriented chains and the higher oriented chains are not taken into account by this analysis. In order to obtain the complete information, lineshape simulations have to performed, e.g. by means of Monte-Carlo methods (Photinos D.; Samulski E. T.; Dardin A., University of Patras, Greece; University of North Carolina at Chapel Hill; Max-Planck-Institut für Polymerforschung Mainz, Germany unpublished data).

Nevertheless, an explanation for the extremely broad parts in the region about ±20 kHz can be given offhand. Assuming completely extended cis-polybutadiene segments which are only allowed to rotate fast around their end-to-end vector (Figure 9) the resulting ^{2}H-NMR spectra would be a reduced Pake pattern scaled by 1/3 compared to the original spectral width of static systems. Thus, the resulting splitting of the singularities would be about 42 kHz.

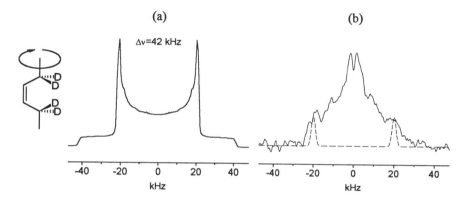

Figure 9. ^{2}H-NMR spectrum for a crankshaft motion of (a) a cis-polybutadiene unit around its end-to-end axis and (b)comparison of the spectrum of an oriented sample with the spectrum of dPB-U35A-4% at λ=19

For an oriented sample (perpendicular with respect to the magnetic field) this spectrum would change to two single absorption lines with a splitting of about 42 kHz. The real sample contains cis- as well as trans-polybutadiene units which would cause the splitting to be narrowed. However, the spectra show evidence for highly oriented network chains besides chains that are less oriented. Owing to the nature of the networks, a distribution of end-to-end vector orientation is conceivable which could arise from the distribution of network chain lengths given by the statistical functionalization with polar groups. Thus, shorter chains would be oriented more than the longer ones as already predicted for similar covalently crosslinked networks (9). But this would be inconsistent with recent results on bimodal networks (27), where short and long chains show the same splitting in the ^{2}H-NMR-spectra which could be explained with Deloche's mean-field concept (5).

Another possibility is a distribution of orientation along a single chain. Since the supramolecular aggregates of U4A units are shown to be more oriented than the polymer backbone (23) in an uniaxially stretched sample, an orientational coupling between the supramolecular aggregates and chain segments seems to be probable. Consequently, these parts of the polymer backbone which are in the neighborhood to the junction zones are most likely more oriented than parts in the 'middle' of the network chains.

Conclusion

The ^{2}H-NMR investigations of thermoplastic elastomers as obtained by the self aggregation of the polar substituted urazole groups (U4A, U35A) show unusual lineshapes as well as extremely large quadrupolar splittings. The attempt to analyze the spectra according to newer theories as developed by Sotta and Deloche or Brereton failed. Here, the use of only a single order parameter is not sufficient to describe the ^{2}H NMR spectra of the stretched samples.

Measurements with variation of the angle between stretching direction and external magnetic field and T_1-inversion-recovery demonstrate clearly that the broad components in the spectra arise from anisotropic averaging, i.e. from chain orientation, and not from incomplete averaging due to limited molecular mobility. An orientation distribution of chain segments has to be invoked to explain the spectra; this distribution could result from either different orientations for short and long network chains or from different degrees of segmental orientation along a single chain (resulting from a coupling to the extended junction zones). However, in this distribution of orientation, the splitting observed in the maxima intensities of the spectra is only a measure of the least oriented chains, and it could be shown that the broad components of the spectra can be explained by competely extended polybutadiene chains.

Acknowledgments

Helpful discussions with Demetri Photinos, University of Patras, Greece, are gratefully acknowledged. A.D. is indebted to Chi-Duen Poon, Chapel Hill, N.C., USA, for many discussions and his help with the spectrometer.

The authors thank the Graduierten Kolleg 'Physik und Chemie Supramolekularer Systeme' (University Mainz) for financial support. R.S. also acknowledges support from the BMFT through grant 03M4074A1.

Literature Cited

1. Spiess, H. W. *Coll. & Polym. Sci* **1984**, *261*, 193.
2. Boeffel, C.; Spiess, H. W. In *Developments in Oriented Polymers*; Ward, I. M., Ed.; Applied Science Publishers, London 1975; 128.
3. Samulski, E. T. *Polymer* **1985**, *Vol.26*, 177.
4. Nakatani, A. I.; Poliks, M. D.; Samulski, E. T.; *Macromolecules* **1990**, *23*, 2686.
5. Sotta, P.; Deloche, B. *Macromolecules* **1990**, *23*, 1999.
6. Gronski, W.; Stadler, R.; Jacobi, M. M. *Macromolecules* **1984**, *17*, 741.
7. Brereton, M. G. *Macromolecules* **1991**, *24*, 6160.
8. Brereton, M. G. *Macromolecules* **1993**, *26*, 1152.
9. Gronski, W.; Emeis, D.; Brüderlin, A.; Jacobi, M. M.; Stadler, R.; Eisenbach, C. D. *British Polymer Journal* **1985**, *17*, 103.

10. Hilger, C.; Stadler, R. *Makromol. Chem.* **1990**, *191*, 1347.
11. Hilger, C.; Stadler, R.; Dräger M. *Macromolecules* **1992**, *25*, 2498.
12. Hellmann, J. *PhD Thesis* **1994**, University Mainz.
13. Hilger, C.; Stadler, R. *Macromolecules* **1990**, *23*, 2095.
14. Hilger, C.; Stadler, R.; de Lucca Freitas, L. *Polymer* **1990**, *31*, 818.
15. Hilger, C.; Stadler, R. *Makromol. Chem.* **1991**, *192*, 805.
16. Hilger, C.; Stadler, R. *Macromolecules* **1992**, *25*, 6670.
17. Dardin, A.; Stadler, R.; Boeffel, C.; Spiess, H. W. *Makromol. Chem.* **1993**, *194*, 3467.
18. Hentschel, D.; Spiess, H. W.; Sillescu, H. *Polymer* **1984**, *25*, 1078.
19. Hentschel, D.; Spiess, H. W.; Sillescu, H.; Voelkel, R.; Willenberg, B. *Magn. Reson. Relat. Phenom., Proc. Congr. Ampere 19th* **1976**, 381.
20. Kuhn, W.; Grün, F. *Kolloid Z.* **1942**, *101*, 248.
21. Gronski, W.; Forster, F.; Stadler, R.; Pyckhout, W.; Frischkorn, K.; Jacobi, M. M. In *Local Orientation and Global Deformation of Network Chains in Strained Elastomeric Networks, Molecular Basiscs of Polymer Networks*; Baumgärtner, A.; Picot, C. E., Eds. Progress in Physics Springer Verlag **1989**, *42*, 120.
22. Jacobi, M. M.; Stadler, R.; Gronski, W. *Macromolecules* **1986**, *19*, 2884.
23. Abetz, V.; Hilger, C.; Stadler, R. *Makromol. Chem., Macromol. Symp.* **1991**, *52*, 131.
24. Poon, C. D.; Samulski, E. T.; Nakatani, A. I. *Makromol. Chem., Macromol. Symp.* **1990**, *40*, 109.
25. Kornfield, J. A.; Chung, G. C.; Smith, S. D. *Macromolecules* **1992**, *25*, 4442.
26. Poon, C. D.; Samulski, E. T. *J. Non-Cryst. Solids* **1991**, *131-133*, 509.
27. Chapellier, B.; Deloche, B.; Oeser, R. *J. Phys. II*, **1993**, *3*, 1619.

RECEIVED January 25, 1995

Chapter 14

Monte Carlo Simulations of End-Grafted Polymer Chains Under Shear Flow

Pik-Yin Lai

Institute of Physics, National Central University, Chung-li, Taiwan 32054, Republic of China

Polymer chains with one end grafted on a surface and exposed to a shear flow in the $+x$ direction are modeled by a non-equilibrium Monte Carlo method using the bond-fluctuation model. In the dilute case of an isolated chain, the velocity profile is assumed to increase linearly with the distance from the surface, while for the case of polymer brushes the screening of the velocity field is calculated using a parabolic density profile for the brush whose height is determined self-consistently. Linear dimensions and orientations of isolated chains are obtained over a wide range of shear rates $\dot{\gamma}$, and the deformation of the coil structure by the shear is studied in detail. For brushes it is found that the density profile differs only little from the shear-free case, while the monomer distribution in the flow direction parallel to the wall is strongly modified. It is shown that the average scaled chain trajectory is a universal function independent of shear rate, and the results are discussed in terms of appropriate theories.

It is well-known that complex fluids can exhibit interesting non-Newtonian behavior, an example of which is shear thinning[1] in which the viscosity of a (non-grafted) polymer solution or melt confined in parallel plates becomes smaller at high shear rates. The behavior of two surfaces grafted with polymer in shear motion[2-4] has been much less investigated despite its important applications in friction control and adhesion. A crucial aspect of this application is the dynamic response of the polymer layers to the flow of the fluid in the process involved in those applications. It would also provide some understanding on the fundamental mechanism of interfacial friction and adhesion[2-4] on a molecular level. While much theoretical effort has been devoted towards the understanding of the static properties of polymers grafted at surfaces (see references [5-7] for a review), the understanding of the dynamics of such layers both in equilibrium[8-11] and exposed to shear flow is very limited. Most of the experimental and theoretical studies concerning the flow induced behavior of polymers[12] in the past has concentrated on non-grafted polymer chains under elongational flow and the associated coil-stretch transition[13] and dynamic response. In recent years, the structure and dynamics of end-functionalized grafted polymers under the action of flow fields have drawn much research interest both experimentally*(14-20)* and theoretically*(21-24)*. One major reason for the growing research interests on such end-grafted polymers layers is that the interaction between these grafted surfaces can be modified in a more or less controlled way. With current technologies, by coating the surfaces with the appropriate grafted polymer layers, surfaces with desired properties can almost be custom made. These end-grafted polymer chains have numerous applications in technologies such as polymeric stabilization of colloidal dispersions*(25,26)*, biocompatibility*(27)*, chromatographic separation*(28,29)*, lubrication and adhesion*(30)*, etc. Furthermore, such grafted polymer layers are of great fundamental interests in polymer physics since the stretched configurations of these chains give rise to very unique physical properties.

The deformation of flexible polymer coils in bulk solution under a flow field is already a formidable problem and under current theoretical discussion*(31-35)*. Since we are far from a complete theoretical understanding of sheared polymer layers, and hence the interpretation of recent experiments where the dynamic response of such layers to shear directly measured by surface force apparatus techniques*(2-4)* is somewhat incomplete, it would be desirable to address these questions by computer simulation. While some important issues of the dynamics of polymers in melts such as the reptation mechanism*(32,36)*, could be clarified by both Molecular Dynamics (MD)*(37)* and Monte Carlo (MC)*(38)* simulations, those treatments are simplified since hydrodynamic backflow interactions*(31,32)* are neglected. If these hydrodynamic interactions are included, even for the simplest bead-spring models of polymer chains, the MD technique can currently handle, at best, the relaxation of a single chain of moderate chain length *(39)*. Consequently, previous approaches to terminally attached chains at surfaces under shear flow applying Brownian Dynamics techniques*(14,18-20)*, neglected these hydrodynamic interactions throughout. This approximation is particularly severe since in the extremely dilute case considered (response of a single chain)*(20)*, there is very little hydrodynamic screening. Even under these circumstances, one is limited to very short chains and considerable statistical errors are a problem. To avoid the above difficulties, we have decided to choose an entirely different approach using a non-equilibrium Monte Carlo (NEMC) method. Inspired by MC studies of driven lattice gases*(40)*, in this model the flow is represented in terms of asymmetric jump rates of monomers on the lattice. The jump rate of effective monomers to neighboring lattice sites is smaller against the flow direction than in the flow direction with the assumption that this difference in jump rates is proportional to the local velocity of the flowing fluid. This idea has also been used successfully to simulate polymers near boundaries in shear*(41)*. With this simple idea we can take advantage of the huge computational advantage of very fast MC algorithms for polymer simulation such as the bond fluctuation model*(38,42,43)*. Although such coarse-grained models of polymers*(44)* clearly are even more idealized than bead-spring type models used in MD*(10,37,39)* or Brownian Dynamics*(14,18-20)* work, it has been demonstrated*(11,41-44)* that the bond fluctuation lattice model can approximate many properties of polymers in continuous space rather closely. Moreover, it has a distinct advantage that detailed information on equilibrium properties of "bond-fluctuation"-polymer brushes is available over a wide range of chain lengths, grafting density, $\sigma(11)$ and solvent quality*(45)*. In addition, for rather dense (multi-chain) layers, hydrodynamic interactions are screened out*(31,32,36,46,47)*, and thus their neglect is better justified, at least over length scales of the same order as average distances between grafting sites in the polymer brush. Of course, at the same time, one must consider the reaction of the polymers on the flow: the flow field itself is damped when penetrating into the polymer layers*(23)*, thus making a fully self-consistent treatment of these flow effects difficult.

Bond-Fluctuation Model and the NEMC Method

The bond-fluctuation model*(38,42,43)* for macromolecular chains is used in the simulations. Previous simulations using this model*(11,45,48,50)* have been very successful in exploring the physical behavior of grafted polymer layers. In the bond-fluctuation model, each monomer occupies a cube of eight lattice sites on a simple cubic lattice and excluded volume effects are modeled by the requirement that no two monomers can share a common site. The 108 allowed bond vectors connecting two consecutive monomers along a chain are obtainable from the set $\{(2,0,0), (2,1,0), (2,1,1), (2,2,1), (3,0,0), (3,1,0)\}$ by the symmetry operations of the cubic lattice. We consider a polymer brush in a good solvent and there is no monomer-monomer interaction other than self-avoidance. The Monte Carlo procedure starts by choosing a monomer at random and trying to move it one lattice spacing in one of the randomly selected directions: $\pm x, \pm y, \pm z$. The move will be accepted only if both self-avoidance is satisfied and the new bonds still belong to the allowed set. Monodisperse polymer chains of N monomers are placed inside a box of linear dimensions $L \times L \times M$ where one surface of size $L \times L$ (in the xy plane) is chosen as a grafting surface, and the other $L \times L$ surface is a large distance M away so that the configurations of the grafted chains are not restricted. We choose periodic boundary conditions in the x and y directions, while the two other boundaries in the z direction (at $z = 1$ and $z = M$, respectively) are treated as hard impenetrable walls. For the dimensions of the box, we choose $L = 32$ and $M = 3N + 1$. The chain lengths cover a range of $10 \leq N \leq 60$. For the simulation of grafted layers, several suitable grafting densities are chosen so that the systems are well into the brush regime from our previous experiences*(11,48)*. The anchoring site of the chains in the brush are randomly chosen but strictly self-avoiding grafting at the wall. To model the effect of shear solvent flow in the $+x$ direction (with shear rate $\dot{\gamma}$), we consider the monomer as acted upon by a viscous force $F^x \propto v(z)$ in the $+x$ direction where $v(z)$ is the velocity difference between the solvent and the monomer at a distance z from the grafting wall (see Figure 1). This net viscous force physically arises from the impact on the monomer by the random (but with a net flow in the $+x$ direction) motions of the solvent molecules. Denoting the probabilities of a monomer moving in the $+x$ and $-x$ directions by p_{+x} and p_{-x} respectively, one has

$$p_{+x} - p_{-x} \propto F^x \propto v(z) \tag{1}$$

The effect of shear is assumed not to affect the probabilities of monomers moving in the y and z directions. Hence the shear flow can be modeled once the velocity profile $v(z)$ is known. In practice the constant of proportionality in equation 1 is chosen such that even for the longest chain in its fully extended configuration, $p_{+x} < 1/3$ still holds. The velocity profile would be $v(z) = \dot{\gamma} z$ if the polymers were absent. The presence of polymers should change the flow profile and one expects that some screening occurs which diminishes the velocity of flow inside the brush. Following reference *(23)*, the grafted layer is treated as a porous medium with pore size given by the hydrodynamic screening length ξ_H. In a semi-dilute polymer solution one can approximate*(47)* $\xi_H \approx$ order of the blob size ξ. Then, the velocity profile of the solvent flow is obtained by solving the Brinkman equation*(51)* at constant pressure,

$$\frac{d^2 v}{dz^2} = \frac{v}{\xi^2(z)} \tag{2}$$

But, unlike in reference *(23)* which ignored the effect of swelling and only obtained the solution near the free brush end, we shall take into account the swelling effects which are important in the present good solvent case, and solve for the velocity profile across the entire brush. According to scaling theory*(36)* in the semi-dilute regime, one has $\phi(z) = (a/\xi(z))^{4/3}$ where ϕ is the volume fraction and a is the monomer size. For very strong shear (Pincus regime)*(52)* the blob size depends on the strength of the force (shear) that is stretching the chain rather than the concentration. Here we assume that the brush is not under extremely strong shear flow. Knowing $\phi(z)$, $v(z)$ can be solved from equation 2. Our previous studies*(11,48)*, for the case of no shear, showed that $\phi(z)$ can be quite accurately represented by the SCF parabolic density profile*(53,54)*. In order to allow for an analytical calculation of $v(z)$, we take $\phi(z)$ to be of parabolic form,

$$\phi(z) = \frac{3Na\sigma}{2h}\left[1 - \left(\frac{z}{h}\right)^2\right] \tag{3}$$

where a, σ and h are the monomer size, surface coverage and brush height respectively. Instead of assuming that the brush height is unaltered by the flow, h is regarded as a parameter to be determined self-consistently from our Monte Carlo data as follows: Starting from an arbitrary value of h, $\phi(z)$ is calculated by MC, resulting in an improved estimate of the brush height. The improved brush height is in turn used as the new input in the next simulation and this procedure is repeated until the input brush height is consistent with the output estimate. In practice, we find that only very few iterations are needed to obtain the self-consistent h. Using the parabolic $\phi(z)$ and changing to the convenient variable $u \equiv z/h$, equation 2 reads

$$\frac{d^2v}{du^2} = Bv(1 - u^2)^{3/2} \tag{4}$$

where $B \equiv (h/a)^{1/2}(3N\sigma/2)^{3/2}$. The boundary conditions are

$$v(z=0) = 0 \quad \text{and} \quad v(z=h) = \dot{\gamma}h \tag{5}$$

Equation 4 is solved by the series method and we obtain

$$v = u\sum_{j=0}^{\infty} A_j u^{2j} \quad \text{with}$$

$$A_j = \frac{B}{2j(2j+1)} \sum_{k=0}^{j-1} \frac{3}{4\pi} \frac{\Gamma(k-\frac{3}{2})}{k!} A_{j-k-1} \tag{6}$$

The initial constant A_0 is determined by matching the boundary condition $v(z=h) = \dot{\gamma}h$. A simple dimensional analysis yields

$$v = \dot{\gamma}h F(B,u) \tag{7}$$

for some function F. The series sum of v in equation 6 is evaluated up to $\mathcal{O}(u^{40})$ for numerical calculations. Figure 2 shows the result of the velocity profile $v(z)$. The kink at $z = h$ arises from the abrupt vanishing of monomer density at the brush end in a parabolic density profile. In principle, $v(z)$ can be determined

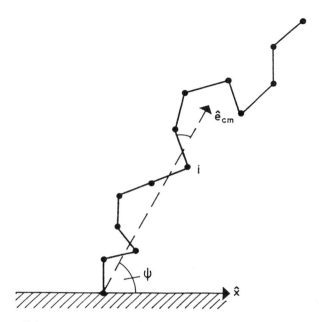

Figure 1. Schematic picture of a grafted chain under shear. $\hat{e}_{cm}$ is the direction of the center of mass vector (dashed line). ψ is the angle between the center of mass direction and the x-axis. (Reproduced with permission from ref. 24 Copyright 1993 American Institute of Physics.)

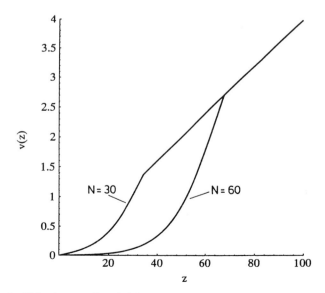

Figure 2. Velocity profile $v(z)$ for a polymer brush under shear flow as obtained from equation 6. At $\dot{\gamma} = 0.04$ for polymer brushes with $\sigma = 0.0625$ and $N = 30, 60$. (Reproduced with permission from ref. 24 Copyright 1993 American Institute of Physics.)

self-consistently by using an arbitrary $\phi(z)$ at first, and the improved $\phi(z)$ can be obtained from the MC data which is in turn used as the input for the next simulation until a self-consistent $\phi(z)$ is obtained. In this case, $\phi(z)$ will have a smooth tail (due to the finite chain lengths in a simulation) and the kink in Figure 2 will not appear. Such an approach requires a series expansion of the $\phi(z)$ obtained from the MC data, and equation 2 is solved for $v(z)$ in a similar manner. However, as we shall see, the structure and especially $\phi(z)$, depends very little on the fine details of $v(z)$.

The effects of shear are characterized by the inclination $\langle \cos \psi \rangle$ which is defined as

$$\langle \cos \psi \rangle = \langle \hat{e}_{cm} \bullet \hat{x} \rangle \qquad (8)$$

where $\hat{e}_{cm}$ is the unit vector in the direction from the anchor point to the center of mass of the chain, $\hat{x}$ is the unit vector in the x direction and ψ is the angle between $\hat{e}_{cm}$ and $\hat{x}$ (see Figure 1). Under no shear, the inclination is zero while for extremely strong shear the chain is tilted in a direction almost parallel to $\hat{x}$ with $\langle \cos \psi \rangle \approx 1$.

Dilute Case: Single Grafted Chain

For the case of a single grafted polymer which corresponds to the dilute limit, the screening effect can be ignored with $\xi \to \infty$ and the velocity profile is linear, $v(z) = \dot{\gamma}z$. Consider an end-grafted chain consisting of N monomers under shear flow in the x direction. The chain is expected to stretch under the shear flow. Analogous to the description of the coil-stretch transition, the degree of stretching can be measured by the stretching parameter (or reduced root-mean-square radius of gyration) $\langle R_g^2 \rangle^{1/2}/N$. The coil state is characterized with a vanishing stretching parameter (in the $N \to \infty$ limit) while its value in the stretch state is of order unity. Figure 3 is a semi-log plot for the variation of the stretching parameter over a wide range of the shear rates ($\dot{\gamma}$ in arbitrary unit) for various chain lengths. As the shear rate increases, the chain transforms from the coil state to the stretched state and the stretching parameter saturates at very high $\dot{\gamma}$ indicating that the chain is fully stretched. The x, y and z components of the radius of gyration are also monitored. Both $\langle R_{gz}^2 \rangle$ and $\langle R_{gy}^2 \rangle$ decrease monotonically as $\dot{\gamma}$ increases. At very high shear rate this decrease levels off at the value $\approx a^2$, indicating that the chain is almost lying flat parallel to the wall and is in the direction of flow, as can also be observed from the data of $\langle R_{gx}^2 \rangle$. A qualitatively similar variation of the components of the radius of gyration is also reported in a recent Brownian dynamics simulation of a single grafted chain of freely joined bead-rods*(20)*. We attempt to look for a scaling form of $\langle R_{gx}^2 \rangle$ as a function of $\dot{\gamma}$ and N. For our computer model, the relaxation time τ of a single chain should obey the Rouse prediction $\tau \sim N^{1+2\nu}$ where $\nu \approx 0.59$. By considering the dimensionless quantity $\tau\dot{\gamma} \sim \dot{\gamma}N^{1+2\nu}$, we formulate the scaling relation

$$\langle R_{gx}^2 \rangle = N^{2\nu} f(\dot{\gamma}N^{1+2\nu}) \qquad (9)$$

where $f(x) \sim$ constant for $x << 1$. This scaling is tested in Figure 4. The data show a rough collapse for moderate shear rates. For large $\dot{\gamma}$, the data scatter significantly, which is expected because the chain approaches its fully stretched limit at such high shear rates and the Rouse relaxation scaling does not hold.

Figure 5 shows $\langle \cos \psi \rangle$ vs. $\dot{\gamma}$. At high shear rate, the chain is essentially lying flat on the wall and a further increase in shear merely extends the chain towards its fully stretched configuration. This can also be observed in Figure 5 for the variation of the mean square bond length $\langle \ell^2 \rangle$ vs. $\dot{\gamma}$. The bond length

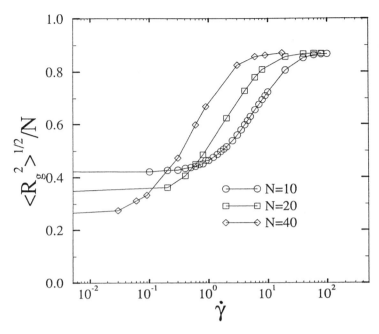

Figure 3. $\langle R_g^2 \rangle^{1/2}/N$ vs. $\dot{\gamma}$ for single grafted chain under shear flow in the x-direction.

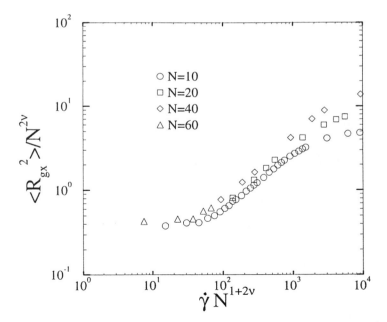

Figure 4. Scaling plot of single grafted chain under shearflow in the x-direction as suggested in equation 9.

is essentially constant at low shear rate, indicating that the increase in shear in this regime has the effect of unfolding the chain segments. At higher $\dot\gamma$, $\langle\ell^2\rangle$ increases indicating the chain is elongated towards its fully extended limit. In our study on the grafted layer, we will not consider situations of such strong shear, but rather concentrate on the regime of weaker shear rate in which the chains are still quite flexible.

Polymer Brush Under Shear

For the case of polymer brushes, we consider the case where the shear rates are moderate such that the derivation for the result in section 2 still holds. Figure 6 shows $\langle\cos\psi\rangle$ vs. $\dot\gamma$ for different values of N for the same grafting density. We find that $\langle\cos\psi\rangle$ increases with the shear rate as expected. Furthermore, at the same $\dot\gamma$, longer chains are tilted more by the shear flow. The fluctuation about the mean orientation of a chain decreases for higher $\dot\gamma$, suggesting that the grafted polymer chains become less flexible as the shear rate increases. This is related to the fact that the chains are end-anchored and the shear flow has the effect of stretching out the chain segments.

Figure 7 shows the monomer volume fraction at a distance z from the grafting plane, $\phi(z)$, at various shear rates. The shape of $\phi(z)$ depends very weakly on the shear rate. Even under strong shear ($\dot\gamma = 0.2$ and $\langle\cos\psi\rangle \approx 0.7$ in Figure 6) the brush thickness only decreases by about 2% as compared to the case of no shear. For weaker shear rates (which are the cases we are interested in), $\phi(z)$ is practically independent of $\dot\gamma$. This indicates that the response of the grafted chain to the shear flow is to incline in the direction of flow, but at the same time unfolds the chain segments such that the end-to-end distance increases and the brush height remains essentially unaltered. The span of a grafted chain in the direction of flow can also be characterized by the average x position of the end monomer relative to the grafting site, $\langle x_N\rangle$. For given N and σ, $\langle x_N\rangle$ varies linearly with $\dot\gamma$, which can be understood from the following scaling argument. In the case of no shear, SCF theory predicts the average position of the ith monomer along a chain is given by

$$\langle z_i\rangle = \langle z_N\rangle \sin\frac{i\pi}{2N} \qquad (10)$$

From our MC result, this relation is practically unaltered even under quite strong shear flow. At equilibrium, the elastic deformation force in the x direction must be balanced by the viscous force due to the shear flow. The elastic free energy can be taken to be Gaussian, since under the condition of not very strong shear, a linear force law still holds. Denoting the x position of the n-th monomer by $x(n)$, one thus has (with continuum notation)

$$\frac{d^2 x(n)}{dn^2} \sim F^x \propto v(z) \qquad (11)$$

Under the condition of not too strong shear flow, the z position of the monomer is assumed to be unaltered (this is justified by our MC data), i.e. equation 10 (its continuum analog) holds, and $h \sim N\sigma^{1/3}$, as in the case of no shear. Using (10) and the scaling form of v as given in equation 7, one deduces that $x(n)$ has the functional form

$$x(n) = \dot\gamma N^3 \sigma^{1/3} G(N^2\sigma^{5/3}, z(n)/z_N) \qquad (12)$$

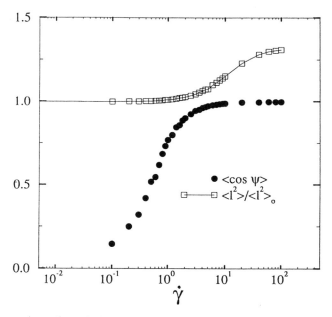

Figure 5. $\langle\cos\psi\rangle$ and the normalised mean square bond length vs. $\dot{\gamma}$ for a single grafted chain under shear. $\langle\ell^2\rangle_o$ is the mean square bond length for the shear-free case.

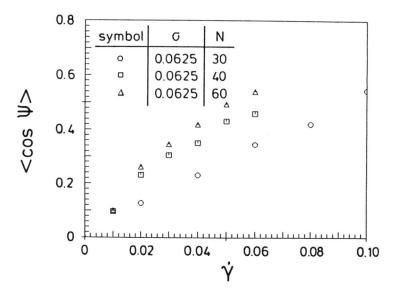

Figure 6. $\langle\cos\psi\rangle$ vs. $\dot{\gamma}$. For grafted layers with $\sigma = 0.0625$ and $N = 30, 40, 60$. (Reproduced with permission from ref. 24 Copyright 1993 American Institute of Physics.)

for some function G. Hence $\langle x_N \rangle \propto \dot{\gamma}$ for given N and σ. The proposed scaling form in equation 12 is also consistent with our MC data *(24)*. The average "real space trajectory" of a chain is obtained by plotting $\langle x_i \rangle / \langle x_N \rangle$ vs. $\langle z_i \rangle / \langle z_N \rangle$, which is shown in Figure 8. As suggested by equation 12, for a given value of N and σ, data at different shear rates collapse onto a master curve.

Discussion

Our findings should be compared with the recent theoretical and experimental studies on polymer brushes under shear. The surface force experiments by Klein et al.*(2,3)* which measured the force between two surfaces with end-grafted polymers and sliding parallel to each other, indicated an increase in repulsion when the relative speed of sliding exceeded some critical value. This result has been interpreted as an increase in layer thickness when solvent flow passed through the brush. The theoretical calculation by Rabin and Alexander (RA)*(21)*, however, indicated that for a single brush under shear, the layer thickness h is pinned down at the equilibrium value of no shear, h_o. Finally Barrat*(22)* revisited the RA calculation and claimed that their model is capable of predicting an increase in layer thickness. These two theoretical results are purely static in nature and the hydrodynamics of the solvent are not accounted for. On the other hand, our MC results indicate that the layer thickness decreases very slightly even when the inclination of the chains is quite large. There are good reasons to expect some disagreement between the present simulation and corresponding theories *(21,22)*. The theories take into account the excluded-volume interaction of a chain with itself and the way in which both this interaction and the interchain one are modified by the applied shear. Under strong deformations, these excluded-volume effects give rise to a nonlinear response of the chain to the applied shear and result in increased osmotic repulsion between the polymers in the grafted layer (swelling of its thickness). This goes beyond the SCF model, which assumes that the chains are Gaussian, and therefore the deformations of polymers in one direction does not affect their dimensions along orthogonal directions. The price paid in *(21)* is that one cannot calculate the flow profile inside the brush, and one has to assume that the hydrodynamic force on the brush is applied only at its external surface. In our calculation, on the other hand, a self-consistent calculation is performed, using the SCF model in order to calculate the velocity profile inside the brush. Since in our approach strong penetration of flow into the brush is observed, especially for the shorter chains, one expects different results concerning the shear-induced swelling of the layers, as indeed is the case.

We should point out that our results for the grafted layers apply only to moderate shear rates where the scaling rules for the blob picture in the semi-dilute regime still hold. Under very strong shear, the blob size ξ depends on the shear force and not on ϕ and thus would produce a non-linear response*(52)* that may couple to the z-direction affecting strongly the brush height. One can identify two regimes of weak and strong shear in which different physical laws govern the structure of the grafted layer under shear: for the shear-free case, the chain configuration can be schematically represented in Figure 9a which consists of blobs of size ξ with the brush height of the order $h_o \propto \xi(N/N_{blob})$ (where N_{blob} is the number of blobs), but at the same time the chain has a random walk like structure in the direction perpendicular to the z- direction covering a region of dimension $\sim \xi\sqrt{N/N_{blob}}$ (this has been verified by MC simulations in reference *(11)*). The blob size is determined by the volume fraction with $\xi \simeq a\phi^{-3/4}$, and blobs near the brush end should have larger sizes but they are drawn with the same size in Figure 9 for convenience. When a weak shear is applied to the brush in the x-direction (see Figure 9b), it has the effect of

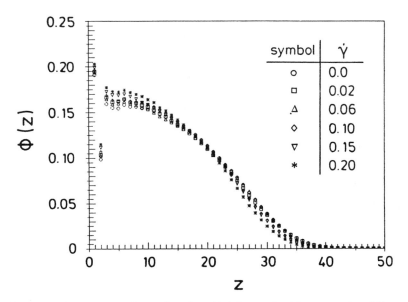

Figure 7. Monomer volume fraction $\phi(z)$ for grafted layers under different shear rates. $\sigma = 0.0625$ and $N = 30$. (Reproduced with permission from ref. 24 Copyright 1993 American Institute of Physics.)

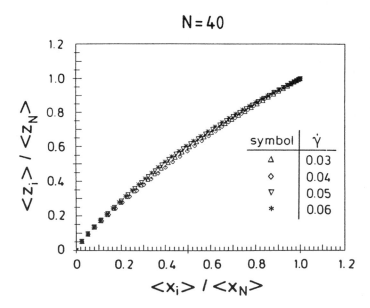

Figure 8. Scaled trajectory of a grafted chain in a brush. $N = 40$ with $\sigma = 0.0625$ at various shear rates. (Reproduced with permission from ref. 24 Copyright 1993 American Institute of Physics.)

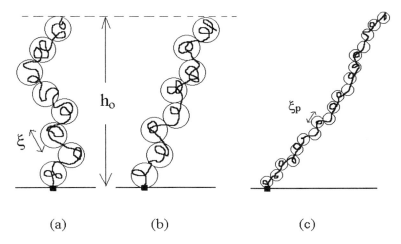

Figure 9. Schematic illustrations of the blob pictures for the configurations of end-grafted chains in the (a) no shear (b) weak shear and (c) strong shear regimes.

tilting the chain of blobs and aligning them in a certain direction but the blob size remains unaltered. The chain of blobs is tilted along the flow direction and at the same time the blobs are stretched by unfolding themselves in such a way that the brush height is essentially unchanged. ξ is still determined by the concentration and a linear elastic force law still holds. As the shear rate becomes stronger, the chain of blobs becomes more fully aligned along the tilted direction, and a further increase in the shear rate enters the chains into the strong shear regime (see Figure 9c). In this regime (Pincus regime), the blob size, $\xi_p(< \xi)$, is determined by the force that stretches the chain and the elastic force is non-linear[52]. Crossover between these two regimes occurs at $\xi = \xi_p$. The experimental observation of swelling of a layer at sufficiently large shear rate may correspond to the strong shear regime. The apparent increase in the thickness of the grafted layer under shear in the force measurement may share the same physical origin as the observation of an increase of effective hydrodynamic thickness for surface bound polymers in some experiments[17]. All theoretical treatments and computer simulations[14-20] on surface-bound polymers *do not* indicate an increase in the thickness. However, none of these studies (including the present simulation) include real hydrodynamic effects (interaction with the explicit solvent molecules, backflow, etc.). Some attempts have been proposed to explain the shear thickening by an elongation component in the flow, due to surface non-uniformity or interference between the chain segments[19]. Our model is expected to give an increase in layer thickness if the velocity profile $\vec{v}(z)$ has an elongation component (say v_z is positive).

Of course, one must bear in mind that our present NEMC technique, where the fluid flow enters via an asymmetry of jump rates and a self-consistent calculation of the velocity profile, involves an approximation as compared to real hydrodynamics. To study the system with real hydrodynamics by computer simulations, explicit solvent molecules should be put into the system, and their interactions with the monomers taken into account[39]. However, the computing power at present is still not adequate to study a multi-chain system incorporating these effects. A comparison of our results for the single chain problem with corresponding Brownian Dynamics simulations (which, however, include neither explicit solvent particles nor hydrodynamic backflow effects) indicates great qualitative similarity. Thus far, the statistical accuracy of the Brownian Dynamics work seems rather limited, and hence it is difficult to extend that technique to polymer brushes. Thus the present NEMC approach seems to be the most convenient route to a study of sheared brushes at this stage. This subject is still under current research and we recently developed a new simulation method which can incorporate the effects of shear for brushes in a more natural way (P.-Y. Lai, in preparation).

Acknowledgment: Part of this research was completed in collaboration with Prof. K. Binder. Support in part by the Bundesministerium für Forschung und Technologie (BMFT) grant no. 03M4040A in Germany and National Council of Science of Taiwan under grand no. NCS 83- 0208-M008-036 are gratefully acknowledged.

Literature Cited

1. Ferry, J.D. *Viscoelastic Properties of Polymers*, Wiley: New York, 1980.
2. Klein, J.; Perahia, D.; Warburg, S. *Nature (London)* **1991**, *352*, 143.
3. Klein, J.; Kamiyama, Y.; Yoshizawa, H.; Israelachvili, J. N.; Fredrickson, G. H.; Pincus, P.; Fetters, L.J. *Macromolecules* **1993**, *26*, 5552.
4. Yoshizawa, H.; Chen, Y.-L.; Israelachvili, J.N. *J. Phys. Chem.*, **1993**, *97*, 11300.
5. Halperin, A.; Tirrell, M.; Lodge, T.P. *Adv. in Polymer Science* **1991**, *100*, 31.
6. Milner, S.T. *Science* **1991**, *251*, 905.
7. Grest, G.S.; Murat, M. in *Monte Carlo and Molecular Dynamics Simulations in Polymer Science*; Binder, K., Ed.; Oxford University Press: Oxford, 1994.
8. Halperin, A.; Alexander, S. *Macromolecules* **1989**, *22*, 2403.
9. Klushin, L.I.; Skvortsov, A.M. *Macromolecules* **1991**, *23*, 1549.
10. Murat, M.; Grest, G.S. *Macromolecules* **1989**, *22*, 4054.
11. Lai, P.-Y.; Binder, K. *J. Chem. Phys.* **1991**, *95*, 9288.
12. *Polymer-Flow Interaction*; Rabin, Y., Ed.; AIP: New York, NY, 1985.
13. de Gennes, P.G. *J. Chem. Phys.* **1974**, *60*, 5030.
14. Lee, J.J.; Fuller, G.G. *Macromolecules* **1984**, *17*, 375.
15. Cohen, Y. *Macromolecules* **1988**, *21*, 494.
16. Gramain, Ph.; Myard, Ph. *Macromolecules* **1981**, *14*, 180.
17. Bagassi, M.; Chauveteau, G.; Lecoutier, J.; Englert, J.; Tirrell, M. *Macromolecules* **1989**, *22*, 262.
18. di Marzio, E. A.; Rubin, R.J. *J. Polym. Sci.* **1978**, *16*, 457.
19. Atkinson, J.; Goh, C.J.; Phan-Thien, N. *J. Chem. Phys.* **1984**, *80*, 6305.
20. Parnas, R.S.; Cohen, Y. *Macromolecules* **1991**, *24*, 4646.
21. Rabin, Y.; Alexander, S. *Europhys. Lett.* **1990**, *13*, 49.
22. Barrat, J.-L. *Macromolecules* **1991**, *24*, 832.
23. Milner, S. *Macromolecules* **1991**, *24*, 3704.
24. Lai, P.-Y.; Binder, K. *J. Chem. Phys.* **1993**, *98*, 2366.
25. Napper, D. *Polymer Stablization of Colloidal Dispersions*; Academic Press: London, 1983.
26. *Polymer Adsorption and Dispersion Stability;* Goddard, E.; Vincent, B., Eds.; ACS Symposium Ser. 240; American Chem. Soc.: Washington, D.C., 1984.
27. *Surface and Interfacial Aspects of Biomedical Polymers;* Andrade, J. D., Ed.; Plenum: New York, 1985.

28. Yau, W. W.; Kirkland, J. J.; Bly, D. *Modern Size Exclusion Chromatography;* John Wiley and Sons: New York, 1979.
29. Lai, P.-Y.; Halperin, A. *Macromolecules* **1992**, *25*, 6693.
30. Wu, S. *Polymer Interfaces and Adhesion*, M. Dekker: New York, 1982.
31. Bird, R. B.; Armstrong, R. C.; Hassager, O. *Dynamics of Polymeric Liquids;* J. Wiley and Sons: New York, 1977.
32. Doi, M.; Edwards, S. *The Theory of Polymer Dynamics;* Oxford University Press: Oxford, 1986.
33. *Dynamics and Patterns in Complex Fluids;* Onuki, A.; Kawasaki, K., Eds.; Spring Verlag: Berlin-Heidelberg, 1990.
34. Doi, M.; Onuki, A. *J. Phys. II (France)* **1992**, *2*, 1631.
35. Helfand, E.; Fredrickson, G.H. *Phys. Rev. Lett.* **1989**, *62*, 2468.
36. de Gennes, P.G. *Scaling Concepts of Polymer Physics;* Cornell University Press: Ithaca, NY, 1979.
37. Kremer, K.; Grest, G. S. *J. Chem. Phys.* **1990**, *92*, 5057.
38. Paul, W.; Binder, K.; Heermann, D. W.; Kremer, K. *J. Chem. Phys.* **1991**, *95*, 7726.
39. Dünweg, B.; Kremer, K. *Phys. Rev. Lett.* **1991**, *66*, 2996.
40. Katz, S.; Lebowitz, J. L.; Spohn, H. *J. Stat. Phys.* **1984**, *34*, 497.
41. Duering, E.; Rabin,Y. *Macromolecules* **1990**, *23*, 2232.
42. Deutsch, H.-P.; Binder, K. *J. Chem. Phys.* **1991**, *94*, 2294.
43. Deutsch, H.-P.; Dickman, R. *J. Chem. Phys.* **1990**, *93*, 8983.
44. Binder, K. *Macromol. Chem. Macromol. Symp.* **1991**, *50*, 1.
45. Lai, P.-Y.; Binder, K. *J. Chem. Phys.* **1992**, *97*, 586.
46. Muthukumar, M.; Edwards, S. F. *Polymer* **1982**, *23*, 345.
47. Richter, D.; Binder, K.; Ewen, B.; Stühn, B. *J. Phy. Chem.* **1984**, *88*, 6618.
48. Lai, P.-Y.; Zhulina, E. B. *J. Phys. II (Paris)* **1992**, *2*, 547.
49. Lai, P.-Y.; Zhulina, E. B. *Macromolecules* **1992**, *25*, 5201.
50. Lai, P.-Y. *J. Chem. Phys.* **1993**, *98*, 669.
51. Brinkman, H. C. *Appl. Sci. Res.* **1947**, *A1*, 27.
52. Pincus, P. *Macromolecules* **1976**, *9*, 386.
53. Milner, S. T.; Witten, T. A.; Cates, M. E. *Macromolecules* **1988**, *21*, 2610.
54. Skvortsov, A. M.; Gorbunov, A. A.; Pavlushkov, I. V.; Zhulina, E. B.; Borisov, O. V.; Priamitsyn, V.A. *Polymer Science USSR* **1988**, *30*, 1706.

RECEIVED February 15, 1995

BLOCK COPOLYMERS
AND MULTICOMPONENT POLYMER SYSTEMS

Chapter 15

Effect of Shear on Self-Assembling Block Copolymers and Phase-Separating Polymer Blends

M. Muthukumar

Polymer Science and Engineering Department and Materials Research Laboratory, University of Massachusetts, Amherst, MA 01002

> The theoretical efforts to understand the effect of shear on block copolymers and polymer blends are briefly reviewed. The known analytical results are only perturbative, and the status of theory is at a primitive stage in comparison with the recent advances made in the experimental front. Much more theoretical and computational work is necessary to fully understand the rich phenomenological experimental results.

The effect of shear on the relative stability and features of different morphologies formed by block coplymers and on the spinodal decomposition of polymeric mixtures has been an active research area. Although the experimental work has produced rich phenomenological results of tremendous intrigue, theoretical progress has been rather modest. Whatever little we know based on theory is reviewed here and much more theoretical and computational work is required to even begin to understand the various phenomenological results. Although the fact that the composition fluctuations are suppressed by shear is obvious, the calculation of the extent of this effect is technically very complicated. Only the linearized theory has been studied so far, thus restricting the applicability of theoretical predictions to the general experimental conditions. Even the verification of predictions of linear theory by experiment is not yet available for the appropriate conditions. The general theoretical framework is based on the pioneering paper (*1*) by Onuki and Kawasaki. The results presented below are essentially contained in the original theory of Onuki and Kawasaki.

Formulation

We consider a volume element **r** in an $A-B$ diblock or $A-B-A$ triblock copolymer or in a polymer blend $(A+B)$ under shear. Based on the conservation of mass and momentum, the equation of motion for the local volume fraction $\phi(\mathbf{r}, t)$ of component A at time t obeys (1)

$$\frac{\partial \phi}{\partial t} = -\nabla \cdot J - \mathbf{v} \cdot \nabla \phi + f \tag{1a}$$

$$\frac{\partial \mathbf{v}}{\partial t} = \eta_0 \nabla^2 \mathbf{v} - \left(\nabla \phi \frac{\delta F}{\delta \phi}\right)_\perp + (\mathbf{f}')_\perp. \tag{1b}$$

Here J is the local current of the A component,

$$\mathbf{J} = \mathbf{J}_d + \mathbf{J}_e + \mathbf{J}_c. \tag{2}$$

The diffusive flux $\mathbf{J}_d$ caused by the chemical potential is given by

$$\mathbf{J}_d = -\frac{\Lambda[\phi]}{k_B T} \nabla \frac{\delta F}{\delta \phi} \tag{3}$$

where Λ is the Onsager coefficient, $k_B T$ is the Boltzman constant times the temperature, and F is the free energy functional appropriate to a particular polymer system. The elastic flux $\mathbf{J}_e$ corresponds to the transport of material due to gradients in strain inside the system. $\mathbf{J}_c$ is the convective flux due to the imposed flow field. We take the shear flow of the form,

$$\mathbf{u}(\mathbf{r}) = \dot{\gamma} y \, \mathbf{e}_x \tag{4}$$

where $\mathbf{u}(\mathbf{r})$ is the average velocity at $\mathbf{r}$, $\mathbf{e}_x$ is the unit vector along x-axis and $\dot{\gamma}$ is the shear rate. $\mathbf{v}$ is the deviation of the local velocity from its average $\mathbf{u}$. η_0 is the shear viscosity of the system. $(\cdots)_\perp$ denotes taking the transverse part. f and $\mathbf{f}'$ are noises.

In general, the free energy F is a Taylor series in the order parameter $\phi(\mathbf{r})$. Usually, the series is truncated at the quartic term. Substituting the functional derivative $\delta F/\delta \phi$ in equations 1a and 1b, we get a coupled set of nonlinear equations. While this set of equations should be solved numerically to obtain the results given by the continuity equations, we (2-8) have so far considered only the linear regime in view of the analytical tractability. If we are interested in only the linear terms of $\phi(\mathbf{r})$ in equation 1a then equations 1a and 1b become decoupled with $\mathbf{v}$ replaced by $\mathbf{u}$.

Making this linear approximation and assuming further that the velocity gradients are weak due to high viscosities of the polymeric material and that the elastic contribution to the free energy is negligible, we obtain

$$\frac{\partial \phi}{\partial t} = \nabla \frac{\Lambda}{k_B T} \nabla \frac{\delta F}{\delta \phi} - \dot{\gamma} y \frac{\partial \phi}{\partial x} + f \tag{5}$$

where only the term of ϕ^2 is kept in F.

$$\frac{F}{k_B T} = \frac{1}{2} \int d\mathbf{r} \int d\mathbf{r}' \, \phi(\mathbf{r}) \, T_2(\mathbf{r} - \mathbf{r}') \, \phi(\mathbf{r}') + \cdots. \tag{6}$$

Assuming further that the Onsager coefficient Λ is local, and using the fluctuation-dissipation theorem for the noise f,

$$\langle f(\mathbf{r},t) f(\mathbf{r}',t') \rangle = -2\Lambda \delta(\mathbf{r} - \mathbf{r}') \, \delta(t - t'), \tag{7}$$

the Fourier transform of equation 5 becomes

$$\frac{\partial \phi_k}{\partial t} = -\Lambda \, k^2 \, T_2(k) \, \phi_k + \dot{\gamma} \, k_x \, \frac{\partial \phi_k}{\partial k_y} + f_k \tag{8}$$

and the experimentally observed scattered intensity $S_\mathbf{k}(t)$ at wave vector $\mathbf{k}$ and at time t is

$$S(\mathbf{k}, t) = \langle \phi_k^2(t) \rangle \tag{9}$$

which follows from the above equations to be given by

$$\frac{\partial}{\partial t} S(\mathbf{k}, t) = 2\Lambda k^2 - 2\Lambda \, k^2 \, T_2(k) \, S(\mathbf{k}, t) + \dot{\gamma} \, k_x \, \frac{\partial S(\mathbf{k}, t)}{\partial k_y}. \tag{10}$$

The role of flow can be parametized by considering the $\Lambda = 0$ case. Now

$$\frac{\partial S}{\partial t} = \frac{\partial \mathbf{k}}{\partial t} \cdot \nabla_k S = \dot{\gamma} \, k_x \, \frac{\partial S}{\partial k_y} \tag{11}$$

leads to "time-dependent" scattering vector

$$\mathbf{k}(t) = \mathbf{k} + \dot{\gamma} \, t \, k_x \, e_y \tag{12}$$

which satisfies the second equality of equation 11. It then follows for $\Lambda \neq 0$ that

$$S(\mathbf{k}, t) = e^{f(\mathbf{k}, t)} S(\mathbf{k}(t), 0) + 2\Lambda \int_0^t ds \, e^{f(\mathbf{k}, t) - f(\mathbf{k}, s)} \, k^2(s) \tag{13}$$

where

$$f(\mathbf{k}, t) = -2\Lambda \int_0^t ds \, k^2(s) \, T_2(\mathbf{k}(s)). \tag{14}$$

In the steady state, equation 10 yields

$$S(\mathbf{k}) = 2\Lambda \int_0^\infty dt\, k^2(t)\, e^{-2\Lambda \int_0^t ds\, k^2(s)\, T_2(\mathbf{k}(s))} \tag{15}$$

where $\mathbf{k}(s)$ and $\mathbf{k}(t)$ are given by equation 12. If $\dot\gamma$ is small enough, the integrals in equation 15 can be performed as a perturbation series in $\dot\gamma$ to get

$$S(\mathbf{k}) = \frac{1}{T_2(k)} + c_1 \dot\gamma\, k_x k_y + c_2 \dot\gamma^2\, k_x^2 k_y^2 + \cdots \tag{16}$$

where c_1 and c_2 are coefficients which can be readily calculated (5) for the specific polymer system. Even in the linear regime, it is necessary to perform the integrals in equations 13 and 15 numerically to obtain the full $\dot\gamma$ and t dependencies of $S(\mathbf{k}, t)$. Since this has not been done yet, we give below only the leading terms which can be obtained perturbatively.

Effect on Block Copolymers

The free energy F of a collection of n block copolymer chains is given by

$$e^{-\frac{F}{k_B T}} \equiv Z = \int \prod_{\alpha=1}^n \mathcal{D}[\mathbf{R}]\, e^{-H_E} \tag{17}$$

where H_E is the Edwards Hamiltonian appropriate to the particular block copolymer chain and $\int \prod_\alpha^n \mathcal{D}[\mathbf{R}]$ denotes the sum over all configurations of all chains. As we are interested in F as a functional of an order parameter, say $\phi(\mathbf{r})$, which may be determined experimentally, we impose a prescribed spatial variation of $\phi(\mathbf{r})$. The configuration sum in equation 17 is performed in two steps: first, sum over all chain configurations which are compatible with the prescribed ϕ ; second, sum over all possibilities of ϕ. This allows us to write Z of equation 17 as

$$Z = \int \mathcal{D}[\phi]\, \exp(-H_{LOK}[\phi]) \tag{18}$$

where $H_{LOK}[\phi]$ is the Leibler-Ohta-Kawasaki Hamiltonian (9-11) given by

$$H_{LOK}[\phi] = \frac{1}{2!} \int_\mathbf{k} T_2(\mathbf{k}, -\mathbf{k})\, \phi(\mathbf{k})\, \phi(-\mathbf{k}) +$$

$$\frac{1}{3!} \int_{\mathbf{k}_1} \int_{\mathbf{k}_2} T_3(\mathbf{k}_1, \mathbf{k}_2, -\mathbf{k}_1 - \mathbf{k}_2) \, \phi(\mathbf{k}_1) \, \phi(\mathbf{k}_2) \, \phi(-\mathbf{k}_1 - \mathbf{k}_2) +$$

$$\frac{1}{4!} \int_{\mathbf{k}_1} \int_{\mathbf{k}_2} \int_{\mathbf{k}_3} T_4(\mathbf{k}_1, \mathbf{k}_2, \mathbf{k}_3, -\mathbf{k}_1 - \mathbf{k}_2 - \mathbf{k}_3) \cdot$$

$$\phi(\mathbf{k}_1) \, \phi(\mathbf{k}_2) \, \phi(\mathbf{k}_3) \, \phi(-\mathbf{k}_1 - \mathbf{k}_2 - \mathbf{k}_3) + \cdots. \tag{19}$$

The explicit forms of the coefficient functions are well-known in the literature (9) as obtained using the random-phase approximation in the incompressibility limit. Since the order parameter ϕ is a fluctuating quantity,

$$\phi = \phi_0 + \psi \tag{20}$$

where ϕ_0 is the mean and ψ is the fluctuating part. Substituting this in equation 18, we get (11)

$$Z = \exp\left(-H_{LOK}(\phi_o)\right) \int \mathcal{D}[\psi] \, \exp(-\tilde{H}[\phi_0, \psi]) \tag{21}$$

so that the free energy F is given by

$$F = H_{LOK}(\phi_0) + F_f \tag{22}$$

where F_f is the fluctuation part of the free energy,

$$F_f = -\ln \int \mathcal{D}[\psi] \, \exp(-\tilde{H}[\phi_0, \psi]). \tag{23}$$

Explicit expressions for $\tilde{H}$ can be readily derived (11). The mean field part of F, $H_{LOK}(\phi_0)$ contains chain fluctuations due to the explicit wave vector dependence of T_2 at all length scales within a polymer chain. Using the full expression for free energy as given by equation 22 and parameterizing ϕ_0 to suit the density profiles of different morphologies, and then minimizing F with respect to these parameters, the phase diagram of different morphologies of block copolymers can be constructed (11-13).

The static structure factor $S(\mathbf{k})$

$$S(\mathbf{k}) = \langle \psi(\mathbf{k}) \psi(-\mathbf{k}) \rangle \tag{24}$$

is an infinite series using the Hamiltonian, $\tilde{H}[\phi_0, \psi]$. Applying the random phase approximation to enable a truncation of this series yields

$$\frac{1}{S(\mathbf{k})} = T_2(\mathbf{k}, -\mathbf{k}) +$$

$$\frac{1}{2}\int_{\mathbf{k}_1} T_4(\mathbf{k}, -\mathbf{k}, -\mathbf{k}_1, \mathbf{k}_1)\{\phi_0(\mathbf{k}_1)\psi_0(-\mathbf{k}_1) + S(\mathbf{k}_1)\}. \tag{25}$$

Near the disorder-order transition, the exact expression for $T_2(\mathbf{k}, -\mathbf{k})$ can be written in the Brazovskii form (14)

$$T_2 \simeq [\tau + (k - k^*)^2] \tag{26}$$

where $k^* \sim \frac{1}{\sqrt{N}}$ and $\tau \sim (X_s - X)$. Here N is the chain length, X is the chemical mismatch parameter between the A and B type blocks, and X_s is the spinodal value of X based on the mean field part of F. The various coefficients of equation 26 depend on whether the block copolymer is diblock or triblock. The coeffients are known in the literature. Since the order parameter ϕ_0 is very small near the order-disorder transition, it is usually assumed to be zero in equation 25. Furthermore, it is usual to approximate $T_4(\mathbf{k}, -\mathbf{k}, -\mathbf{k}_1, \mathbf{k}_1)$ to be a constant T_4. Now the $\mathbf{k}_1$-integral of equation 25, depicting the fluctuation correction to T_2,

$$\int_{\mathbf{k}_1} \frac{1}{\tau + (k - k^*)^2} \sim \frac{(k^*)^2}{\sqrt{\tau}} \tag{27}$$

diverges as $\tau \to 0$. In view of this we let

$$S(k) = \frac{1}{r_0 + (k - k^*)^2} \tag{28}$$

and determine r_0 self-consistently from equation 25. Using the above mentioned several approximations, we get (14)

$$r_0 = \tau + \frac{(k^*)^2}{4\pi} \frac{T_4}{\sqrt{r_0}}. \tag{29}$$

This result was originally derived by Brazovskii. This shows that fluctuations increases the X_t, the value of X corresponding to the order-disorder transition. If X is inversely proportional to T, then the order-disorder transition temperature is depressed by the fluctuations.

In the presence of shear, the steady state structure factor is similarly given self-consistently by

$$\frac{1}{S(\mathbf{k})} = T_2(\mathbf{k} - \mathbf{k}) + \frac{T_4}{2} \int_{\mathbf{k}_1} S(\mathbf{k}_1) \tag{30}$$

where $S(\mathbf{k}_1)$ is now given by equation 16. Again replacing τ of $T_2(k)$ by the unknown r_0, the $\mathbf{k}_1$-integral becomes from equations 16 and 30.

$$\int_{\mathbf{k}_1} \left\{ \frac{1}{r_0 + (k - k^*)^2} + c_1 \dot{\gamma} k_x k_y + c_2 k_x^2 k_y^2 + \cdots \right\}. \tag{31}$$

The self-consistent determination of r_0 from equation 30 gives the shear rate dependence of the order-disorder transition X, X_t. The result is

$$\frac{X_s - X_t(\dot\gamma)}{X_s - X_t(0)} = 1 - \alpha \left(\frac{\dot\gamma N^{5/2}}{\Lambda}\right)^2 + \cdots \tag{32}$$

where α is a known positive coefficient of order unity. The actual value of α depends on the architecture of the block copolymer (whether diblock or triblock) and the particular morphology of the ordered state (whether lamellar or cylinder). If X is proportional to $1/T$, then the shear rate dependence of the order-disorder transition temperature T_{ODT} is given by

$$\frac{T_{ODT}(\dot\gamma)}{T_{ODT}(0)} = 1 + \beta \left(1 - \frac{T_{ODT}(0)}{T_s}\right) \left(\frac{\dot\gamma N^{5/2}}{\Lambda}\right)^2 + \cdots \tag{33}$$

where β is a positive coefficent related to α. Therefore shear suppresses the composition fluctuations and raises the T_{ODT}. The result of equation 33 is only a perturbative result. It can be shown that when $\dot\gamma$ is very high, T_{ODT} becomes the mean field value of Leibler (9).

Although some experimental results begin to be available, even the functional form of the perturbative term, $\dot\gamma N^{5/2}/\Lambda$, has not yet been verified.

Effect on Spinodal Decomposition in Polymer Blends

We first give explicit forms for the free energy F and the Onsager coefficient Λ.

Free Energy. We take the free energy to be the generalized Flory-Huggins free energy to account for spatial inhomogeneities,

$$\frac{F}{k_B T} = \int d\mathbf{r} \left[\frac{\phi(\mathbf{r})}{N_A} \ln \phi(\mathbf{r}) + \frac{1-\phi(\mathbf{r})}{N_B} \ln(1-\phi(\mathbf{r})) + \right.$$
$$\left. \chi \phi(\mathbf{r})(1-\phi(\mathbf{r})) + \kappa_0^2 (\nabla \phi(\mathbf{r}))^2 \right] \tag{34}$$

where N_A and N_B are the numbers of effective segments in a chain of type A and B respectively. κ_0 contains two parts: one part arising from the range of interactions responsible for the χ parameter, analogous to the Cahn-Hilliard term, and the other part arising from the entropy of the chain. Using the random phase approximation for polymer blends and the Cahn-Hilliard theory for small molecular systems,

$$\kappa_0^2 = \frac{1}{3}\left(\frac{R_A^2}{\phi N_A} + \frac{R_B^2}{(1-\phi)N_B}\right) + \chi \ell^2 \tag{35}$$

where R_i^2 is the mean square radius of gyration of i and ℓ is the Kuhn length.

Onsager Coefficient. Using the standard procedure in nonequilibrium thermodynamics, the Onsager coefficient Λ is given by

$$\Lambda = \phi(1-\phi)\left[N_A D_A (1-\phi) + N_B D_B \phi\right] \tag{36}$$

where D_i are the self-diffusion coefficients of component i. For the symmetric blend, $N_A = N_B = N$ and $D_A = D_B = D$, we get

$$\Lambda = DN\phi(1-\phi) \tag{37}$$

For the asymmetric case, $N_A = N \gg 1$ and $N_B = 1$, Λ can be approximated by

$$\Lambda = D_0 \phi(1-\phi) \tag{38}$$

where the Rouse law, $D_A = D_0/N$ is assumed in addition to taking D_B to be D_0. For arbitrary asymmetry in the molecular weight the full form of equation 36 must be used.

The noises **f** and **f'** in equation 1 are specified in terms of their respective fluctuation-dissipation theorem. The general description is nontrivial and the specific instances are discussed below.

Under the experimental conditions where the velocity gradients are weak due to high viscosities of the blends and molecular weights of the polymers not too high so that the elastic contribution to the free energy is weak, equation 1a reduces to

$$\frac{\partial \phi}{\partial t} = \nabla \cdot DN\phi(1-\phi)\nabla\left[\frac{1}{N}\ln\frac{\phi}{(1-\phi)} + \chi(1-2\phi)\right.$$
$$\left. + \frac{\ell^2(1-2\phi)}{36\phi^2(1-\phi)^2}(\nabla\phi)^2 - \frac{\ell^2}{18\phi(1-\phi)}\nabla^2\phi\right] - \dot{\gamma}\frac{\partial \phi}{\partial x} + f. \tag{39}$$

We call this equation as the mean field equation since the coupling of ϕ to the velocity field is ignored. We assume that the enthalpic part of the square gradient term is negligible in comparison with the entropic part. This approximation has been established previously to be a good approximation for the polymer problems. Although the above equation is written for the symmetric blend, the general case can be readily considered.

Linear Analysis. Taking the local concentration ϕ to be a minor perturbation from the average value,

$$\phi = \phi_0 + \delta\phi \tag{40}$$

and ignoring higher order fluctuations of $0\,(\delta\phi)^2$, we get from equation 39,

$$\frac{\partial \delta\phi}{\partial t} = DN\,\phi_0\,(1-\phi_0)\,\nabla^2 \left[-\frac{\ell^2 \nabla^2}{18\,\phi_0\,(1-\phi_0)} + 2(\chi_s - \chi)\,\delta\phi - \dot\gamma\,y\,\frac{\partial \delta\phi}{\partial x} + f \right. \tag{41}$$

where

$$\chi_s = 1/2\,N\,\phi_0\,(1-\phi_0) \tag{42}$$

and the fluctuation-dissipation theorem for the noise f is

$$\langle f(\mathbf{r},t)\,f(\mathbf{r}',t') \rangle = -2DN\,\phi_0\,(1-\phi_0)\,\delta\,(\mathbf{r}-\mathbf{r}')\,\delta\,(t-t'). \tag{43}$$

The experimentally observed scattered intensity $I_\mathbf{k}(t)$ at wave vector $\mathbf{k}$ and at time t is the Fourier transform of the mean square fluctuations in the order parameter, $\langle(\delta\phi)^2\rangle$. It follows from equation 41 that

$$\frac{\partial I_\mathbf{k}(t)}{\partial t} = 2L_0\,k^2 - 2L_0\,k^2\,(Kk^2 - \kappa^2)\,I_\mathbf{k}(t) + \dot\gamma\,k_x\,\frac{\partial I_\mathbf{k}(t)}{\partial k_y} \tag{44}$$

where

$$L_0 = DN\,\phi_0\,(1-\phi_0)$$
$$\kappa^2 = 2\,(\chi - \chi_s)$$
$$K = \ell^2/18\,\phi_0\,(1-\phi_0). \tag{45}$$

Although the form of equation 44 is identical to that of Imaeda, Onuki and Kawasaki (*8*) (IOK), the analysis in this report contains notable differences from IOK. The role of flow can be parametrized by considering the $L_0 = 0$ case. Now

$$\frac{\partial I}{\partial t} = \frac{\partial \mathbf{k}}{\partial t} \cdot \nabla_\mathbf{k} I = \dot\gamma\,k_x\,\frac{\partial I}{\partial k_y} \tag{46}$$

leads to "time-dependent" scattering vector

$$\mathbf{k}(t) = \mathbf{k} + \dot\gamma\,t\,k_x\,\mathbf{e}_y \tag{47}$$

which satisfies the second equality of equation 46. It then follows for $L_0 \neq 0$ that

$$I_{\mathbf{k}}(t) = e^{f(\mathbf{k},\,t)} I_{\mathbf{k}(t)}(0) + 2L_0 \int_0^t ds\, e^{f(\mathbf{k},\,t) - f(\mathbf{k},\,s)} k^2(s) \qquad (48)$$

where

$$f(\mathbf{k},\,t) = 2L_0 \int_0^t ds [\kappa^2 k^2(s) - K k^4(s)] \qquad (49)$$

For $\dot{\gamma} t > 1$, $k^2(t)$ is approximated by

$$k^2(t) = (k_y + \dot{\gamma} t k_x)^2 + k_z^2 \qquad (50)$$

Using the change of variables,

$$\ell^2 = k^2/\kappa^2$$

$$\tau = L_0 \kappa^4$$

and

$$\Im(\mathbf{l},\,\tau) = \kappa^2 I_{\mathbf{k}}(t) \qquad (51)$$

we obtain

$$\Im_{\mathbf{l}}(\tau) = \Im_{\mathbf{l}(\tau)}(0) e^{F(\mathbf{l},\,\tau)} + 2 \int_0^\tau du\, l_\perp^2(u) e^{F(\mathbf{l},\,\tau) - F(\mathbf{l},\,u)} \qquad (52)$$

where

$$F(\mathbf{l},\,\tau) = 2 \int_0^\tau du\, [l_\perp^2(u) - K l_\perp^4(u)]$$

$$l_\perp^2(u) = l_\perp^2 + 2u l_x l_y + u^2 l_x^2 \qquad (53)$$

with $l_\perp^2 = l_y^2 + l_z^2$.

Results. We now proceed to look at the limiting behaviors of $\Im$. Since $F(\mathbf{l},\,\tau)$ is complicated, we ignore the cross term $2u l_x l_y$ in equation 53. The asymptotic results obtained below are independent of this approximation. Now,

$$F(\mathbf{l},\,\tau) \simeq 2(l_\perp^2 - K l_\perp^4)\tau + \frac{2}{3} l_x^2 (1 - 2K l_\perp^2) \tau^3 - \frac{2}{5} K l_x^4 \tau^5 \qquad (54)$$

1). It follows from equation 54 that fluctuations with $l_\perp > 1/\sqrt{K}$ cannot be enhanced. Therefore we may assume that $l_\perp < 1$. Now, for longer times, $F(1, \tau)$ goes essentially as $-\tau^5$.

2). For $l_x = 0$, $\Im(1, \tau)$ grows indefinitely.

3). For $l_x > 1$, $l_\perp < 1$: The second integral in equation 52 can be approximated as

$$\int_0^\tau l_\perp^2(u) e^{-F(1,u)} \simeq \int_0^\tau du\, u^2\, l_x^2\, \exp(\frac{2}{5} l_x^4 u^5)$$

$$= \frac{1}{2}\left(\frac{2}{5l_x}\right)^{\frac{2}{5}} \int_0^\tau 'dx\, x^{-\frac{2}{5}} \exp(x)$$

$$= \frac{1}{2}\left(\frac{2}{5l_x}\right)^{\frac{2}{5}} (-1)^{-\frac{3}{5}} \left[\Gamma(\frac{3}{5}) - \Gamma(\frac{3}{5}, -\tau')\right] \quad (55)$$

where $\Gamma's$ are gamma functions and $\tau' = \frac{2}{5} l_x^4 \tau^5$. Substituting equation 55 into eauation 52 yields for $\tau > l_x^{-4/5}$

$$\Im(1, \tau) = l_x^{-2} \tau^{-2} + \cdots. \quad (56)$$

Therefore the intensity decays with time instead of reaching a constant value as claimed by IOK.

4). $l_\perp < 1$, $l_x < 1$: Now the second integral in equation 52 can be approximated as

$$\int_0^\tau du\, l_\perp^2(u) e^{-F(1,u)} \simeq 2\,|\,l_x\,|^{-\frac{2}{3}} \int_0^\infty dp\, p^2 \exp\left(\frac{2}{3}\,|\,l_x\,|^{-\frac{2}{3}} p^3 - \frac{2}{5} p^5\right)$$

$$= \sqrt{2\pi}\,|\,l_x\,|^{-\frac{1}{2}} \exp\left(\frac{4}{15}\,|\,l_x\,|^{-1}\right). \quad (57)$$

Substituting equation 57 into equation 52 yields for $\tau > 1/\,|\,l_x\,|$

$$\Im(1, \tau) = \sqrt{2\pi}\,|\,l_x\,|^{-\frac{1}{2}} \exp\left(\frac{4}{15}\frac{1}{|\,l_x\,|} - \frac{2}{5} l_x^4 \tau^5\right) \quad (58)$$

so that the scattered intensity decays exponentially with τ^5. This result also is in disagreement with the theoretical prediction of IOK but in agreement with the results of Nakatani et al (*16*).

Conclusions

The status of theoretical predictions on the effects of flow on polymeric systems is still in a primitive state. Even in the simplest cases, the problem is technically and computationally intense although the methodology is clear. There are many issues which need to be addressed by theoretical and computational methods before we can successfully predict the phase behavior of polymers under flow. Some of this issues are the following:

1. Calculation of nonlinear effects even for the mean field case.
2. Numerical solution of coupled equations between the order parameter and velocity field.
3. Derivation of correct coupled equations including the contributions from elasticity of the various components.
4. Incorporation of topological constraints, as associated with bridges and loops of $A-B-A$ triblock copolymers in self-assembled morphologies for example, and coupling with external flow fields.
5. Generalization of the coupled equations when there are many order parameters.

Acknowledgments

I am grateful to the polymer blend group at NIST for making critical comments on this problem. Acknowledgment is made to the NSF grant DMR-9221146001 and to the Materials Research Laboratory at the University of Massachusetts.

Literature Cited

1. Onuki, A.; Kawasaki, K. *Ann. Phys.* **1979**, *121*, pp. 456.
2. Frederickson, G.H.; *J. Chem. Phys.* **1986**, *85*, pp. 5306.
3. Onuki, A.; *J. Chem. Phys.* **1987**, *87*, pp. 3692.
4. Cates, M.E.; Milner, S.T. *Phys. Rev. Lett.* **1989**, *62*, pp. 1856.
5. Marques, C.M.; Cates, M.E. *J. Phys. (Paris)* **1990**, *51*, pp. 1733.
6. Frederickson, G.H.; Larson, R.G. *J. Chem. Phys.* **1987**, *86*, pp. 1553.
7. Douglas, J. *Macromolecules.* **1992**, *25*, pp. 1462.
8. Imaeda, T.; Onuki, A.; Kawasaki, K. *Prof. Theor. Phys.* **1984**, *71*, pp. 16.
9. Leibler, L. *Macromolecules.* **1980**, *13*, pp. 1602.
10. Ohta, T.; Kawasaki, K. *Macromolecules.* **1986**, *19*, pp. 2621.
11. Muthukumar, M. *Macromolecules* **1993**, *26*, pp. 5259.
12. Melenkevitz, J.; Muthukumar, M. *Macromolecules.* **1991**, *24*, pp. 4199.

13. Lescancec, R.L.; Muthumkumar, M. *Macromolecules.* **1993**, *26*, pp. 3908.
14. Brazovskii, S.A. *Sov. Phys. JETP* **1979**, *91*, pp. 7228.
15. Frederickson, G.H.; Helfand, E. *J. Chem. Phys.* **1987**, *87*, pp. 697.
16. Hobbie, E.K.; Hair, D.W.; Nakatami, A.I.; Han, C.C. *Phys. Rev. Lett.* **1982**, *69, pp. 1951.*

RECEIVED February 15, 1995

Chapter 16

Shear-Induced Changes in the Order—Disorder Transition Temperature and the Morphology of a Triblock Copolymer

C. L. Jackson[1], F. A. Morrison[2], A. I. Nakatani[1], J. W. Mays[3], M. Muthukumar[4], K. A. Barnes[2], and C. C. Han[1]

[1]Polymers Division, National Institute of Standards and Technology, Building 224, Room B210, Gaithersburg, MD 20899
[2]Department of Chemical Engineering, Michigan Technological University, 1400 Townsend Drive, Houston, MI 49931—1295
[3]Department of Chemistry, University of Alabama, 901 South 14th Street, Birmingham, AL 35294
[4]Polymer Science and Engineering Department, University of Massachusetts, 701 Lederle Graduate Tower, Amherst, MA 01003

An overview of ongoing research on a triblock copolymer subjected to steady-shear flow is presented. The copolymer is composed of polystyrene-polybutadiene-polystyrene (23 wt% polystyrene-d_8) with cylindrical morphology in the quiescent state. The order-disorder transition (ODT) was studied by in-situ SANS in a couette geometry as a function of temperature and shear rate. We observed that the ODT temperature shifted upward by 20 °C as the shear rate was increased from ~ 0.1 to 25 s^{-1}. Sheared samples were also quenched and studied by transmission electron microscopy. A unique shear-induced morphology was found and tentatively identified as a transformation from the original p31m space group to a p3m1 space group of lower symmetry, similar to a martensitic transformation in metals.

The study of polymer melts in a flow field is a subject of great relevance to the industrial processing of these materials via extrusion, injection molding and other methods. It is also a difficult area of research, because of the rapid changes that may occur in the system as a function of temperature, pressure and shear rate and the inherent non-equilibrium nature of a flowing system. In this work, the in-situ study of a triblock copolymer under steady shear flow is presented. Block copolymers are materials used in many industrial applications such as blend compatibilizers, surfactants and thermoplastic elastomers (1-2).

There have been other reports on the effect of deformation on the morphologies of block copolymer melts. The main effect of shear is to orient lamellar and cylindrical phases. Studies of diblock (*3-8*) and triblock (*9-12*) copolymers have included extrusion through a tube (*9*) steady shearing (*5,11*), oscillatory shearing (*5,10,12*) and roll-casting (*6*). Most of these studies have relied on quenched specimens, where the effect of the quench on the final structure is difficult to assess. More recently, in-situ SANS studies on diblock copolymers have shown an increase in the isotropic-to-lamellar transition temperature in a reciprocating shear field (*3*) and various intermediate phases have been observed by SANS between the lamellar and cylindrical phases on heating and shearing through the transition region (*4*). The added complexity of structure in triblock copolymers, which can adopt either a chain bridging or looping conformation in equilibrium, has complicated the development of theories to interpret data on these technically important materials.

This chapter summarizes results from a series of papers (*13-14*) and work still in progress (*15*) on the structure of a polystyrene-d_8/polybutadiene/polystyrene-d_8 (SBS) block copolymer (23 wt % styrene-d_8, cylindrical morphology) in a steady shear field, as studied by small-angle neutron scattering (SANS)(*13-15*) and transmission electron microscopy (TEM) (*14*). The in-situ measurements were made in a couette geometry as a function of shear rate and temperature, near the rheologically determined order-disorder transition (ODT) temperature of 116 °C. The TEM studies were conducted on a sheared and quenched specimen of the same copolymer, where the samples were prepared in a cone-and-plate rheometer under conditions similar to the SANS couette cell.

Experimental Procedures

Materials. The SBS copolymer with a block architecture of $8 \times 10^3/54 \times 10^3/8 \times 10^3$ g/mole ($M_w/M_n = 1.05$) was synthesized by coupling living diblock under vacuum using dichlorodimethylsilane, followed by fractionation to remove residual diblock polymer (*13*). The overall composition of the polymer is 23 wt.% polystyrene, and the ordered phase of this sample is a hexagonally packed cylindrical array. The ODT of this material was 116 ± 5 °C, by dynamic mechanical measurements. The glass transition temperature of the polystyrene phase in this sample is estimated to be 81 °C (*16*).

Small-Angle Neutron Scattering. SANS experiments were performed at the Cold Neutron Research Facility (CNRF) at the National Institute of Standards and Technology (NIST) using an incident wavelength of 0.6 nm for the in-situ melt studies. The SANS shear cell used has been described previously (*17*). In the standard configuration, the shear gradient is oriented so that the incident beam is parallel to the gradient axis (y-direction), and scattering is monitored in the xz plane (x is the flow direction and z is the neutral direction) called the through-view. We have also translated the apparatus so the beam enters the couette cell tangentially and refer to this as the yz or tangential-view. SANS measurements (at $\lambda=0.9$ nm) were also made on the quenched disks from the cone-and-plate rheometer in the xz plane (*14*), which is equivalent to the through-view for the in-situ studies.

Ultramicrotomy and Electron Microscopy. The quenched disk was sectioned with the surface normals both parallel and perpendicular (transverse) to the flow direction, x. This corresponds to the through and tangential view in the SANS patterns, respectively. Ultra-thin sections cryomicrotomed at -100 °C had a nominal section thickness of 50-70 nm. The polybutadiene blocks were stained with osmium tetroxide vapors by exposing the grid to a 4% aqueous solution at room temperature for 4 h (*14*). A Philips 400T (*18*) model TEM was used at 120 kV to obtain the images. The spacings on the TEM negatives were measured using an optical bench with a He/Ne laser (λ=632.8 nm) and calibrated with the image of a grating.

Results and Discussion

In-Situ SANS During Shear Flow. Three basic types of SANS behavior are seen in-situ as a function of shear rate and temperature. In Figure 1, we summarize the results with typical data taken at 116 °C, in the through-view. In Figure 1a, an isotropic ring of scattered intensity is seen at $q_{max} = 0.29$ nm^{-1}, where q is the scattering vector, $q = (4\pi/\lambda)\sin(\theta/2)$, λ is the incident wavelength, and θ is the scattering angle. This scattering pattern could either be from an isotropic, disordered melt or from a multigrain, randomly oriented, cylindrical phase. In Figure 1b, a single set of peaks is observed normal to the flow direction at $q_{max1} = 0.29$ nm^{-1}. This scattering pattern is the result of orienting the grains of a microphase-separated cylindrical phase. The spacing observed is $d_{(100)} = (2\pi/q) = 21.7$ nm, which corresponds to the spacing expected for the equilibrium cylindrical structure of this triblock copolymer (*14*). In Figure 1c, increasing the shear rate produces a second set of peaks at higher q, normal to the flow direction at $q_{max2} = 0.50$ nm^{-1} (d = 12.5 nm). The origin of this second set of peaks is the main discussion of this work and will be described in detail below. A scattering pattern obtained in the tangential geometry (yz plane) at 117 °C is shown in Figure 2, to confirm the hexagonally packed structure of the PS cylinders. The SANS intensity contour plots are normalized to the maximum intensity of the image, thus only the relative peak intensities within a given image can be compared.

The dynamics of the triblock copolymer during shear are evident in the evolution of the scattering patterns with time. The temporal evolution of the through-view scattering patterns following the inception of shear is shown in Figure 3 for shear rate and temperature conditions that are representative of data above and below T_{ODT}. The times were converted to strain units (strain, $\gamma = \dot{\gamma}t$) to compare the different experimental conditions. For this analysis, sector averages of the scattering intensity were studied. For the sample at 100 °C and 0.12 s^{-1}, the development of the high-q peak (q_2) coincides with the diminishment of the low-q peak (q_1) for all shear rates studied. This implies that the fraction of the sample which is responsible for the low-q peak (q_1) is reduced as the fraction of the sample which scatters at high q is formed. Since both the peak intensities and areas may vary with time, caution must be exercised when relating these quantities to the population of a certain structure, due to the multigranular nature of the sample and the orientation relative to the neutron beam. We will present the analysis of the anisotropic scattering data in the context of liquid crystalline order parameters elsewhere (*15*).

In-Situ SANS Upon Cessation of Shear. The ratio of the q_{max} values for the two sets of peaks shown in Figure 1 is 1:1.7, which is consistent with reflections from the (100) and (110) planes of a hexagonally packed cylindrical array, with a unit cell parameter, $a = 25$ nm. This interpretation proves to be too simple, however.

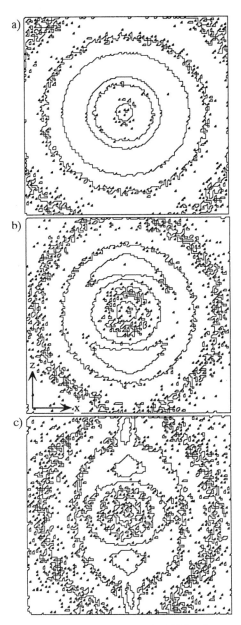

Figure 1. Shear rate dependence of in-situ SANS in the couette shear cell at 116 °C in the through-view (xz plane) shown by intensity contour plots: (a) 0 (b) 1.3 (c) 6.5 s^{-1}. (Adapted from ref. 13)

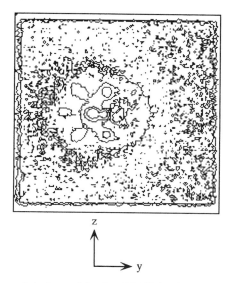

Figure 2. Tangential-view of in-situ SANS in the couette shear cell at 117 °C in the yz plane at a shear rate of 1.3 s^{-1}

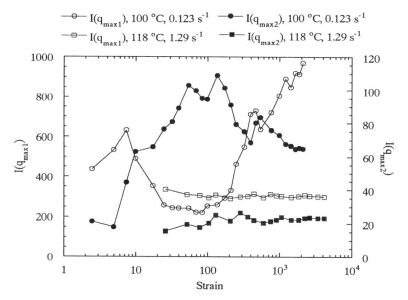

Figure 3. Strain dependence of neutron scattering intensity normal to flow around T_{ODT}: ○ I_{max1} and ● I_{max2} at 100 °C, 0.12 s^{-1} and □ I_{max1} and ■ I_{max2} at 118 °C, 1.3 s^{-1}.

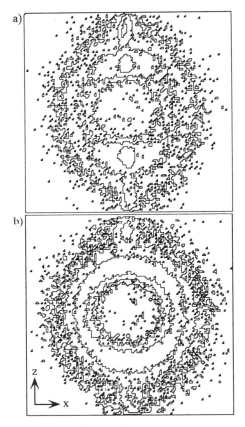

Figure 4. Time dependence of scattering patterns after cessation of shear (6.5 s^{-1}) at 116 °C at (a) 0 s (b) 120s after shear stops. (Adapted from ref. 13)

The time-dependence of the scattering behavior after cessation of shear provides strong evidence that the two sets of peaks originate from two distinct morphologies. As shown in Figure 4, the low-q peak was observed to broaden azimuthally and become isotropic in a short time (120 s), while the high-q peak remained sharp over the same time period (13). The shape of the high-q peak was very different from the low-q peak, also suggesting a different origin.

Morphology of Sheared and Quenched Specimen by TEM. The structure of the sheared, quenched block copolymer (0.59 s^{-1}, 100 °C) parallel to the flow direction is seen in Figure 5. The larger spacing, labelled 1, is measured to be 21 ± 2 nm. A second structure labelled 2, with a much smaller spacing of 12 ± 2 nm, is also evident in Figure 5. In the region shown, a large area of the second structure is shown for purposes of illustration, but in general, the overall fraction of this structure in the sample is low. SANS on this specimen confirmed this observation (14). The two spacings are evident in the optical diffractogram in the inset, which illustrates that they are oriented in the same direction. It is not possible that the smaller spacing is the (110) spacing of the primary structure in this type of cylindrical block copolymer, because the diameter of the cylinders is very large (d = 12 nm) within the unit cell.

The large errors on the d-spacings reflect the approximate nature of the TEM measurement, especially since the cylinders and matrix have an undulating interface due to the shear (14,19). The image in Figure 5 was chosen to demonstrate the presence of both structures in the same field of view, but quantitative estimates of the amount of each structure present in the sample should not be made from such a small area. The domain sizes shown also may not reflect the average domain size. A significant portion of poorly ordered material was also present in the sheared sample but is not shown in detail here.

The morphology of the sheared, quenched block copolymer (0.59 s^{-1}, 100°C) in the transverse view is shown in Figure 6. This confirms that the primary structure is a hexagonally packed array of cylinders, with an approximate spacing, $d_{(100)}$ = 22 ± 2 nm. Based on the SANS results obtained on this specimen, we show the cylinders to be oriented with the (100) planes parallel to the shear plane, or xz plane. Another region is shown in Figure 6 where the best image of the small cylinders, end-on, is identified with an arrow. The cylinder diameters are approximately 8-9 nm, which is significantly smaller than seen in the primary structure. They are fairly well resolved in a small region. The central domain of each hexagonal unit cell is absent for the region identified, and the relative intensity of the spots in the small hexagon alternate in brightness, appearing as either white or gray. This is an indication that the PS concentration in the gray cylinders is slightly lower than in the white cylinders. The small cylinders are very difficult to image due to their small diameter (~8-9 nm) relative to the sample thickness of 50-70 nm, and the low volume fraction of this component.

Possible artifacts that might produce the observed results, including deformation during cutting, tilt effects, thickness, staining and focussing conditions, have been given careful consideration and were insufficient to explain the two spacings observed (14, 19-22). Finally, although there is some uncertainty in size measurements made by TEM, we have found fairly good agreement between the measurements in the parallel and transverse cases. The appearance of both

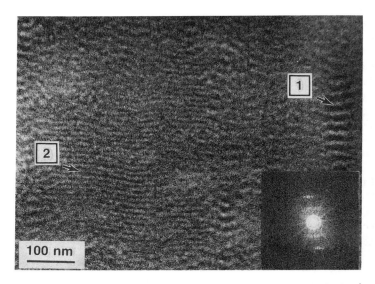

Figure 5. TEM micrograph of the sheared and quenched SBS (0.59 s^{-1} at 100 °C), parallel-view (*xy* plane, flow direction approximately horizontal). The butadiene matrix has been stained dark with osmium tetroxide. The image shows regions of a large spacing, ~ 22 nm, labelled 1, coexisting with regions of a smaller spacing, ~ 12 nm, labelled 2. The optical diffractogram of the TEM negative is inset. (Reproduced with permission from ref. 14. Copyright 1995 American Chemical Society).

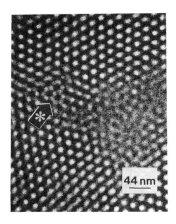

Figure 6. TEM micrograph of the sheared and quenched SBS (0.59 s^{-1} at 100 °C), transverse-view (*yz* plane, flow direction normal to page), showing the cylinders end-on. A region of the primary structure of the large PS cylinders of spacing, ~ 22 nm, coexist with a region of the smaller PS cylinders of spacing, ~ 12 nm, as designated by the arrow. (Reproduced with permission from ref. 14. Copyright 1995 American Chemical Society).

structures in the same field of view for both geometries directly demonstrates the relative size difference in the two structures.

Shear-Induced Martensitic-like Transformation. The experimental observation of an additional cylindrical structure forming and coexisting with the original cylindrical morphology under shear is unique and has not been previously reported in the field of block copolymers. We propose that the shear-induced morphology may be related to the original hexagonal microstructure as a "superlattice", with analogy to other crystalline materials. The unit cells and corresponding lattice structures are shown in Figure 7, where the positions of the cylinders in the unit cells correspond to the atomic positions in a crystallographic lattice, and the two lattices are drawn to scale relative to each other. In block copolymers, the crystals are two-dimensional since the cylinders extend along the c-axis. The unit cell of the structure in Figure 7a has the same symmetry as the p31m crystallographic space group while the unit cell of the structure in Figure 7b belongs to the p3m1 space group (23). The white circles represent the PS cylinders, while the shaded circles in the p3m1 structure represent cylinders of some intermediate composition. This corresponds to the alternating pattern of white and gray cylinders observed experimentally in Figure 6. The dark matrix is the PB-rich phase. The reflection observed at high-q is identified as the (110) spacing, $d_{(110)} \sim 12$ nm, from Figure 7b and is oriented with the (110) planes at approximately 30° to the (100) planes of the primary structure. The crystallographic spacing of the shear-induced phase is directly related to the original unit cell of the quiescent phase by $\sqrt{3}$ for the symmetry reasons outlined in Figure 7. For simplicity we refer to the p31m structure as "triangular hexagonal" and the shear-induced, p3m1 structure as "simple hexagonal".

The transformation of the p31m unit cell of dimension $a = 25$ nm to a hexagonal microstructure of lower (p3m1) symmetry, also with $a = 25$ nm, is analogous to martensitic transformations known in solid-state materials such as iron (24,25). Martensitic transformations are defined as shear-dominant, lattice-distortive, diffusionless processes which occur by nucleation and growth (24). The kinetics and morphology of these transformations are dominated by strain energy. In metals, these transformations are also observed in deformation processes and properties such as hardness are improved (25).

Shear-Induced Shift in the ODT Temperature by In-Situ SANS. Three distinct regimes of shear rate response for the triblock copolymer are identified in Figure 8. The strain-dependence of the high-q peaks is very significant, and the presence of two sets of peaks at any time during the experiment for a given shear rate and temperature was the criterion used to construct this plot. From Figure 8, we have identified two critical shear rates. The first is where the transition from an isotropic to an ordered cylindrical morphology occurs, called $\dot{\gamma}_{C1}$, denoted by the upper curve. The second critical shear rate, called $\dot{\gamma}_{C2}$, is denoted by the lower curve. This is where a morphological change (14), or an order-order transition, occurs in the sample, as observed by TEM on a sheared and quenched specimen.

The SANS data presented in Figure 8 proved a useful alternative means for identifying T_{ODT}, a measurement that is sometimes ambiguous for block copolymers by conventional rheological measurements (15). Since hexagonal symmetry exists for a single, anisotropic peak (Figures 1-2), we can extrapolate the $\dot{\gamma}_{C1}$ line to zero

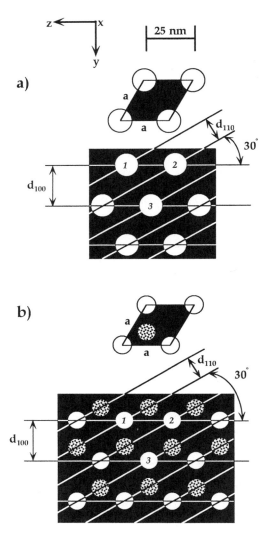

Figure 7. Hexagonal lattice projected along the c-axis. (a) p31m space group and related "triangular hexagonal" structure (b) p3m1 space group and related "simple hexagonal" structure. The white circles represent the cylinders of PS, the shaded circle represents an intermediate concentration of PS and the black area represents the PB matrix. (Adapted from ref. 14)

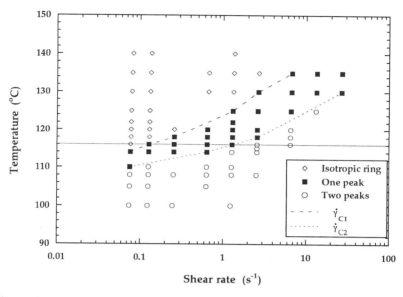

Figure 8. Temperature versus shear rate, showing the increase in the ODT temperature with increasing shear rate based on the appearance of through-view SANS patterns.

shear rate and obtain an estimate of T_{ODT} of 110 °C. We believe the discrepency between this value and the rheological value for ODT of 116 ± 5 °C, represent genuine differences between the two methods.

The shift in ODT is shown as a function of shear rate in Figure 9, where the log of reduced T_{ODT} versus log shear rate is plotted for the data obtained in this study. We find a slope of 0.4 from a linear regression analysis. A theory proposed by Cates and Milner (26) predicts a shift in the T_{ODT} with shear rate for diblock copolymers, but this theory may not be appropriate for triblocks under the conditions here. Further progress in the theoretical treatments of flowing block copolymers will provide a better understanding of the observed experimental data.

Model of Flow Behavior and Morphology. Two cylindrical microstructures having different symmetries were found to exist in a polystyrene-d_8/polybutadiene/polystyrene-d_8 (SBS) block copolymer subjected to shear flow. The fact that the unit cell parameters for the two structures are actually the same ($a = 25$ nm) has important ramifications. It is apparent from the symmetry of the two space groups shown in Figure 7 that only two structures with the same unit cell parameters will form, instead of structures with a wide range of apparent spacings. Similarly, the energy required for the transition to occur may not be very large, since the basic unit cell dimensions are unchanged. As outlined in Figure 7, the white PS cylinders at the positions marked 1,2 and 3 remain stationary but reduce in size. Conservation of mass dictates the resultant cylinder size and polystyrene composition of the newly formed cylinders, shown as the shaded circles in Figure 7. It is important to note that the reduced cylinder diameter cannot be due to a simple elongation of the original triangular hexagonal structure, because of the lack of a central cylinder

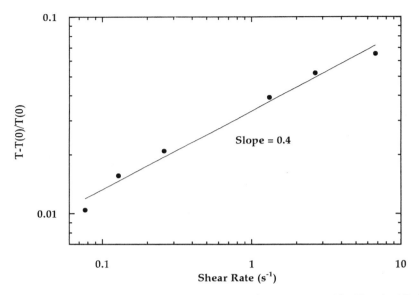

Figure 9. Plot of the reduced shift in ODT ($\Delta T_{ODT} = [\ T_{ODT}(\dot{\gamma}) - T_{ODT}(\dot{\gamma}=0)]\ /\ T_{ODT}(\dot{\gamma}=0)$ as a function of shear rate. The symbols are data from the current study on triblocks and the line is a linear regression fit to the data, slope 0.4.

in the new structure and the alternating composition of the cylinders that is observed. Because of the preferred orientation of the original triangular hexagonal structure relative to the shear field (Figure 2), imposition of a non-symmetric deformation such as simple shear permits the transformation from the triangular hexagonal to the simple hexagonal structure. It is possible that if the triangular hexagonal structure had another orientation relative to the gradient direction, a different shear-induced morphology would be produced.

A theory to describe the effect of shear on the morphology of microphase-separated block copolymers is challenging because the relative stability of the morphology is complicated by the presence of topological constraints. The response of these constraints to shear flow is difficult to predict. Even the simpler problem of diluted polymer solutions subject to shear is imperfectly understood because of our limited understanding of the shear properties of entangled polymer solutions (27). Although the identification of the two space groups allows us to gain understanding of this transformation purely on the basis of symmetry arguments, it is compelling to propose a molecular rationalization of the nature of the transformation. Future experimental and theoretical work will be focussed on this area.

Acknowledgments

The martensitic-like transformation in a triblock copolymer under shear was originally suggested by J. Cahn. We acknowledge his insight into this phenomenon. The authors also thank J. Douglas, R. Briber, R. M. Waterstrat, E. DiMarzio, F. Khoury, and J. Barnes for helpful discussions during the course of this work.

Literature Cited

1) Aggrawal, C. Ed., *Block Copolymers*, Plenum Press, New York, 1979.
2) Brown, R. A.; Masters, A. J.; Price, C. and Yuan, X. F. in *Comprehensive Polymer Science, Vol. 2*; Pergamon Press, Oxford, 1989.
3) Koppi, K. A.; Tirrell, M.; Bates, F. S. *Phys. Rev. Lett.*, **1993**, *70*, 1449.
4) Hamley, I. W.; Koppi, K. A.; Rosedale, J. H.; Bates, F. S.; Almdal, K.; Mortensen, K. *Macromolecules*, **1993**, *26*, 5959.
5) Winey, K. I.; Patel, S. S.; Larson, R. G.; Watanabe, H. *Macromolecules*, **1993**, *26*, 2542.
6) Albalak, R. J.; Thomas, E. L. *J. Polym. Sci.:Polym. Phys. Ed.*, **1993**, *31*, 37.
7) Balsara, N. P.; Hammouda, B.; Kesani, P. K.; Jonnalagadda, S. V.; Straty, G. C. *Macromolecules* **1994**, *27*, 2566.
8) Balsara, N. P.; Dai, H. J.; Kesani, P. K.; Garetz, B. A.; Hammouda, B. *Macromolecules*, **1994**, *27*, 7406.
9) Keller, A.; Pedemonte, E.; Willmouth, F. M. *Koll. Z. Z. Polym.*, **1970**, *238*, 385.
10) Hadziioannou, G.; Mathis, A.; Skoulios, A. *Colloid Polym. Sci.*, **1979**, *257*, 136.
11) Morrison, F. A.; Winter, H. H.; Gronski, W.; Barnes, J. D. *Macromolecules*, **1990**, *23*, 4200.
12) Winter, H. H.; Scott, D. B.; Gronski, W.; Okamoto, S.; Hashimoto, T. *Macromolecules*, **1993**, *26*, 7236.
13) Morrison, F.; Mays, J. W.; Muthukumar, M.; Nakatani, A. I.; Han, C. C. *Macromolecules*, **1993**, *26*, 5271.
14) Jackson, C. L.; Barnes, K. A.; Morrison, F. A., Mays, J. W.; Nakatani, A. I.; Han, C. C. *Macromolecules*, **1995**, *28*, 713.
15) Nakatani, A. I.; Morrison, F. A.; Douglas, J. F.; Jackson, C. L.; Mays, J. W.; Muthukumar, M.; Han, C. C. in preparation.
16) Richardson, J.; Saville, N. G. *Polymer*, **1977**, *18*, 3.
17) Nakatani, A. I.; Kim, H.; Han, C. C. *J. Res. NIST*, **1990**, *95*, 7.
18) Certain commercial materials and equipment are identified in this paper in order to specify adequately the experimental procedure. In no case does such identification imply recommendation or endorsement by the National Institute of Standards and Technology nor does it imply that the material or equipment identified is necessarily the best available for this purpose.
19) Odell, J. A.; Dlugosz, J. ; Keller, A. *J. Polym. Sci.:Polym. Phys. Ed.*, **1976**, *14*, 861.
20) Handlin, D. L.; Thomas, E. L. *Macromolecules*, **1983**, *16*, 1514.
21) Dlugosz, J.; Keller, A.; Pedemonte, E. *Kolloid Z. Z. Polymere*, **1970**, *242*, 1125.
22) Roche, E. J.; Thomas, E. L. *Polymer*, **1981**, *22*, 333.
23) *International Tables for Crystallography*; Hahn, T., Ed.; D. Reidel Publishing Co.: 1983, Vol. A, pp. 78-79.
24) Olson, G. B. In *Martensite*; Olson, G. B.; Owen, W. S., Eds.; ASM International, 1992, Chapter 1.
25) Waterstrat, R. M. *Platinum Metals Rev.*, **1993**, *37 (4)*. 194.
26) Cates, M. E.; Milner, S. T. *Phys. Rev. Lett.*, **1989**, *62*, 1856.
27) Douglas, J. F. *Macromolecules*, **1992**, *25*, 1468.

RECEIVED March 1, 1995

Chapter 17

Phase-Separation Kinetics of a Polymer Blend Solution Studied by a Two-Step Shear Quench

A. I. Nakatani, D. S. Johnsonbaugh, and C. C. Han

Polymers Division, National Institute of Standards and Technology, Building 224, Room B210, Gaithersburg, MD 20899

A polystyrene (PS) and polybutadiene (PB) blend (50:50 by weight) dissolved in dioctyl phthalate (DOP) (8% total polymer by weight) was subjected to a two-step shear quench with a delay time between the two shear rates to examine the phase separation kinetics by light scattering. The time dependence of the peak maximum position, q_m, and the intensity of the peak maximum, $I(q_m)$, were measured. The growth rate after cessation from the second shear rate was unaffected by the delay time. The mechanism for the coarsening parallel to flow was coupled to the extent of coarsening which occurred normal to the flow direction during the second shear rate.

In a prior report on the phase separation kinetics of a polymer blend solution following cessation of shear [1], we observed that a solution which is initially in the two-phase region becomes homogeneous at high shear rates (626 s^{-1}). Upon cessation of shear, phase separation occurred and isotropic domains formed after short times (30 s). At lower shear rates (5.22 s^{-1}), the system remained inhomogeneous, droplet deformation was observed, and upon cessation of shear, the droplets remained deformed for long times (10-15 min). This experimental result raised the question of the relationship between the initial, quiescent two-phase structure and its effect on the subsequent coarsening behavior after shearing, since the results conflicted with the anticipation that a system which experiences a small perturbation should return to equilibrium faster than the same system which has undergone a large perturbation. From the results described above, it appears that orientation induced at high shear rates is shorter lived than the lesser degree of orientation produced at lower shear rates. Hence, the production of oriented two-phase materials may actually be improved by using lower shear rates. For the processing of two-phase materials, this result has important implications because two-phase materials traditionally have been oriented by applying high shear rates to a sample to induce morphological order. We will examine this behavior further by studying the phase separation

This chapter not subject to U.S. copyright
Published 1995 American Chemical Society

kinetics during a two-step, steady shearing process. In these experiments, the sample will be allowed to reach steady state at high shear rate, then "shear-quenched" to a lower shear rate. We have employed a delay time, t_d, between the two shear rates, in order to control the amount of structure development before the second shear rate, $\dot{\gamma}_2$, is applied. By controlling the structure evolution before application of $\dot{\gamma}_2$, we can test the hypothesis which was implied by the initial studies (1) that the existing structure affects the coarsening kinetics after cessation of $\dot{\gamma}_2$. Similar types of experiments have been performed previously by Takebe and Hashimoto (2), except that they have used a delay time of 0 s.

In this work, we will demonstrate that the preexisting structure does not affect the ultimate coarsening behavior but the magnitude of the second shear rate is the variable which controls the ultimate coarsening. The two-step experiments also revealed that the coarsening process is accelerated during the second shear rate. After cessation of the second shear rate, deceleration of the coarsening is observed. The coarsening behavior in the directions parallel and normal to flow also appear to be coupled. When phase separation occurs during $\dot{\gamma}_2$ normal to flow, phase separation parallel to flow only occurs in a q-independent fashion, after cessation of $\dot{\gamma}_2$. Spinodal-like phase separation parallel to flow is only observed when the second shear rate suppresses growth in all directions.

Experimental

The sample used in this study was a 50:50 blend (by weight) of polystyrene-d_8 (PSD, $M_w = 9.0 \times 10^4$ g/mol, $M_w/M_n = 1.03$) and polybutadiene (PB, $M_w = 2.2 \times 10^4$ g/mol, $M_w/M_n = 1.1$) dissolved in dioctyl phthalate (DOP). The total polymer concentration was 8% by weight. The solution exhibits upper critical solution temperature (UCST) behavior and this particular three component mixture has a cloud point of 34 °C (1). The composition is off-critical, with the critical composition being 35:65 PSD:PB. All experiments are performed at room temperature, where the quiescent sample is in the two-phase region.

Shear light scattering measurements were performed on an instrument described previously (3). The instrument has a cone-and-plate geometry with the incident beam parallel to the gradient direction (y-axis). The scattering patterns were obtained in the x-z plane, where the x-axis is the flow direction and the z-axis, perpendicular to the flow direction, is the vorticity direction. The samples were exposed to the three types of shear histories shown in Figure 1: a) **Single step shear** - The relaxation after cessation from a single steady shear rate was measured (Figure 1a). b) **Two-step shear, variable t_d** - The delay time, t_d, was varied (0, 7.5, 15, 22.5 and 30 s) and $\dot{\gamma}_2$ was fixed (5.22 s^{-1}). The data was taken only after cessation of $\dot{\gamma}_2$ so the effect of t_d could be monitored (Figure 1b). c) **Two-step shear, variable $\dot{\gamma}_2$** - The values of $\dot{\gamma}_2$ were varied ($\dot{\gamma}_2$ = 1.31, 5.22, 52.2 and 329 s^{-1}) and the delay time was held constant for various delay times (t_d = 0, 7.5, and 15 s). Data was taken immediately after stopping $\dot{\gamma}_1$ so that the coarsening behavior during the delay, during $\dot{\gamma}_2$, and after cessation of $\dot{\gamma}_2$ were monitored (Figure 1c).

For the two-step experiments, the initial shear rate, $\dot{\gamma}_1$, was constant at 626

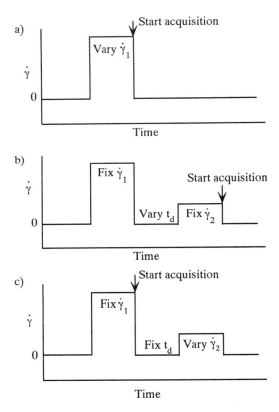

Figure 1. - Schematic of the shear histories imposed on sample: a) Single step shear; b) Two-step shear, variable t_d; c) Two-step shear, variable $\dot{\gamma}_2$.

s^{-1} and the duration of both shear rates was also held constant at 15 s. In all experiments, the data acquisition sequence consisted of 20 frames of data each with an exposure time of 0.1 s at 1.6 s intervals so that the elapsed time for each experiment was slightly longer than 30 s.

Results and Discussion

Single-Step Shear Experiments. The results for the phase separation behavior after a single shear rate were first measured for 626 s^{-1} and 5.22 s^{-1} (Figures 2 and 3 respectively) to verify the previously observed behavior *(1)*. In all scattering patterns the original flow direction is along the vertical axis. Application of the single-step shear was performed after the sample had been at rest for 15 minutes. The intensity contours parallel and normal to the original flow direction were analyzed. From the contours, the temporal evolution of the growth in the scattering intensity at various q values was monitored (where q is the magnitude of the scattering vector = $(4\pi n/\lambda)\sin(\theta/2)$, n is the refractive index of the sample, λ is the incident wavelength (632.8 nm) and θ is the scattering angle). All intensities were corrected for a background scattering contribution which was subtracted and assumed constant for all scattering angles. By plotting the natural logarithm of the corrected intensity versus time, the q-dependent growth rates of the sample can be obtained in the fashion of Cahn and Hilliard *(4)*. The q-dependent growth rates after cessation from a single shear rate, normal and parallel to the flow direction, are shown in Figures 4 and 5 for 626 and 5.22 s^{-1}, respectively. After cessation of 626 s^{-1} (Figure 4), the growth of a spinodal ring structure in the scattering pattern is observed. The peak intensity increases and the peak position shifts to smaller angle with increasing time in both directions. On the other hand, after cessation of 5.22 s^{-1} (Figure 5), the growth patterns are quite different from the spinodal-like growth observed after high shear rates. Parallel to flow the growth rates are nearly independent of q as demonstrated by the set of nearly parallel growth curves, while normal to flow, the low q modes grow slowly and the high q modes decrease in intensity. Here, the quantitative behavior of the q-dependent growth rates is not as important as the general patterns of growth (spinodal-like versus q-independent) exhibited by the two different experimental conditions.

Two-Step Shear, Variable t_d Experiments. The growth rate patterns for the experiments where the delay time was varied and $\dot\gamma_2$ was fixed at 5.22 s^{-1} are shown in Figure 6. As mentioned previously, $\dot\gamma_1 = 626$ s^{-1} is sufficient to homogenize the sample, therefore all prior deformational and orientational history should be erased. The duration time of 15 s for $\dot\gamma_1$ is assumed to be sufficient for the sample to reach steady state. By varying t_d, we are able to control the size of the developing concentration fluctuations which are present before applying $\dot\gamma_2$. The longer t_d is, the greater the extent of coarsening before applying $\dot\gamma_2$. The same growth rate patterns are observed in both directions regardless of the value of t_d. Figure 6 demonstrates that the growth rate behavior after cessation of the second shear rate is not influenced by the prior shear history or degree of coarsening. The slow shear rate, which is ineffective in producing a homogenized state, is sufficient to erase all memory of the prior deformation history as well as any previous coarsening of the sample following

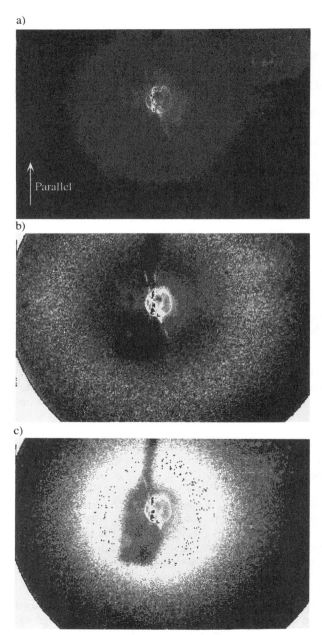

Figure 2. - Temporal evolution of scattering patterns following cessation of a single stage steady shear ($\dot{\gamma} = 626$ s^{-1}, time on = 1 min). a) 0.7 s; b) 10.3 s; c) 24.7 s after cessation.

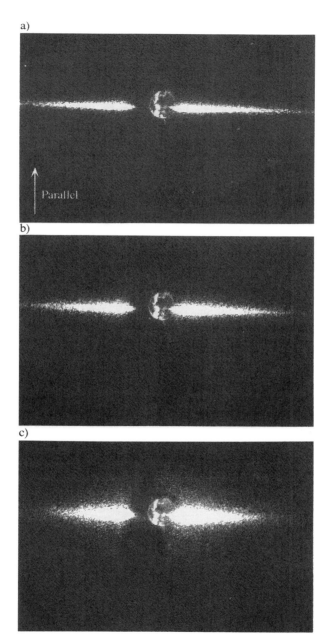

Figure 3. - Temporal evolution of scattering patterns following cessation of a single stage steady shear ($\dot\gamma = 5.22$ s^{-1}, time on = 1 min). a) 0.7 s; b) 10.3 s; c) 24.7 s after cessation.

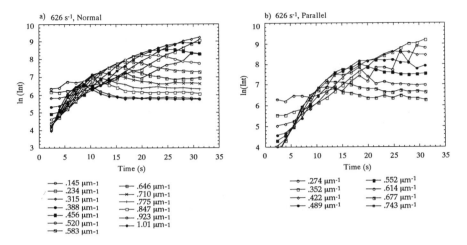

Figure 4. - The q dependent growth rate patterns after cessation of a single-step steady shear ($\dot{\gamma} = 626$ s^{-1}): a) Normal to flow. b) Parallel to flow.

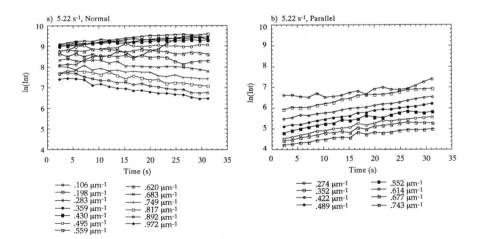

Figure 5. - The q dependent growth rate patterns after cessation of a single-step steady shear ($\dot{\gamma} = 5.22$ s^{-1}): a) Normal to flow. b) Parallel to flow.

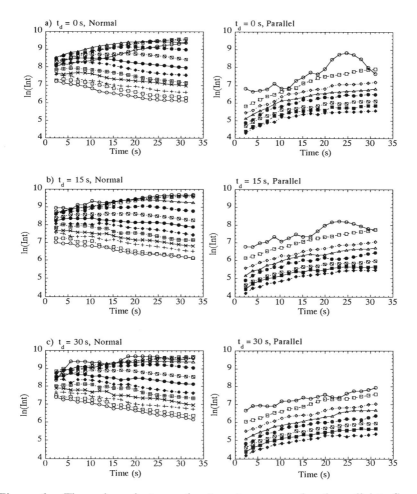

Figure 6. - The q dependent growth rate patterns normal and parallel to flow after cessation of a two stage shear ($\dot{\gamma}_1 = 626$ s^{-1}, $\dot{\gamma}_2 = 5.22$ s^{-1}): a) $t_d = 0$ s. b) $t_d = 15$ s. c) $t_d = 30$ s.

cessation of the first shear rate. From these results the hypothesis, that pre-existing structure within the sample was affecting the coarsening kinetics following cessation of slow shear rates, is not supported.

Two-Step Shear, Variable $\dot{\gamma}_2$ Experiments. Examples of the temporal evolution of the scattering patterns during the delay time and second shear rate are shown in Figures 7 and 8, which contrast the behavior between a low second shear rate ($\dot{\gamma}_2$ = 1.31 s^{-1}) and a high second shear rate ($\dot{\gamma}_2$ = 329 s^{-1}), respectively. Both 1.31 and 329 s^{-1} alone, produce anisotropic scattering patterns at steady state, which are streaks of scattered light along the normal direction to flow, indicating the presence of deformed droplets and a lack of sample homogenization. During the delay time after $\dot{\gamma}_1$ (t_d = 7.5 s), the sample begins to coarsen in the same manner as after cessation of 626 s^{-1}, as expected (Figures 7a and 8a). With the onset of $\dot{\gamma}_2$ = 1.31 s^{-1}, a deformation of the "spinodal ring" is observed as an ellipse (Figure 7b). The long axis of the ellipse is oriented perpendicular to the flow direction indicating deformation of the developing concentration fluctuations along the flow direction. As the patterns continue to develop during $\dot{\gamma}_2$, the peak positions on the major and minor axes of the ellipse shift to smaller angles with increasing time. After cessation of $\dot{\gamma}_2$, the coarsening process continues, with the anisotropy persisting for the duration of the experiment (Figure 7c). In contrast, with the onset of $\dot{\gamma}_2$ = 329 s^{-1}, the coarsening is completely suppressed (Figure 8b), and after cessation of 329 s^{-1}, the sample just begins to coarsen as observed previously after 626 s^{-1} (Figure 8c).

The growth rate patterns for various values of $\dot{\gamma}_2$ with constant delay times, t_d = 0 s and 7.5 s, are shown in Figures 9 and 10, respectively. In the direction normal to flow, the intensities do not always reach a steady state during $\dot{\gamma}_2$. The duration of $\dot{\gamma}_2$ is fixed at 15 s in all cases, but the scattering does not always reach steady state, since lower values of $\dot{\gamma}_2$ do not prevent coarsening in the sample. This may influence the ultimate coarsening behavior after $\dot{\gamma}_2$ and will be discussed later. In Figure 9 (t_d = 0 s), for $\dot{\gamma}_2$ = 1.31 s^{-1} (Figure 9a), the sample coarsens in a spinodal-like fashion in the normal direction while the second shear rate is being applied. Parallel to the flow direction, the growth is somewhat suppressed during $\dot{\gamma}_2$. After cessation of 1.31 s^{-1}, coarsening in the parallel direction proceeds with q-independent growth rates reminiscent of the behavior following cessation of 5.22 s^{-1} only, while the spinodal-like growth continues normal to the original flow direction. For the intermediate case where $\dot{\gamma}_2$ = 52.2 s^{-1} (Figure 9b), during shear, growth is completely suppressed parallel to flow and occurs to some extent normal to flow. After cessation of 52.2 s^{-1}, the coarsening is q-independent parallel to flow while normal to flow a q-dependent, non-spinodal coarsening is observed. For $\dot{\gamma}_2$ = 329 s^{-1} (Figure 9c), the coarsening behavior in both directions is suppressed during $\dot{\gamma}_2$. After cessation of 329 s^{-1}, the coarsening proceeds in a spinodal-like fashion in both directions.

Similar results are shown in Figure 10 for the same shear rates and with t_d equal to 7.5 s. During the delay time, spinodal growth is observed in both directions for all cases as expected. During the lower shear rates (Figures 10a and 10b), coarsening is observed in the normal direction, while scattering is partially suppressed parallel to flow for $\dot{\gamma}_2$ = 1.31 s^{-1} and completely suppressed parallel to flow for all other values of $\dot{\gamma}_2$. After cessation of $\dot{\gamma}_2$, the coarsening parallel to flow proceeds in

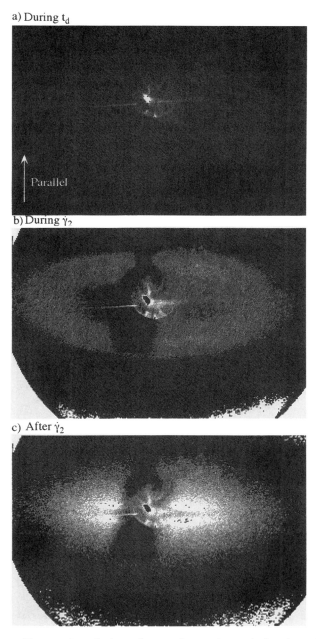

Figure 7. - Temporal evolution of scattering patterns during two-step steady shear ($\dot{\gamma}_1 = 626$ s^{-1}, time on = 15 s; $t_d = 7.5$ s; $\dot{\gamma}_2 = 1.31$ s^{-1}, time on = 15 s): a) 0.7 s; b) 10.3 s; c) 24.7 s after stopping $\dot{\gamma}_1$.

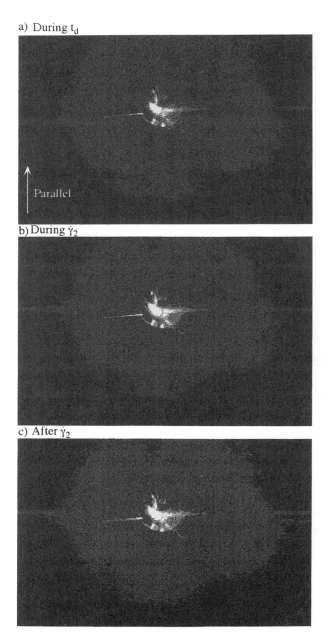

Figure 8. - Temporal evolution of scattering patterns during two-step steady shear ($\dot{\gamma}_1 = 626$ s^{-1}, time on = 15 s; $t_d = 7.5$ s; $\dot{\gamma}_2 = 329$ s^{-1}, time on = 15 s): a) 0.7 s; b) 10.3 s; c) 24.7 s after stopping $\dot{\gamma}_1$.

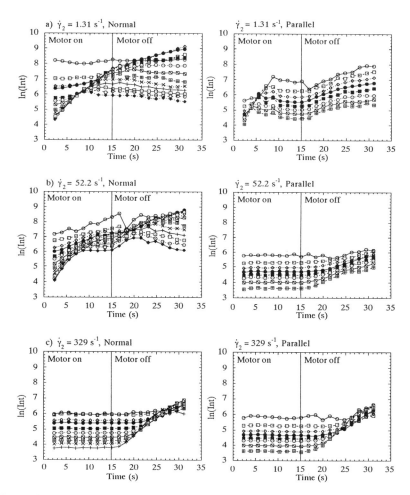

Figure 9. - The q dependent growth rate patterns normal and parallel to flow during a two stage shear process ($\dot{\gamma}_1 = 626$ s^{-1}, $t_d = 0$ s): a) $\dot{\gamma}_2 = 1.31$ s^{-1}; b) $\dot{\gamma}_2 = 52.2$ s^{-1}; c) $\dot{\gamma}_2 = 329$ s^{-1}.

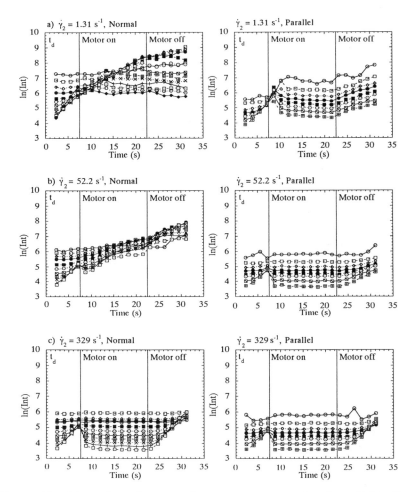

Figure 10. - The q dependent growth rate patterns normal and parallel to flow during a two stage shear process ($\dot{\gamma}_1 = 626$ s^{-1}, $t_d = 7.5$ s): a) $\dot{\gamma}_2 = 1.31$ s^{-1}; b) $\dot{\gamma}_2 = 52.2$ s^{-1}; c) $\dot{\gamma}_2 = 329$ s^{-1}.

a q-independent fashion for the lowest values of $\dot{\gamma}_2$. Spinodal-like growth is only observed after cessation of the highest value for $\dot{\gamma}_2$ (Figure 10c). For $\dot{\gamma}_2$ values of 52.2 and 329 s^{-1} normal to flow, both shear rates are capable of inhibiting the coarsening of the system to some extent (some growth is observed). This is despite the fact that the normal direction is the vorticity axis and these shear rates alone are insufficient to homogenize the sample. From Figures 9 and 10, it appears that parallel to flow, the only condition where spinodal-like growth occurs is when shear has completely suppressed growth in the normal direction (Figures 9c and 10c). Whenever growth in the normal direction occurs, the coarsening process parallel to flow occurs in a q-independent fashion (Figures 9a-b and 10a-b).

For the experimental conditions where peaks in the contours were observed, the peak position, q_m, and the intensity of the peak maximum, $I(q_m)$, were monitored as a function of time. Examples of the time dependence of q_m and $I(q_m)$ are shown in Figures 11 and 12. In Figure 11a, a double logarithmic plot of q_m versus elapsed time after cessation of $\dot{\gamma}_1$ is shown for $\dot{\gamma}_2 = 1.31$ s^{-1}, with t_d equal to 0 and 7.5 s. For reference, the behavior after cessation of 626 s^{-1} alone is also shown ($\blacklozenge$ = Normal, $\lozenge$ = Parallel). Due to the size and orientation of the beamstop, data is limited for the peak position in the direction parallel to flow. The power law exponent for $q_m(t)$ for 626 s^{-1} alone is approximately -1, indicating late stage spinodal behavior. This may indicate that the sample at 626 s^{-1} is not homogeneous at length scales smaller than can be observed by light scattering. However, since the contrast is insufficient to observe scattering at shorter times and higher q ranges are not accessible, we cannot firmly conclude that the sample is slightly inhomogeneous at 626 s^{-1}.

For two-step shear experiments with t_d equal to 0 s, during application of the second shear rate, q_m shifts to lower q values much faster than observed after 626 s^{-1} alone ($\blacksquare$ = Normal during 1.31 s^{-1}, $\boxplus$ = Parallel during 1.31 s^{-1}). This is evident from the steeper slopes observed for the data during $\dot{\gamma}_2$ than the data after 626 s^{-1} alone. The same type of behavior is observed for t_d equal to 7.5 s. During $t_d = 7.5$ s, coarsening occurs in the same fashion as after 626 s^{-1} alone (O = Normal, $\triangle$ = Parallel). For $t_d = 7.5$ s, during $\dot{\gamma}_2$ ($\bullet$ = Normal during 1.31 s^{-1}, $\blacktriangle$ = Parallel during 1.31 s^{-1}), the shift in q_m to lower values is faster than the reference rate after 626 s^{-1} alone. For both t_d values, the second shear rate appears to be driving the system to a new steady state faster than can be achieved after cessation from a single, high shear rate by a spinodal-like process.

After cessation of $\dot{\gamma}_2$, the peak position, q_m, migrates to lower q values more slowly than the reference behavior after 626 s^{-1}. This behavior is more obvious for $t_d = 7.5$ s ($\odot$ = Normal after $\dot{\gamma}_2$, No data in the parallel direction) than for $t_d = 0$ s ($\square$ = Normal after $\dot{\gamma}_2$, No data in the parallel direction). However, a slowing down of the coarsening after cessation of $\dot{\gamma}_2$ is indicated in both cases.

Similarly, $I(q_m)$ grows more quickly in both directions during $\dot{\gamma}_2$, compared to the behavior of $I(q_m)$ after 626 s^{-1} alone (Figure 11b). Although not as obvious as the q_m behavior, the growth rate of $I(q_m)$ also appears to decelerate after cessation of the second shear rate. Therefore, it appears that the effect of $\dot{\gamma}_2$ is to ultimately control the rate of coarsening in the sample after cessation of all shear rates. While application of the second shear rate accelerates the coarsening process, after cessation of the second shear rate, the coarsening rate decelerates. This deceleration is probably related to the original observation that relaxation is slower after slow shear

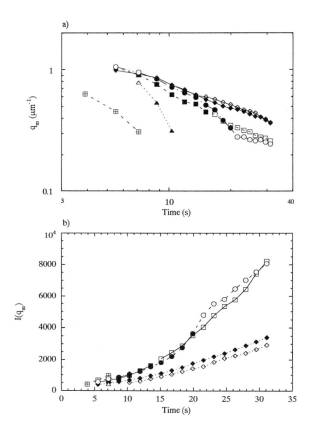

Figure 11. - Time dependence during a two-step deformation process for a) q_m; b) $I(q_m)$. Symbols: ◆ - After 626 s^{-1} only, Normal; ◇ - After 626 s^{-1} only, Parallel; ■ - $t_d = 0$ s, Normal, During $\dot\gamma_2 = 1.31$ s^{-1}; ⊞ - $t_d = 0$ s, Parallel, During $\dot\gamma_2 = 1.31$ s^{-1}; □ - $t_d = 0$ s, Normal, After $\dot\gamma_2 = 1.31$ s^{-1}; ○ - During $t_d = 7.5$ s, Normal; • - $t_d = 7.5$ s, Normal, During $\dot\gamma_2 = 1.31$ s^{-1}; ⊙ - $t_d = 7.5$ s, Normal, After $\dot\gamma_2 = 1.31$ s^{-1}; △ - During $t_d = 7.5$ s, Parallel; ▲ - $t_d = 7.5$ s, Parallel, During $\dot\gamma_2 = 1.31$ s^{-1}.

rates. In both cases, the accelerated coarsening during shear must also be considered, so the domain sizes over the entire deformation history remains similar to the domain sizes after cessation of 626 s^{-1} alone.

For comparison, the complementary data to Figure 11 are shown in Figure 12 for $\dot\gamma_2$ equal to 329 s^{-1}. In this case, it was impossible to determine a peak position while the second shear rate was being applied. Since 329 s^{-1} suppresses all coarsening, the time axis does not represent the elapsed time after cessation of $\dot\gamma_1$ as in Figure 11, but the elapsed time from the onset of coarsening. Obviously all data from the different delay times should superimpose. The q_m and $I(q_m)$ data taken after cessation of 329 s^{-1} show no differences as expected. For $\dot\gamma_2$ values which are sufficient to prevent coarsening of the sample, the phase separation behavior after cessation of $\dot\gamma_2$, is the same as observed after cessation of 626 s^{-1} alone. There is

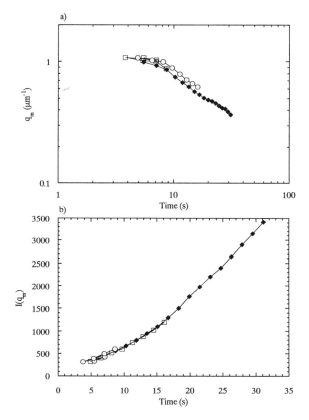

Figure 12. - Time dependence during a two-step deformation process for a) q_m; b) $I(q_m)$. Symbols: ♦ - After 626 s^{-1} only, Normal; □ - $t_d = 0$ s, Normal, After $\dot{\gamma}_2 = 329$ s^{-1}; ☉ - $t_d = 7.5$ s, Normal, After $\dot{\gamma}_2 = 329$ s^{-1}.

currently insufficient data to determine the shear rate dependence of the rates of growth after cessation of $\dot{\gamma}_2$.

Conclusions

Two-step shear experiments have provided some insight into the mechanism of phase separation and coarsening following cessation of shear. Three conclusions may be drawn from this study: 1) The slow relaxation of anisotropic scattering patterns produced by low shear rates to an isotropic pattern was found to be independent of the sample history prior to the deformation. The nature and extent of pre-existing structure in the sample is therefore not responsible for the slow relaxation behavior. 2) The two-step experiments revealed that the coarsening process is accelerated during the second shear rate, then decelerates after cessation of the second shear rate. The effect of $\dot{\gamma}_2$ is to ultimately control the rate of coarsening in the sample after cessation of all shear rates. 3) The coarsening behavior in the directions parallel and normal to flow are coupled. When phase separation occurs during $\dot{\gamma}_2$ normal to flow, phase separation parallel to flow only occurs in a q-independent fashion, after cessation of $\dot{\gamma}_2$. Spinodal-like phase separation parallel to flow is only observed when the second shear rate suppresses growth in all directions.

For the range of shear rates examined in this study, the droplet (or cluster) dimension parallel to flow is always larger than that perpendicular to flow. At low shear rates the droplet is deformed with the long axis parallel to flow. At intermediate shear rates, the string phases proposed by Hashimoto and coworkers (5) align parallel to the flow direction. Finally at the highest shear rates, the sample appears homogenized over the length scales examined by light scattering (> 2 μm^{-1}), the interface between the two phases becomes blurred, and the macroscopic droplets have lost their definition (clusters). One possible interpretation for the apparent coupling between the parallel and normal coarsening mechanisms is given below.

During the two-step shear quench, the q dependent growth rate associated with spinodal decomposition is always observed in the normal direction, however parallel to flow, spinodal type growth behavior was rarely observed. We assume that once the droplets are finely dispersed to the same length scale as the polymer single chain dimensions, then the concentration fluctuations are of the microscopic (or molecular) type, and spinodal decomposition, with the signature q dependent growth rate, occurs after the shear quench. On the other hand, if the droplet dimensions remain macroscopic, (orders of magnitude larger than the polymer chain dimensions), then the q independent droplet coarsening kinetics are observed. Therefore, the first explanation for the coupled behavior is that the sample is more homogeneous in the normal direction rather than the parallel direction during shear. The seemingly related growth mechanisms between the parallel and normal directions may originate from the relative mixing conditions in the two directions, with the normal direction always being better mixed.

Presently, it is uncertain whether changes in behavior at even higher shear rates are possible. It is possible that the clusters may mix to an even finer level so that the cluster dimensions become smaller in the parallel direction than the normal direction as for the mean field systems we have examined by small angle neutron scattering (SANS) (6). This question may be resolved by using SANS to study this same system at higher shear rates and by varying the temperature to approach mean field-like behavior. Since shear tends to increase the dimensions of droplets along the flow direction, experiments at much lower q ranges are also necessary to examine the long wavelength behavior in these systems.

Literature Cited

1) Nakatani, A. I.; Waldow, D. A.; and Han, C. C. *Polymer Preprints* **1991**, *65*, 266.
2) Takebe, T.; Hashimoto, T. *Polymer* **1988**, *29*, 261.
3) Nakatani, A. I.; Waldow, D. A.; Han, C. C. *Rev. Sci. Instr.* **1992**, *63*, 3590.
4) Cahn, J. W. *J. Chem. Phys.* **1965**, *42*, 93.
5) Hashimoto, T.; Matsuzaka, K.; Kume, T.; Moses, E.; *PMSE Preprints* **1994**, *71*, 111.
6) Nakatani, A. I.; Kim, H.; Takahashi, Y.; Matsushita, Y.; Takano, A.; Bauer, B. J.; Han, C. C. *J. Chem. Phys.* **1990**, *93*, 795.

RECEIVED January 25, 1995

Chapter 18

Shear-Induced Structure and Dynamics of Tube-Shaped Micelles

L. E. Dewalt[1], K. L. Farkas[1], C. L. Abel[1], M. W. Kim[2], D. G. Peiffer[2], and H. D. Ou-Yang[1,3]

[1]Department of Physics, Lehigh University, Bethlehem, PA 18015
[2]Exxon Research and Engineering Company, Annandale, NJ 08801

> Tube shaped micelles formed from mixtures of allyl dimethylhexadecylammonium bromide surfactant and sodium salicylate salt show dramatic shear thickening at a critical shear rate. A comparison is made between our experimental data and theories by Wang (*11*) and Bruinsma et al. (*10*). At low concentrations we observe that the critical shear rate decreases with concentration, consistent with both theories. But, moving to concentration near the overlap concentration, we see a crossover to a range where the critical shear rate increases with concentration, unaccounted for by the existing theories. Modeling the temperature dependence of the critical shear rate as an activation process yields activation energies ranging from 12 to 39 k_BT for different surfactant to salt molar ratios. There is no clear relationship between the activation energy and the surfactant to salt molar ratio.

The structure of self-assembling surfactant systems in the aqueous phase has been studied extensively from both theoretical and experimental viewpoints (*1-12*). This is due to their importance in a wide variety of applications and also to fundamental interests such as phase transitions and anomalous behaviors in the nonequilibrium state. One of the striking features that is common to the surfactant micelles is the onset of an abrupt shear thickening even at low solid concentration. Wang and Bruinsma, Gelbart, and Ben-Shaul have developed theories to investigate this shear thickening (*10,11*). Bruinsma et al. use a competition between shear enhanced collisions and shear breaking to predict the critical shear rate for aggregation. Wang shows that shear alignment reduces the shear breaking enough to allow for nonequilibrium self-assembly of the micelles.

Here, we compare shear thickening data obtained from rheological and light scattering measurements to these theories. At low concentrations the results are consistent with both theories, but once the micelle overlap region is reached, effects

[3]Corresponding author

which have not been predicted appear. Investigation of the shear thickening phenomenon as a function of counter-ion concentration yields activation energies which we compare to Cates and Candau's calculations for the scission energy of these micelles (*12*).

Theoretical Background

Recent theoretical work in shear thickening of rigid surfactant micelle solutions follows several different lines of thought. Bruinsma, Gelbart, and Ben-Shaul consider that the shear enhanced collisions of micelles lead to aggregation of the micelles, while shear forces tear apart large micelles (*10*). Their work describes a micellar system in a diagram with four separate regimes (Figure 1) each of which show a typical plot of the effective quasi free-energy V as a function of the micelle length. The value of the dimensionless concentration, C, and the Peclet number, P, (a comparison of the particle flux due to shear with the particle flux from Brownian motion) define which regime the system is in. The dimensionless concentration is defined to be $C \equiv \phi \ell^3$, where ϕ is the micelle number density and ℓ is the micelle length. Note that the overlapping concentration is $C^*=1$ in this notation.

For P<1 (low shear) and C<1 the system is in the equilibrium phase and there is a well defined mean micelle length. Increasing the flow at this concentration allows some aggregation, but there are not enough collisions to overcome the shear breaking. Hence, the system ends up in a distorted equilibrium with a slightly higher steady state micelle length than that of the quiescent state. Large aggregate structures can not form until the C=1 condition is exceeded. For P<1 and C>1, an activated gelation phase occurs where a metastable micelle length exists. However, if the system can overcome the free energy barrier, the system will aggregate and form a gel. One way of overcoming this barrier is to increase the flow, therefore moving the system into the gelation region. Thus this picture suggests a shear induced phase transition.

In contrast, S. Q. Wang has developed a nonequilibrium thermodynamic theory for the self assembly of cylindrical micelles (*11*). He compares the elastic energy of scission in the shear flow with the thermodynamics of self assembly to determine micelle size distributions. Aggregation occurs because flow alignment reduces the tension in the sheared micelle. Wang's theory does not consider collisions and is therefore only valid at low concentrations. Still, he finds a shear induced redistribution of micelle lengths.

Both theories predict that the critical shear rate needed to induce aggregation or growth should be a decreasing function of concentration. The power law dependence is found to have an exponent of -1.34 by Wang and -1.5 by Bruinsma et al. Simple scaling arguments also show that, for C<1, the critical shear rate should decrease with surfactant concentration. Starting with the critical shear rate being related to the characteristic rotational time, τ_r, we see that

$$\dot{\gamma}_c = 1/\tau_r \propto 1/\ell_3 \tag{1}$$

where ℓ is the mean micelle length. Dynamic light scattering data (*13*) show that the hydrodynamic radius increases as a power law with an exponent of 0.57, so one expects

$$\dot{\gamma} \propto C^{-1.71}. \qquad (2)$$

So for low concentration we expect the critical shear rate to decrease with concentration, however, for higher concentration (C>1) it is not clear what the trend would be. An interesting point, however, is that the mean micelle length, and thus the concentration C, is an increasing function of the shear rate.

In this study we carry out a shear study on allyl dimethylhexadecylammonium bromide at different surfactant concentrations, surfactant to counterion ratios, temperatures and also perform a kinetic study by quenching the applied stress. We are interested in the onset of shear thickening under these various conditions and the relaxation from this shear thickened state.

Experimental

Materials. In this paper semi-dilute, aqueous solutions of allyl dimethylhexadecylammonium bromide (ADHAB) and salicylic sodium salt are studied. The molecular structure is given in Figure 2. There are two representative synthesis routes to produce this cationic surfactant monomer. The first procedure involves the reaction of an allyl bromide with the appropriate N, N dimethylalkylamine in a 1:1 stoichiometric ratio (50 °C, 4 hours). The amine of interest in this study is dodecylamine. The product was recrystallized three times from an ethanol-acetone mixture. Alternatively, the alkyl substituted cationic monomer was prepared by dissolving the two reactants in a stoichiometric ratio into ethylacetate. The precipitated product was filtered, recrystallized from ethylacetate and dried under vacuum for 24 hours at 50 °C. Aqueous solutions were prepared by dissolving at room temperature the cationic monomer with an aromatic organic salt (sodium salicylate). Sample concentrations, C, range from 0.01 to 0.75 wt%, while the molar ratio of surfactant to salt vary from 10:1 to 1:10. Samples with a high molar ratio of salt were made by the addition of salt to a 10:1 stock sample at room temperature. Dynamic light scattering and scanning electron micrographs show that, for an equimolar surfactant to salt ratio at a concentration of about 0.2 wt%, rod-like micelles form with mean lengths of 500 Å and monodisperse diameters of 50-60 Å (about twice as large as an individual surfactant monomer) (*13*).

Experimental apparatus. A Zimm viscometer is used to measure the viscosity of the micelle solution. This viscometer has a Couette geometry with a gap size of 1.47 mm. It has a fixed outer cylinder and a free inner cylinder with radius 2.047 cm. Rotational stress is applied to the inner cylinder through magnetic induction and the shear rate is determined by measuring the rotation rate of the inner cylinder. During an experiment a computer controls the external stress which is varied by changing the rotational speed of a U-shaped magnet. The shear rate range is about $0 \rightarrow 600$ s^{-1}. Details of the set up will be published elsewhere (*14*).

Rheological Measurements.

   Critical Shear Rate in Viscosity Measurements. In a typical run, the applied stress slowly increases to a high stress and then decreases in order to observe

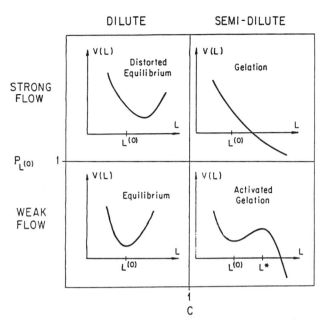

Figure 1. Schematic picture of the effective potentials determining the micelle lengths in different ranges of concentration and applied flow. Reproduced with permission from ref. 10. Copyright 1992 Journal of Chemical Physics.

$$\begin{array}{c} CH_2 \quad\quad (CH_3)_2 \\ \parallel \quad\quad\quad\quad | \\ CH-CH_2-\overset{+}{N}-(CH_2)_{15}-CH_3 \\ \overline{Br} \end{array}$$

Figure 2. Chemical structure of allyl dimethylhexadecylammonium bromide.

any hysteresis effects. The experimentally obtained shear stress - shear rate relation is converted to a viscosity - shear rate curve as shown in Figure 3. This plot shows typical viscosity vs. shear rate relationships for several different concentrations at a 1:1 molar ratio of surfactant to salt at 24 °C. In the two lower concentration samples, 0.05% and 0.15% surfactant concentrations, Newtonian behavior is seen at low shear rates with viscosities very close to that of the solvent. However, a dramatic shear thickening occurs beyond a well defined point, for example point A on the 0.20% wt sample in Figure 3. We define this point to be the critical shear rate. Throughout the shear thickening region, large fluctuations in shear rate are observed at a constant shear stress. Reducing the applied stress, the system begins to relax, but with an apparent hysteresis loop. Increasing the stress beyond that used in Figure 3, the system is rather unstable and has been seen to go into a regime of shear thinning or additional shear thickening.

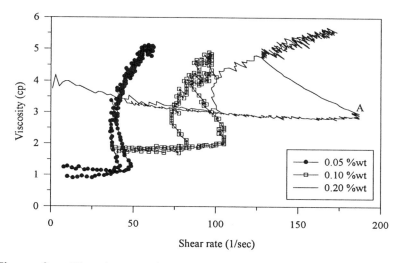

Figure 3. Viscosity vs. shear rate data for three different surfactant concentrations as shown with a surfactant to salt molar ratio of 1:1 at a temperature of 24 °C. The critical shear rate is defined to be the point at which the shear thickening starts to occur while ramping up in shear stress, i.e. the point marked A for the 0.20 % wt sample.

At high concentrations, it is common to see also a slight shear thinning at low shear rates, as seen in the 0.20 wt% sample in Figure 3, indicative of the polymeric nature of the micelle. The large fluctuations and the hysteresis in the shear thickening regime indicate the occurrence of a first order phase transition and that slow kinetic processes are involved. The shear thickening behavior is studied as a function of temperature and concentrations of surfactant and counterion. These results will be discussed later.

Kinetic Measurements. The hysteresis loops which appear in Figure 3 indicate that very slow kinetics are involved in the structural phase transition. In order to study the time scales of the structural rearrangement, a quench-relaxation experiment is designed to follow the kinetics. A shear stress quench is applied to a system in the shear thickened regime (as shown schematically in Figure 4). To perform these experiments, we allow the system to reach steady state in the shear thickening regime, then we quickly quench to a low applied stress. In the short time of the quench, the micelles do not have time to relax, and therefore maintain their rheological properties, but the mechanical viscometer system must follow a "Newtonian-like" path (point A to B in Figure 4) to the lower stress. As the micelle aggregates relax from B to C, a computer records the shear rate at the constant stress. Fitting an exponential function to the inverse shear rate vs. time data, we obtain the characteristic relaxation times. Since the inverse shear rate is proportional to the viscosity, this quench procedure measures the transient relaxation from the shear induced structure at A to that at C. The relaxation time obtained from the fitting characterizes structural relaxation. Results will be discussed in the next section.

Results and Discussion

Figure 3 shows viscosity vs. shear rate data for our samples. Here we see the shear thickening at a well defined shear rate, however, the magnitude of the critical shear rate appears to be surprisingly low. If we assume that the critical shear rate is $1/\tau_c$, and using the mean micelle length, we expect the critical shear rate to be about 10^4 s^{-1}, much higher than the observed critcal shear rate. So this implies that the critical shear rate is sensitive to the large micelles in the distribution of lengths. Investigating the concentration dependence of the critical shear rate, Figure 5, we find that at low concentrations the critical shear rate decreases with concentration as predicted. However, there is a crossover into a range where the critical shear rate increases with concentration. This positive slope region is not predicted by existing theories, and this may indicate that the concentration is above the overlap concentration. Evidence for the concentrations being around C=1 is apparent in the zero shear viscosity data, also shown in Figure 5. Here, the zero shear viscosity is close to the solvent viscosity until it reaches the crossover region, where the zero shear viscosity increases dramatically. This increase is an indication that collisions between micelles occur at low shear rates due to an overlap in the effective rotational volume of the rods. The concentration at this overlap defines C*; however, note that C* is a function of the micelle length and, therefore, a function of the shear rate. This means that, increasing the micelle length by shearing, we move C* to lower concentrations, therefore allowing gelation of samples initially below C*. So, the zero shear viscosity data indicate that the crossover region is very close to the center of Figure 1, where C=C*. When just below C* at low shear rates, the system is in equilibrium as shown in the figure's bottom left corner. Entering the distorted equilibrium region by increasing the shear rate allows the micelle size to lengthen. However, this increase in micelle length may allow the micelles to overlap, in effect allowing C to now be above C*, inducing gelation. For low shear and at concentrations at or above C*, the system is in the activated gelation region of figure 1. Here, a small amount of shear overcomes the "potential" barrier before gelation occurs. However, at higher concentrations, where there is strong overlapping as suggested by the zero shear viscosity, it is uncertain why the critical shear rate increases.

The positive slope may also be caused by complicated effects due to the flexibility of these micelles. Electric birefringence measurements reveal that these rods are semi-flexible with about 10 persistence lengths (7). However, neutron scattering data by others find that these rods are much stiffer (15).

If the shear thickening can be modeled as an activation process, the Arrhenius behavior of the critical shear rate with temperature will furnish the activation energy. Table I shows the activation energies for several surfactant concentrations and surfactant to salt ratios. These activation energies are of the same order of magnitude as Cates and Candau's estimate for the scission energy of a micelle, which is 24 $k_B T$

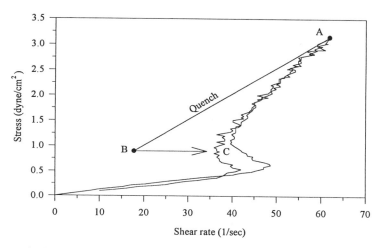

Figure 4. Schematic plot of the stress quench experiment. The process from A to B is Newtonian-like and we expect the structure at B to be the same as at A because of the short quench time. The system will follow the micelle relaxation from B to C.

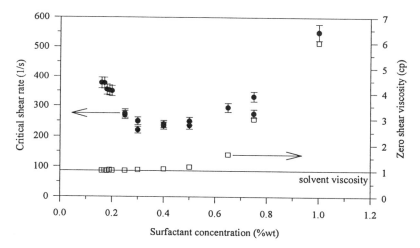

Figure 5. Critical shear rate (●) and zero shear viscosity (□) vs. concentration data for a sample with a surfactant to salt molar ratio of 2:1 at 20 °C.

(*12*). We do not see any clear trends in the activation energies with either concentration or surfactant to salt ratio. In Table I there are no data available at low salt concentrations (i.e. 5:1 and 10:1) because the shear thickening is no longer

Table I. Activation Energies

	Surfactant to salt ratio				
	2:1	1:1	1:2	1:5	1:10
0.01 %wt				23 k_BT	
0.05 %wt		35 k_BT	13 k_BT	12 k_BT	18 k_BT
0.10 %wt			25 k_BT	25 k_BT	
0.25 %wt	35 k_BT				

apparent even at high surfactant concentration (1% wt) and low temperatures.

In the shear thickening regime, observation of the relaxation from a non-equilibrium state yields information about the kinetics of aggregation and break-up. Quench tests as in Figure 4 provide this information. The depth of the quench has a large effect on the relaxation times. Small quenches, such as those done in a typical viscosity run, reach steady state within seconds. However, deeper quenches yield relaxation times of 10^2 to 10^4 seconds. This change in relaxation times with the quench depth indicates that, by adding energy to the system (through mixing in the shear field), the system reaches steady state more quickly. Light scattering during shear also indicates that aggregated structures and slow kinetics occur in this system. Here, in the shear thickened state, the forward scattering intensity is much stronger than in the quiescent state. This implies that more massive objects exist under high shear consistent with the gelation picture. Quenching the shear down to a quasi-quiescent state (enough shear to homogenize the sample), the relaxation of the scattered intensity was found to be an exponential decay with a characteristic decay time of about 10^4 seconds. The very slow kinetics is also consistent with the Bruinsma et al. picture, since the micelles have aggregated in a deep "potential" well. Overall, the shape of the hysteresis curve combined with the large fluctuations in the transition regime and slow relaxation suggest the existence of a first order phase transition. We note that Rehage, Wunderlich and Hoffmann indicate that a phase transition of first order may occur in a similar system (*16*). In both the Bruinsma et al. and Wang pictures, there is a point where the two phases (quiescent micelle and aggregated micelle) coexist, which would be consistent with a first order phase transition.

Summary

Low concentration behavior of self-assembling surfactant micelles qualitatively agrees with theories by Bruinsma et al. and Wang; however, a new trend in the concentration dependence develops when the micelles reach the overlap concentration. At the overlap concentration, the critical shear rate for aggregation begins to increase with the surfactant concentration. At this point we do not have a model to explain this trend. Modeling the shear thickening as an activation process, we obtain activation energies in the range of 12-35 $k_B T$ for several different counterion concentrations. The counterion concentration has a dramatic effect on the shear induced thickening, but there are no clear trends seen in the activation energies. Kinetic measurements by quenching the stress on the shear thickened state reveals that the relaxation is intrinsically slow. But, the accompanying shear appears to accelerate the relaxation process.

Acknowledgments

We would like to thank the Exxon Education Foundation for supporting part of this work. L. E. D. is a fellowship student supported by U.S. Department of Education. K. L. F. and C. L. A. are supported by NSF Research Experiences for Undergraduates scholarships. H. D. O. appreciates insightful discussions with W. M. Gelbart, R. Bruinsma and S. Q. Wang. We would also like to thank D. Pine and P. Dixon at Exxon Research for running the preliminary light scattering measurements.

Literature Cited

1. Missel, P.; Mazer, N.; Benedek, A.; Carey, M. *J. Phys. Chem.* **1984**, *87*, 1264 and references therein.
2. Candau, S. J.; Hirsch, E.; Zana, R.; Adam, M. *J. Coll. Interface Sci.* **1988**, *122*, 430 and references therein.
3. Hoffmann, H.; Plat, G.; Rehage, H.; Schorr, W. *Adv. Coll. Int. Sci.* **1982**, *17*, 275.
4. Safran, S.; Turkevich, L.; Pincus, P. *J. Phys. Let.* **1984**, *45*, L-69.
5. Halle, B.; Landgren, M.; Jonsson, B. *J. Phys. France* **1988**, *49*, 1235.
6. Ulminus, J.; Lindman, B.; Lindblom, G.; Drakenberg, T. *J. Coll. Interface Sci.* **1978**, *65*, 88.
7. Wu, X. L.; Yeung, C.; Kim, M. W.; Huang J. S.; Ou-Yang, H. D. *Phys. Rev. Lett.* **1992**, *68*, 1426.
8. Hoffmann, H.; Löbl, M.; Rehage, H.; Wunderlich, I. *Physikalische Chemie* **1985**, *22*, 290.
9. Wunderlich, I.; Hoffmann, H.; Rehage, H. *Rheo. Acta* **1987**, *26*, 532.
10. Bruinsma, R.; Gelbart, W. M.; Ben-Shaul, A. *J. Chem. Phys.* **1992**, *96*, 7710.
11. Wang, S. Q. *J. Phys. Chem.* **1990**, *94*, 8381.
12. Cates, M. E.; Candau, S. J. preprint.
13. Kim, M. W.; Liu, S-N.; Peiffer, D. G. unpublished.

14. Dewalt, L. E.; Farkas, K. L.; Abel, C. L.; Kim, M. W.; Peiffer, D. G.; Ou-Yang, H. D.; in preparation.
15. Lin, M. private communication.
16. Rehage, H.; Wunderlich, I.; Hoffmann, H. *Progr. Coll. Polym. Sci.* **1986,** 72, 51.

RECEIVED February 3, 1995

LIQUID-CRYSTALLINE POLYMERS

Chapter 19

The Texture in Shear Flow of Nematic Solutions of a Rodlike Polymer

Beibei Diao[1], Sudha Vijaykumar, and Guy C. Berry

Department of Chemistry, Carnegie Mellon University, Pittsburgh, PA 15213

Rheological and rheo-optical studies on the evolution of texture in flow and on cessation of flow are reported on nematic solutions of the rodlike poly(1,4-phenylene-2,6-benzobisthiazole). Measurements include the strain and optical transmission during creep, steady-state flow, and recovery on cessation of flow. Changes of the texture are monitored by the transmission of light through the sample in parallel plate torsional deformation. Although changes in textural features appear to scale with certain strain parameters, consistent with some theoretical proposals for the behavior of non-shear aligning materials, the same is not true of the recoverable strain. Many of the rheological features appear to be qualitatively similar to behavior reported for isotropic polymer solutions containing filler particles, with the strain in creep or recovery scaling with the stress in creep for behavior in slow flow, or for deformation under a recently small strain. It is suggested that the behavior observed is related to the effects of curvature elasticity acting over a scale that is both much smaller than the macroscopic dimensions and dependent on the deformation history. Theoretical treatments have not yet captured the range of behavior reported here.

Earlier studies on the rheological and rheo–optical properties of nematic solutions of rodlike polymers from our laboratory have comprised two principal protocols: (1) Steady state rheological studies on textured samples prepared by steady deformation at high shear rate prior to the initiation of rheological measurements; and (2) Transient rheo–optical measurements on a defect-free

---

[1]Permanent address: DuPont Corian, Buffalo, NY 14207

(monodomain) preparation under conditions of a homogeneous distortion of the director field (1-7). In a so-called *monodomain*, the director $\tilde{n}(r)$ of the nematic field is everywhere the same, and the fluid is a uniform, defect–free preparation. With nematic solutions of high molecular weight rodlike chains, the local thermally induced orientation fluctuations are largely suppressed, diminishing the scattering of light, and giving the monodomain the appearance of a normal isotropic fluid when viewed with natural (unpolarized) light without the use of a polarization analyzer (3-9). By comparison, orientation fluctuations are larger for monodomains of small molecule nematics, giving them a turbid appearance (10,11). When viewed between crossed polars, a monodomain is observed to exhibit a sharp extinction for light polarized either along or orthogonal to the optic axis (e.g., the chain axis for most rodlike molecules), and may also exhibit dichroism. Without special preparation, polymeric nematics will typically exhibit a *mottled* texture, in which the transmission of light between crossed polarizers varies in an apparently random fashion on a length scale much smaller than the sample dimensions (e.g., the sample thickness). It should be noted that this texture is not an equilibrium one, or even meta–stable, but that its evolution toward an equilibrium state will be exceedingly slow in many cases for polymeric nematics, including the materials studied here (3-6).

Rheo–optical studies on monodomains of nematic preparations of rodlike or semiflexible polymers have shown that the shear flow tilts the director in a non-flow aligning direction for materials studied to date (4-6,12-16), indicating that the two Leslie viscosity coefficients α_2 and α_3 have different signs (10,11,17). Studies of the transient response starting with a monodomain provide an example of the importance of the Ericksen number **Er** in slow flow (6,7,16). Here, **Er** is defined as the ratio of the viscous stress $\sigma^{(V)}$ to the elastic stress $\sigma^{(E)}$, with $\sigma^{(V)} \propto \langle\alpha\rangle v/d$, and $\sigma^{(E)} \propto \langle K\rangle/d^2$, respectively (10,17), where $\langle\alpha\rangle$ and $\langle K\rangle$ are, respectively, appropriately averaged Leslie viscosity coefficients and Frank curvature elasticities, and v is the flow velocity and d is the dimension along the shear gradient:

$$\mathbf{Er} = \frac{\sigma^{(V)}}{\sigma^{(E)}} = vd\frac{\langle\alpha\rangle}{\langle K\rangle} \qquad (1)$$

Consequently, viscous stresses dominate for large **Er**, but curvature elasticity is dominant for small **Er**. As elaborated below, a homogeneous distortion of the director field was observed in flow between parallel plates over a range of plate separation d and relative velocity v, provided the apparent strain vt/d was smaller than a certain value, with t the time after onset of flow (6,7,16); for this flow, the appropriate Ericksen number is $\mathbf{Er} = vd|\alpha_2\alpha_3|^{1/2}/K_S$, with K_S the curvature elasticity for a splay distortion (7,16). This behavior is related to the prediction (18-22) that shear flow instability is expected in steady flow if **Er** exceeds a certain critical value Er_c for a material with $\alpha_2\alpha_3 < 0$. With continued shear flow for $Er > Er_c$ and if vt/d exceeds a certain value, the sample may exhibit a phase grating before the texture becomes chaotic, developing a mottled appearance. This behavior has been discussed in terms of the development of "roll–cell instabilities" of the form giving rise to an observed phase grating, but with the appearance of rolls over a wide distribution of dimension and orientation (23,24). The mottled texture scatters light strongly, and is characterized by transmissions $T_\psi^{||}$ and $T_\psi^\perp$ that are small (< 10%), about equal, and independent of the angle ψ between the orientation of the polarization of the incident light and the flow direction (3,5); here, $T_\psi^{||}$ and $T_\psi^\perp$ are the transmissions between parallel and crossed polarizers, respectively. The latter behavior suggests that whereas $\tilde{n}(r)$ may be approximately uniform, with a nearly equilibrium order parameter S, over a small dimension $\ell << d$, corresponding to a small volume $V_S \propto \ell^3$, the average of $\tilde{n}(r)$ over macroscopic dimensions gives $S = 0$. The sample retains the original orientation of $\tilde{n}(r)$ near the surfaces, and will eventually reestablish a defect–free monodomain texture, but that process is typically very slow, requiring days or weeks with the materials of interest here, depending on the sample and its dimensions (3-6).

In studies on textured preparations, with small V_S (i.e., $\ell << d$), it has been found that both the steady state viscosity $\eta(\dot\gamma)$ and the steady-state recoverable compliance $R^{(S)}(\dot\gamma)$ determined with such an initial state exhibited many similarities to the behavior of polymeric fluids containing a low level of solid particles (1-3,25). Here, $\eta(\dot\gamma) = \sigma_{ss}/\dot\gamma_{ss}$, with σ_{ss} and $\dot\gamma_{ss}$ the stress and shear rate in steady flow, respectively, and $R^{(S)}(\dot\gamma) = \gamma_R(\infty)/\sigma_{ss}$, with $\gamma_R(\infty)$ the total recoverable strain on cessation of steady state flow. Thus, the dependence of the steady state viscosity $\eta(\dot\gamma)$ on the shear rate $\dot\gamma$ was described in terms of three regimes of response (1-3,26,27): (1) anomalous slow flow at very low $\dot\gamma$, with $\eta(\dot\gamma)$ increasing with decreasing $\dot\gamma$, but with the recoverable compliance $R^{(S)}(\dot\gamma)$ independent of $\dot\gamma$; (2) slow flow at somewhat larger $\dot\gamma$, with $\eta(\dot\gamma)$ about independent of $\dot\gamma$, equal to a plateau value η_P, over a region of $\dot\gamma$ for which $R^{(S)}(\dot\gamma)$ remains independent of $\dot\gamma$; and (3) fast flow at still larger $\dot\gamma$, for which both $\eta(\dot\gamma)$ and $R^{(S)}(\dot\gamma)$ decrease with increasing $\dot\gamma$. These three regimes have been denoted regions I, II and III in some cases (26,27). With the exception of the anomalous slow flow regime, the behavior is qualitatively similar to the nonlinear rheological behavior of isotropic materials, with the onset of the fast flow behavior occurring for shear rates with $\eta_P R^{(S)}(0)\dot\gamma_{ss} > 1$. It may be noted that in slow flow, with both $\eta(\dot\gamma)$ and $R^{(S)}(\dot\gamma)$ independent of $\dot\gamma$, and $\gamma_R(\infty) = R^{(S)}(\dot\gamma)\eta(\dot\gamma)\dot\gamma_{ss} \approx R^{(S)}(0)\eta_P\dot\gamma_{ss}$ increases with $\dot\gamma_{ss}$. That behavior is inconsistent with certain scaling theories (27,28), for which both $\gamma_R(\infty)$ and $\eta(\dot\gamma)$ are predicted to be constant over a range of $\dot\gamma_{ss}$. For the behavior reported in well–developed fast flow $R^{(S)}(\dot\gamma) \propto \dot\gamma_{ss}^{-\beta}$ and $\eta(\dot\gamma) \propto \dot\gamma_{ss}^{-\alpha}$. Thus, $\gamma_R(\infty)$ tends to be a constant in this regime if $1 - \alpha - \beta \approx 0$, as is sometimes observed, see below.

The objective in this study is to further study the questions of texture and scaling of rheological parameters in the deformation of nematic solutions of a rodlike polymer, with the solutions having a texture with small V_S. Data will be reported on the transient strain under constant stress (creep), the recoverable strain following steady flow, and the corresponding behavior for $T_\psi^{||}$ and $T_\psi^\perp$.

Experimental

Materials Poly (1,4-phenylene-2,6-benzobisthiazole), PBZT, was used as an example of a rigid, rodlike lyotropic polymer:

PBZT

The samples of PBZT used, and their specifications are given in Table I. Polymers used were dried under vacuum prior to use. Nematic solutions were prepared by addition of the appropriate quantities of dry polymer and (99% pure) methane sulfonic acid (Aldrich Chemical Company) in a 35 ml centrifuge tube, containing a Teflon coated magnetic stirring bar. Slow stirring action was achieved by rotating the sample tube, between the poles of a horseshoe magnet for about 2 weeks. The tubes were stored in a vacuum dessicator for at least a month before being used.

Table I. Parameters for Nematic Solutions of PBZT

Weight Fraction w	Contour Length L_w /nm	η_{SS}/Pa·s (Slow Flow)	$\dfrac{c}{c_{NI}}$	Deformation History (See text)
0.040	140	2700	1.21	1
0.040	140	2400	1.21	2
0.050	140	25000	1.52	2
0.041	153	60000	1.40	2
0.043	155	3000	1.38	2
0.050	155	1000	1.61	2
0.058	155	150	1.87	2
0.068	155	30	2.19	2
0.049	140	2000	1.47	3
0.053	155	20000	1.71	3

Apparatus The two rheometers used in this study have been described in prior work: a rheometer with a wire suspension of the cone in a cone and plate geometry (1,29), and a rheometer with an air–bearing suspension of one platen in a parallel plate geometry (25,30). Both permit imposition of a constant torque M on one of the sample platens, using eddy–current torque transducers. The wire–suspension apparatus is used in either of two modes: (a) the rotation of the cone relative to the plate is monitored as constant torque is applied to the cone, with the recoverable strain determined following flow by reducing the torque to zero, and (b) the plate is rotated at constant angular velocity Ω, using a feedback circuit to control the torque to make the cone stationary, with recoverable strain determined following flow by reducing both Ω and M to zero. The air–bearing instrument is used only in the first mode. The apparatus is equipped with parallel pyrex plates to facilitate measurement of $T_\psi^{||}$ and $T_\psi^\perp$ using a plane polarized He-Ne laser (Melles Griot) as a light source (15 mW, λ = 632.8 nm). The light beam is perpendicularly incident on the lower plate, steered to the desired radial position using a movable mirror. A lens placed about 5 mm above the sample focused the transmitted beam on a photodiode. Visual observations were made using the expanded beam from a tungsten lamp; these were recorded using a charge–coupled device camera. Both rheometers are equipped to control the temperature, and are sealed against intrusion by atmospheric moisture, with the interior filled with dry nitrogen under a small positive pressure. In addition, a perimeter seal around the fixture described previously was used to reduce possible contamination by moisture (1).

The samples are installed in different ways with the two rheometers. With the wire–suspension apparatus, the sample is placed in the center of the plate, and the cone and plate are gradually brought to their operating positions over a period of 6-7 hr, while rotating the plate with the cone held fixed. With the parallel plate geometry, the plate separation is adjusted to about three times the desired value, and the sample is extruded into the gap from a port in the center of the upper (stationary) plate, with the lower plate fixed. The plate separation is then adjusted to the operating value, with no relative rotation of the two surfaces.

In the following, the shear strain γ and the stress σ are calculated, respectively, from the torque M and the angular displacement Θ using the usual relations for these geometries when used with homogeneous materials (31): $\sigma = K_\sigma M/\pi R^3$ and $\gamma = K_\gamma \Theta$, where R is the radius of the fixture; K_σ is 3/2 for the cone and plate and $2r/R$ for the parallel plate (see below), and K_γ is α for the cone and plate and r/d for the parallel plate, with α the angle between the surfaces of the cone and plate, r the radial position in the parallel plate and d the separation of the parallel plates. With the rheometers used, $\alpha = 2°$, R is 1.90 cm and 2.40 cm for the cone and plate and parallel plate fixtures, respectively, and d is usually in the range 0.03-0.07 cm. The assumption that γ may be estimated from the relations for a homogeneous flow follows approximations usually assumed in treating nonstationary flows with nematic fluids (28). With the parallel plate platens, the shear rate is not independent of r, and the stress (torque) corresponds to an average over shear rates from zero to $R\Theta$ /d. With rheologically nonlinear materials, the shear stress corresponding to the rate of strain $\dot{\gamma}$ = $R\Theta$ /d at radius R may be calculated with $K_\sigma = \{3 + \partial \ln M/\partial \ln \Theta\}/2$ (32). The deviation from $K_\sigma = 2$ for the linear response is at most 15%, and has not been applied here. Consequently, in the experiments at constant applied torque (creep), the true stress may not be strictly constant.

In rheo–optical measurements, the photodiode response $V_\psi^{\| \, 0}$ for separation with solvent in the gap was determined to verify independence of $V_\psi^{\| \, 0}$ on ψ, and to compute the transmissions $T_\psi^{\|} = V_\psi^{\|}/V_\psi^{\| \, 0}$ and $T_\psi^{\perp} = V_\psi^{\perp}/V_\psi^{\| \, 0}$ from the photodiode responses $V_\psi^{\|}$ and $V_\psi^{\perp}$. Weak absorption at λ = 632.8 nm limits the transmission to ≈ 0.95 for the samples used, in the absence of scattering or other effects. Solutions of PBZT absorb light increasingly for wavelengths less than about 600 nm, with a maximum in the absorbance at about 440 nm. Aligned nematic solutions are markedly dichroic for wavelengths with strong absorption, permitting qualitative evaluation of the dichroism in the parallel plate rheometer by visual observation with a white light source.

Results

Three deformation histories were used as protocols to prepare samples for rheological measurements:
1. Steady flow at 0.75 s⁻¹ (ca. 15 min) and recovery for 1 hr prior to each deformation and recovery cycle;
2. Steady flow at 0.75 s⁻¹ and recovery for 12 hr, before cycling through a series of deformations over a range of stresses, starting at low stress, with the recoverable strain determined following cessation of steady flow;
3. Sample allowed to relax for 24 hours after installation prior to the initiation of sequential creep and recovery cycles, starting at low stress, with sequential creep deformations interspersed with determination of the recoverable strain.

History 1 has been used in our prior studies on nematic solutions of PBZT (1). With the parallel plate rheometer, the radial flow of the sample between the fixed parallel plates during the insertion of the sample imposes a preferential radial alignment, see below.

For the samples used, both $T_\psi^{\|}$ and $T_\psi^{\perp}$ are small ($\approx$ 0.05-0.08 for the sample thickness used) during and following the deformation at shear rate 0.75 s⁻¹ used in histories 1 and 2, implying a mottled texture, with small V_S. Examples of the steady viscosity as a function of shear rate determined with the cone and plate instrument by increasing $\dot{\gamma}$, without recovery between successive determinations of the stress at each

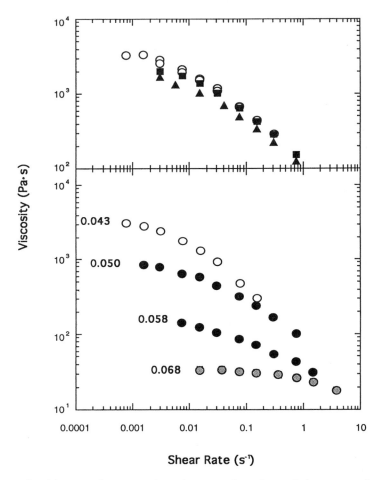

Figure 1. The steady state viscosity as a function of shear rate for nematic solutions of PBZT in methane sulfonic acid (30 °C, using cone and plate platens).
Upper: Behavior under different deformation histories: squares, circles and triangles are for histories 1, 2 and 3, respectively. (L_W = 140 nm; w = 0.040).
Lower: Data under history 2 for the indicated concentrations (weight fraction). (L_W = 140 nm).

constant $\dot\gamma$, are shown in Figure 1 for several solutions (history 1), along with examples of the $\eta(\dot\gamma)$ determined for a single sample under the three deformation histories. Anomalous slow flow (region I) behavior was not observed with these solutions, but based on prior studies (1,5), it is presumed that behavior would occur for shear rates smaller than those utilized here. The decrease of η_P observed in slow flow (region II) with increasing polymer concentration normally observed for concentrations not too much larger than that for the isotropic to nematic phase transition may be noted (2). The creep and recoverable strain behavior observed with the cone/plate and parallel platens was qualitatively similar, despite the different strain rate profiles expected with these two geometries (31,32). Corresponding examples of the transient recoverable strain $\gamma_R(t)$ determined with history 2 for one of the samples are given in Figure 2 in two forms: $\log\{\gamma_R(t)/b_\sigma\}$ versus $\log t$, and $\log \gamma_R(t)$ versus $\log \{h_\gamma t\}$, where b_σ and h_γ are parameters chosen to superpose as nearly as possible the data for different levels of stress or apparent steady state shear rate $\dot\gamma_{ss}$, with emphasis on the data at small t. As may be seen, neither scaling provides superposition of the data over all of the stress studied. The scaling of $\gamma_R(t)/\sigma_{ss}$ with t is similar to that reported in prior work on PBZT solutions, with the data found to superpose well for short times for any stress level, but with $\gamma_R(t)/\sigma_{ss}$ reaching increasingly smaller asymptotic values for larger values of σ_{ss} in the fast flow regime (1,2,5). Such behavior is typical of that normally observed with high molecular weight materials in linear viscoelastic behavior (33). In some cases the scaling of $\gamma_R(t)$ with $h_\gamma t$ does superpose data in the fast flow regime, for which the viscosity $\sigma_{ss}/\dot\gamma_{ss}$ decreases with increasing $\dot\gamma_{ss}$. The parameters b_σ and h_γ determined for several samples are given in Figure 3 as functions of σ_{ss} and $\dot\gamma_{ss}$, respectively. As may be seen, $b_\sigma/\sigma_{ss} \approx$ constant, whereas $h_\gamma/\dot\gamma_{ss}$ increases with increasing $\dot\gamma_{ss}$.

Comparisons of the creep and recovery for histories 2 and 3 are given in Figures 4 and 5, in the form of the compliances $J(t) = \gamma(t)/\sigma_{ss}$ and $R(t) = \gamma_R(t)/\sigma_{ss}$; the data shown were obtained with the parallel plate platens, but similar data have been obtained with the cone and plate platens. The stress scaling provides a reasonable superposition of the data on both $J(t)$ and $R(t)$ for history 2 in the lower stress range, with nonlinear behavior evident at higher stress levels. The behavior with history 3 is seen to be more complex, as might be anticipated. As shown in Figure 1, the steady state viscosity was found to be somewhat smaller at low shear rate for experiments with history 3 than with history 2; here, the viscosity is calculated as $\sigma_{ss}/\dot\gamma_{ss}$. Qualitatively similar results were obtained for the samples with L ≈ 153 and 155 nm. Visual observations of the dichroism by the transmission of white light, polarized along and perpendicular to the radius of the plate (without an analyzer) shows a weak preference for orientation along the radius of the sample one day after the sample is injected radially into the plate-plate gap (with no plate rotation), and a stronger preference for orientation perpendicular to the

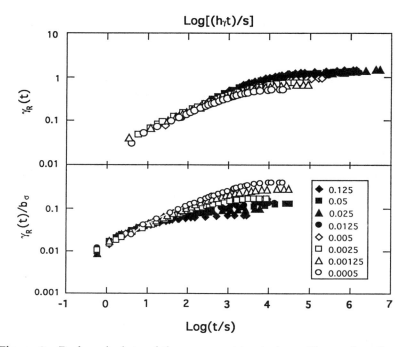

Figure 2. Reduced plots of the recoverable strain $\gamma_R(t)$ as a function of time following cessation of steady flow at the indicated shear rates (s^{-1}) in the order applied (deformation history 2). The upper and lower panels present log $\gamma_R(t)$ versus log $(h_\gamma t)$ and log $\{\gamma_R(t)/b_\sigma\}$ versus log t, respectively (L_W = 155 nm; w = 0.041).

radius following the completion of the creep/recovery cycle shown in Figure 4. Thus, the sample texture was substantially modified during the creep/recovery cycle, resulting in the pronounced failure of the data on $J(t) = \gamma(t)/\sigma_{ss}$ and $R(t) = \gamma_R(t)/\sigma_{ss}$ to superpose over a range of σ_{ss}.

The transmissions $T_\psi^{||}$ and $T_\psi^\perp$ during creep and recovery in history 2 or 3 were found to be essentially independent of ψ and about equal to each other, except as noted below. Typical data on $T_{\pi/4}^\perp(t)$ ($\psi = \pi/4$) in history 3 are shown in Figures 6 and 7, along with the corresponding behavior for $J(t) = \gamma(t)/\sigma_{ss}$ and $R(t) = \gamma_R(t)/\sigma_{ss}$. Similar data for history 2 are given in Figures 8 and 9. In each case, it is seen that $T_{\pi/4}^\perp(t)$ decreases markedly with increasing time for the larger values of σ_{ss}, returning to its initial value during recovery, but more slowly than $\gamma_R(t)$ reaches its limiting value $\gamma_R(\infty)$. As may be seen in Figure 10, the decrease of $T_{\pi/4}^\perp(t)$ during creep appears to scale with the strain $\gamma(t)$ during creep, for both histories 2 and 3; significant decrease in $T_{\pi/4}^\perp(t)$ requires $\gamma(t) > 10$. The slower return of $T_{\pi/4}^\perp(t)$ to its initial value than that of the recoverable strain $\gamma_R(t)$ to its limiting value is shown explicitly in Figure 11 for data in history 2. By contrast, the behavior given in Figures 12 and

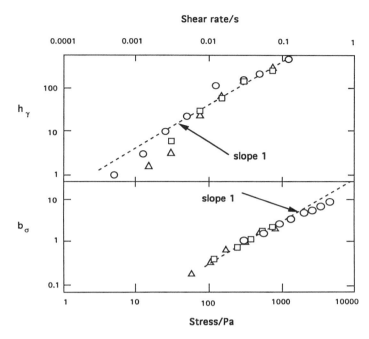

Figure 3. The factors h_γ and b_σ versus shear rate $\dot{\gamma}_{ss}$ and shear stress σ_{ss}, respectively, for three samples: circles, $L_w = 155$ nm, $w = 0.041$ (history 2); squares, $L_w = 140$ nm, $w = 0.040$ (history 2); triangles, $L_w = 140$ nm, $w = 0.040$ (history 1).

13 for the same data shows that during recovery $T_{\pi/4}^{\perp}(t)$ tends to scale with the product $\dot{\gamma}_{ss}t$ for a range of σ_{ss} for which the viscosity $\sigma_{ss}/\dot{\gamma}_{ss}$ decreases with increasing $\dot{\gamma}_{ss}$. Following creep under the larger stresses used, for which $T_{\pi/4}^{\perp}(t)$ dropped to a low level, $T_\psi^{\parallel}$ and $T_\psi^{\perp}$ depended weakly on ψ, and $T_{\pi/4}^{\parallel}(t)$ was about two to three times larger than $T_{\pi/4}^{\perp}(t)$, indicating a weak preferential alignment in the flow direction, consistent with the weak dichroism discussed above. Although the behavior will not be discussed futher, we note that experiments in progress give qualitatively similar results, with no tendency for increased alignment in flow, for the sample with the lowest viscosity in Figure 1 (i.e., $L_w = 155$ nm; $w = 0.068$).

Discussion

Theoretical treatments of the rheological behavior of nematic polymers are based on a constitutive relation for the stress comprising the Ericksen–Leslie (EL) constitutive relation for the viscous stress $\sigma^{(v)}$ and an elastic stress $\sigma^{(E)}$ involving the Frank curvature elasticities. These may be summarized by the expressions (in Cartesian coordinates, and with the Einstein summation convention) (10,11,17):

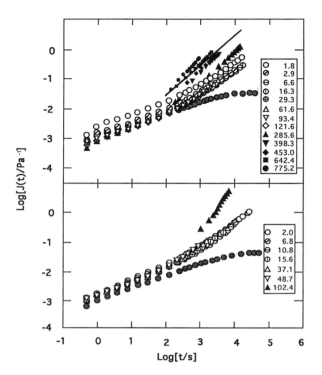

Figure 4. The creep compliance J(t) as a function of time during creep for deformation histories 3 and 2 in the upper and lower panels, respectively, for the indicated stresses (Pa) in the order applied (parallel plate rheometer; L_W = 140 nm; w = 0.049 (upper), 0.050 (lower)). The shaded circles give the recoverable compliance R(t) for the lowest stress level in each panel, see also Figure 5.

$$\sigma_{ij}^{(v)} = 2\mu_0 D_{ij} + 2\mu_1 n_i n_j n_\alpha n_\beta D_{\alpha\beta} + 2\mu_2 (n_i D_{j\alpha} n_\alpha + D_{\alpha i} n_\alpha n_j) + (\alpha_2/\gamma_1) n_j h_j + (\alpha_3/\gamma_2) h_i n_j \quad (2)$$

$$\sigma_{ij}^{(E)} = -\partial_j n_\alpha \left[\frac{\partial F_d}{\partial n_{\alpha,i}}\right] - p\delta_{ij} \quad (3)$$

where the D_{ij} are the symmetric components of the rate of deformation tensor, the μ_v are the Ericksen viscosities (v = 0-2), related to the Leslie viscosity parameters α_i (i = 1–6) by the expressions $2\mu_0 = \alpha_4$, $2\mu_1 = [\alpha_1 - \lambda(\alpha_2 + \alpha_3)]$, and $2\mu_2 = [\alpha_5 + \lambda\alpha_2]$, with $\lambda = -\gamma_2/\gamma_1$, $\gamma_2 = \alpha_6 - \alpha_5 \approx \alpha_3 + \alpha_2$, $\gamma_1 = \alpha_3 - \alpha_2$, and the molecular field $\mathbf{h(r)}$ depends on the distortion free energy $F_d(\mathbf{r})$ of the nematic field (10,11):

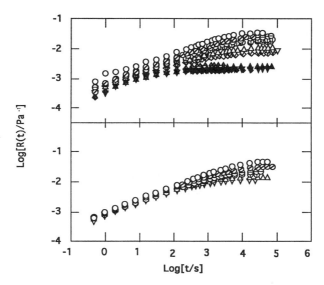

Figure 5. The recoverable compliance R(t) as a function of time following cessation of the creep shown in Figure 4.

$$h_i = \partial_\alpha \left[\frac{\partial F_d}{\partial n_{i,\alpha}} \right] - \frac{\partial F_d}{\partial n_i} + X_i \qquad (4)$$

Here, $\partial_\alpha = \partial/\partial x_\alpha$, $n_{i,\alpha} = \partial_\alpha n_i$, and $\tilde{X}(r)$ is an external field (e.g., an external magnetic field). For weak distortions of the director field $F_d(r)$ is expressed in terms of the Frank curvature elasticities K_μ, with μ = S, T, and B for splay, twist and bend distortions, respectively (10,11,34,35):

$$2F_d(r) = K_S[\text{div } \tilde{n}(r)]^2 + K_T[\tilde{n}(r) \cdot \text{curl } \tilde{n}(r)]^2 + K_B[\tilde{n}(r) \times \text{curl } \tilde{n}(r)]^2 \qquad (5)$$

In theoretical treatments, it is frequently assumed that all of the curvature elasticities are equal, say to a constant K_d, with $2F_d(r) = K_d\{(\text{div } \tilde{n}(r))^2 + (\text{curl } \tilde{n}(r))^2\}$. This single constant approximation approximation is not accurate for nematic solutions of PBZT (8,9). It may be noted that $\tilde{h}(r)$ is zero and $\sigma^{(E)} = -p\delta$ if the director field is undistorted, but that will not usually be the case in the flow of polymeric nematic fluids.

The EL constitutive relation is linear, with a status for the nematic fluid comparable to that of the Newtonian fluid for isotropic fluids. Consequently, although it is generally inadequate to describe the flow behavior of nematic polymers fluids, it is expected to apply under appropriate conditions. For example, in prior studies on defect–free, fully aligned (monodomain) preparations of nematic solutions of PBZT,

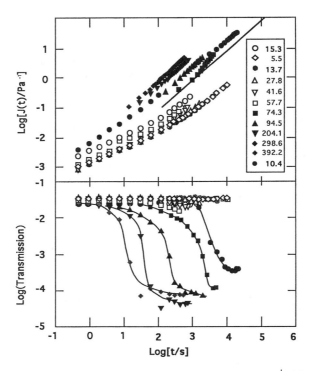

Figure 6. The creep compliance J(t) and transmission $T^{\perp}_{\pi/4}(t)$ during creep as a function of time for the indicated stresses (Pa) in the order applied (parallel plate rheometer, history 3). The transmission was measured with crossed polars oriented 45 degrees from the flow direction. (L_W = 155 nm; w = 0.053; 600 µm thickness).

it has been found that certain predictions of this constitutive relation apply to the distortion of the director field with increasing shear strain starting from rest (7,16). As remarked above, it is known that slow shear flow does not lead to flow alignment for this system, i.e., that $\varepsilon = \alpha_3/\alpha_2 < 0$. Nevertheless, the elastic stress arising from the constraints of the director field at the planar surfaces prevents inhomogeneous distortion of the director field in the initial stages of a shear flow starting from rest. In the experiments cited, an analysis for a slow flow with relative velocity v between plates separated by distance d, leads to the result (7,16)

$$\varphi(z,t) = \varphi_{mid}(\mathbf{Er})\Phi_{vis}(z/d)\, u_{vis}(\mathbf{Er},\, vt/d) \qquad (6)$$

for the orientation φ of the director at distance z from the mid-plane for

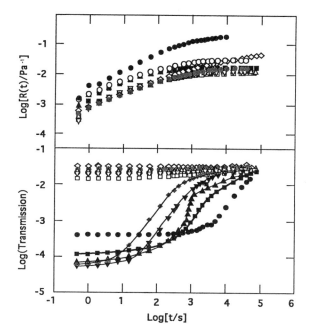

Figure 7. The recoverable compliance R(t) and transmission $T^{\perp}_{\pi/4}(t)$ as a function of time following cessation of the creep shown in Figure 6. The transmission was measured with crossed polars oriented 45 degrees from the flow direction.

small $\mathbf{Er} = vd |\alpha_2\alpha_3|^{1/2}/K_S$ ($\varphi \gtrless 0$ indicating a tilt against the shear gradient), $\varphi_{mid}(\mathbf{Er}) \approx \text{sgn}(\varepsilon) |\varepsilon|^{1/2} \mathbf{Er}/8$, $\Phi_{vis}(z/d) \approx \cos(\pi z/d)$ (20,36-37), and

$$u_{vis}(\mathbf{Er}, vt/d) \approx 1 - \exp[-\alpha_{vis} vt/d] \qquad (7)$$

where $\alpha_{vis} = [A^2 + B^2]^{1/2} \eta_T/|\alpha_3\alpha_2|^{1/2}$, with $B = 8/\mathbf{Er}$, and $A^2 = 4 \text{ sgn } \varepsilon$ (see reference 16 for a complete expression). For $\varepsilon < 0$, the elastic stress is unable to constrain to a homogeneous flow for **Er** less than a certain value $\mathbf{Er}_c$, unless vt/d is smaller than some critical value. The cited results showed that for small strain vt/d, the distortion $\langle\varphi\rangle$ (averaged through the slab) of the director scaled with $\langle\varphi\rangle/vd \propto \langle\varphi\rangle/\mathbf{Er}$ a function of $t/d^2 \propto (vt/d)/\mathbf{Er}$, showing that in that range the elastic stress acted across the sample dimension d to stabilize the shear deformation. With increasing t, the conoscopic interference figures used to determine $\langle\varphi\rangle$ were lost at a strain vt/d that depended weakly on v. This behavior is attributed to incipient formation of a phase grating, with periodic distortion of the director out of the shear plane as the elastic stress is no longer sufficient to stabilize the director field across the sample dimension (5,6,16). Related ef-

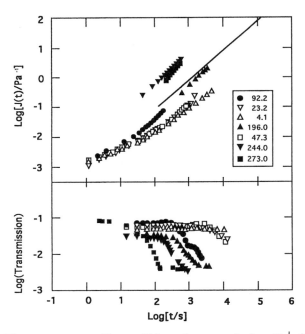

Figure 8. The creep compliance J(t) and transmission $T^{\perp}_{\pi/4}(t)$ during creep as a function of time for the indicated stresses (Pa) in the order applied (parallel plate rheometer, history 2). The transmission was measured with crossed polars oriented 45 degrees from the flow direction. (L_W = 155 nm; w = 0.050; 330 µm thickness).

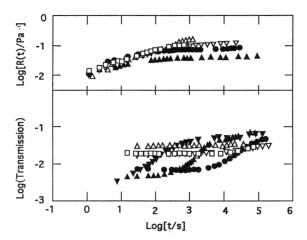

Figure 9. The recoverable compliance R(t) and transmission $T^{\perp}_{\pi/4}(t)$ as a function of time following cessation of the creep shown in Figure 8. The transmission was measured with crossed polars oriented 45 degrees from the flow direction.

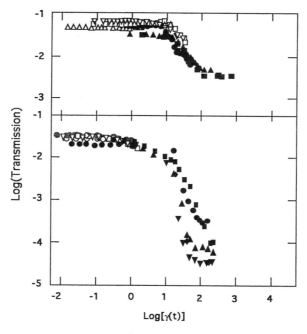

Figure 10. The transmission $T_{\pi/4}^{\perp}(t)$ during creep as a function of the strain $\gamma(t)$ for the data given in Figures 6 (lower) and 8 (upper). The transmission was measured with crossed polars oriented 45 degrees from the flow direction.

fects have been reported for the distortion of an originally homeotropically aligned monodomain of poly(γ–benzyl-L–glutamate) solutions (38).

As emphasized in the preceding, the stress arising from curvature elasticity can be important under appropriate circumstances, even though the Ericksen viscosities may be large. In some theoretical studies, directed principally to an explanation of negative normal stresses noted in steady shear flow over certain ranges of shear rate, an expression for the evolution of the orientation distribution with neglect of curvature elasticity due to Doi (39) was employed (27,40-42). This treatment leads to the appropriate form of the EL constitutive relation for slow flows with a spatially uniform director field, providing estimates of the Leslie viscosities (43), but also provides a means (with negelct of curvature elasticity) to examine behavior in flows strong enough to influence the order parameter. Different treatments for a nematic with $\alpha_3/\alpha_2 < 0$ concluded that non-flow aligning slow flow should become shear aligning for large enough shear rates in the fast flow regime, for which $\eta(\dot{\gamma})$ decreases continuously with increasing $\dot{\gamma}$ (27,40-42). Further, the calculations indicate that flow of nematics may be shear aligning for any $\dot{\gamma}$ as the order parameter S falls below some value, reflecting a change in the sign of α_3/α_2 for low S (27,42). Behavior with flow alignment at

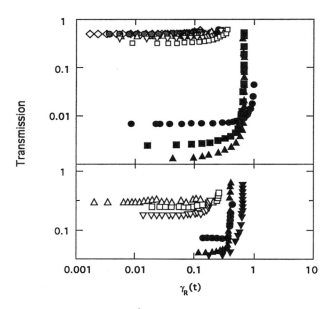

Figure 11. The transmission $T_{\pi/4}^{\perp}(t)$ during recovery as a function of the recoverable strain $\gamma_R(t)$ for the data given in Figures 7 (upper) and 9 (lower). The transmission was measured with crossed polars oriented 45 degrees from the flow direction.

high shear rate has been reported for certain nematic polymer solutions (27,44), but is not observed in the studies reported here. As would be anticipated, the creep behavior is more systematic for samples with histories 1 or 2 than with history 3. Similar effects of the deformation history on rheological properties have been reported with other polymeric nematic solutions, reflecting textural changes in flow that remain for a long period on cessation of flow (45-47). However, as shown in Figure 10, with either history 2 or 3 the transmission $T_{\pi/4}^{\perp}(t)$ (or $T_{\pi/4}^{\parallel}(t)) \approx T_{\pi/4}^{\perp}(t)$) exhibits a systematic dependence on the strain $\gamma(t)$ during creep, with only small decrease in $T_{\pi/4}^{\perp}(t)$ if $\gamma(t) < 10$, and a marked decrease in $T_{\pi/4}^{\perp}(t)$ with larger $\gamma(t)$, with behavior under different stresses superposing, that is, during creep:

$$T_{\pi/4}^{\perp}(t) \approx T_{\pi/4}^{\perp}(0)\, T_C(\gamma(t)) \tag{8}$$

where $T_C(0)$ is unity. The low values of $T_{\pi/4}^{\perp}(t)$ observed for $\gamma(t) > 100$, and the relative independence of $T_\psi^{\parallel}$ and $T_\psi^{\perp}$ from ψ, are inconsistent

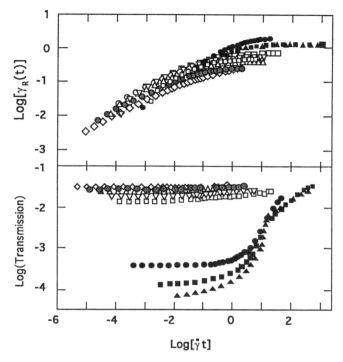

Figure 12. The recoverable strain $\gamma_R(t)$ and transmission $T_{\pi/4}^{\perp}(t)$ as a function of $\dot{\gamma}_{ss}\, t$, with $\dot{\gamma}_{ss}$ the shear rate prior to cessation of flow for the data in Figure 7. The transmission was measured with crossed polars oriented 45 degrees from the flow direction.

with any shear flow alignment, even though the texture appears smooth to the unaided eye. The strain required to effect a substantial decrease in $T_{\pi/4}^{\perp}(t)$ is not achieved at small shear stress within an experimentally accessible time. Nevertheless, as shown in Figures 4–6, the rheological response changes markedly following the first creep deformation for a sample with history 3 in the parallel plate rheometer. As remarked above, the method of extruding the sample radially in the fixed gap in the parallel plate rheometer tends to give a weak radial alignment to the director. Shear flow in this alignment is found to be very unstable with solutions of PBZT (16), as predicted theoretically (21,22). It is likely that the first creep deformation induces a weak tangential alignment. In addition, in creep under higher stresses, it is sometimes found that the logarithmic creep rate $\partial \ln \gamma(t)/\partial \ln t$ exceeds unity, which is not possible unless some structural change occurs in the material. This effect may be noted for certain of the creep deformations under the larger stresses used with a sample with history 3 in Figure 6, and similar effects may be noted in Figure 8 for a sample with history 2 as the stress is increased to large

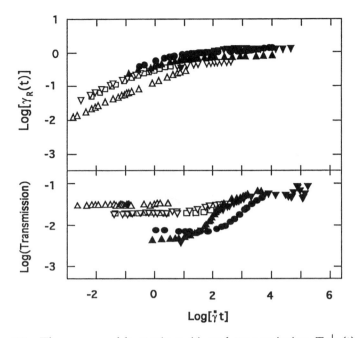

Figure 13. The recoverable strain $\gamma_R(t)$ and transmission $T_{\pi/4}^{\perp}(t)$ as a function of $\dot{\gamma}_{ss}\,t$, with $\dot{\gamma}_{ss}$ the shear rate prior to cessation of flow for the data in Figure 9. The transmission was measured with crossed polars oriented 45 degrees from the flow direction.

values in the final stages of the creep/recovery cycle. The observations reinforce the conclusions that there is no unique textural state of reference, and that the rheological behavior is sensitive to the texture. The strategy of preparing a sample by steady flow at a shear rate larger than any used in subsequent studies will lead to systematic behavior only if further deformations are carried out on a time scale and in a sequence that produces systematic changes in the texture.

It is possible that the high molecular rigidity of PBZT in comparison with other nematogenic polymers studied may play a role in this behavior, producing a relatively high order parameter (at least over regions of volume V_S) under all conditions of shear rate so far studied. An interesting, but so far unexplored, possibility is that in the flow alignment reported for large $\dot{\gamma}$, local order has developed, but that it is rather poorly defined, with a low order parameter. This might correspond to the range of the order parameter predicted to give positive α_3/α_2 in slow flow with the EL constitutive relation (27,42), permitting flow alignment, even though the flow regime might be considered to be too fast for strict application of the EL constitutive relation. This postulate is consistent with a report that the order (birefringence) tended to improve initially on cessation of slow flow (i.e., flow with $\eta(\dot{\gamma}) \approx \eta_P$) (48). This tendency would be frustrated with a rigid polymer that inherently tends to exhibit a high value of S, and promoted for less rigid structures.

The tendency to form roll-cells, or periodic distortions of the director out of the shear plane, leading to phase gratings noted above in slow flow of monodomains, has been known theoretically for some time (18-20,36,37). A detailed treatment applied to polymeric nematics suggests that this tendency may proliferate coexisting cells with a wide range of diameters and orientations with increasing Er, possibly leading to a kind of "director turbulence" on a length scale ℓ that results in appreciable light scattering (23). This behavior may be consistent with the results obtained here. The flow of the mottled nematic preparations studied here does not appear to involve disclinations in general (though they may be implicated in certain slow flow behavior on well aligned monodomain samples (6,15,16)), perhaps because the time constants in the formation and relaxation of disclinations are so much longer (e.g., by several orders of ten) than relevant rheological times for nematic solutions of PBZT (7). The formation of less well defined, but generally smoothly connected texture, for which the director field exhibits high order parameter over a scale $\ell < d$, with essentially zero order parameter over a large scale, seems more representative of the texture encountered with nematic PBZT solutions in flow. A theoretical model with this sort of texture has been constructed on the basis of the EL constitutive relation, including curvature elasticity (in the single constant approximation) (27,28). Motivated by the postulate that a balance of viscous and elastic stresses will occur on a length scale $\ell << d$ for non-shear aligning nematics ($\alpha_3/\alpha_2 < 0$), rather than over the macroscopic dimension d, the stress is computed over regions of scale ℓ, and then macroscopically averaged to compute the shear stress. Thus, the macroscopic Ericksen number in the form $Er = \dot{\gamma} d^2 \langle \alpha \rangle / \langle K \rangle$ is replaced by a textural Ericksen number $Er_{TEX} = \dot{\gamma} \ell^2 \langle \alpha \rangle / \langle K \rangle$ on the scale ℓ. On the presumption that the initial texture is characterized by a scale $\ell_o << d$, the calculation requires an expression for the decrease of ℓ with increasing $\dot{\gamma}$. By analogy with the relaxation kinetics of loop disclinations observed with small molecules nematics (49), it is assumed that (27,28):

$$-\frac{\partial \ell n\, Er_{TEX}}{\partial (\dot{\gamma} t)} = k_1 - k_2/Er_{TEX} \qquad (9)$$

where k_1 and k_2 are constants, sometimes taken to be unity. In this form it is apparent that Er_{TEX} is expected to be a constant in steady state shearing flow, so that in steady flow $\ell \propto \dot{\gamma}_{ss}^{-1/2}$ (27). A certain difficulty is associated with the application of this relation to nematic solutions of

PBZT since it has been found that loop disclinations observed in the late stages of the annealing process do not exhibit the relaxation kinetics found with small molecule nematics, and relax on a time scale several orders of magnitude slower than times of interest rheologically (7). The treatment, which is intended to apply only under conditions for which $\eta(\dot\gamma)$ is independent of $\dot\gamma$ (i.e., slow flow, or Region II), affords a scaling relation for the recoverable strain on cessation of steady flow at shear rate $\dot\gamma_{ss}$ (28):

$$\gamma_R(t) = \gamma_R(\infty) H(\dot\gamma_{ss} t) \tag{10}$$

where $\gamma_R(\infty)$ is expected to be independent of $\dot\gamma_{ss}$. As may be seen in Figures 12 and 13, this scaling is not obeyed for the solutions studied here for shear rates in slow flow (region II), for which $\eta(\dot\gamma)$ is independent of $\dot\gamma_{ss}$, as shown in Fig. 13, as expected theoretically, but $\gamma_R(t)$ does tend to scale with $\dot\gamma_{ss} t$ in recovery following fast flow (region III), for which $\eta(\dot\gamma)$ decreases with $\dot\gamma_{ss}$ (e.g., $\dot\gamma > 0.01$ for the data in Figure 13). By contrast, as seen in Figures 12 and 13, the recovery of the transmission $T_{\pi/4}^\perp(t)$ appears to scale with the product $\dot\gamma_{ss} t$ during recovery, with the recovery of $T_{\pi/4}^\perp(t)$ to its value prior to creep requiring a much longer time than the time required for $\gamma_R(t)$ to reach its limiting value $\gamma_R(\infty)$, so that

$$T_{\pi/4}^\perp(t) \approx T_{\pi/4}^\perp(0) T_R(\dot\gamma_{ss} t) \tag{11}$$

where $T_R(\infty)$ is unity, and $T_{\pi/4}^\perp(0)$ is the transmission prior to the creep and recovery cycle. It does not appear that the strain recovery is associated directly with the measure of texture provided by $T_{\pi/4}^\perp(t)$, in contrast with some suppositions that couple the recoverable strain with the effects of stored curvature elasticity (27). This is observed even with samples with creep under a small stress, for which the recovery proceeds rapidly enough that $\gamma_R(t) \approx \gamma_R(\infty)$ before the transmission exhibits much recovery towards its initial value. It may be that a substantial part of the effect of curvature elasticity occurs before the change in $T_{\pi/4}^\perp(t)$, but that would be unexpected if the elasticity effects are coupled with a change in the scale ℓ.

Finally, we note again that many of the rheological features observed are similar to those found with a polymeric solution filled with spheres, including the three regime behavior for $\eta(\dot\gamma)$, and the scaling behavior found for $\gamma_R(t)/\sigma_{ss}$ (25). The behavior of the filled polymeric solution was attributed to a tendency for the spheres to organize into a space-spanning network, with a weak yield stress. Although we have observed the lack of yield stress behavior for the nematic solutions studied here, the similarity of the steady-state and transient rheological responses may not be entirely coincidental: the inhomogeneous character of the nematic fluid on a scale of dimension $\ell \ll d$ may result in behavior similar to that for the filled system (5,27). That possibility has been the focus of some recent attempts to use "two-phase" models to describe the flow behavior of nematics, especially in the anomalous slow flow regime, and with reference to polydomain models (27,50). The effects of curvature elasticity over distances of scale ℓ would then play the role served by the bead-bead interactions in the filled system. The rheo-optical data obtained here show that for the PBZT solutions studied, the flow never results in a monodomain texture, but rather that an apparently smooth, but still inhomogeneous texture, results in a strong flow. On consideration of the time-scales of the flow, and that required to form true line defects (7), it does not seem likely that line defects can be important in the strong flows studied. It may even be that the expression for $F_d(r)$ is inadequate to describe the effects of gradients in the director field in strong flow (i.e., ℓ may be too small). Theoretical treatments that capture the behavior found in this study will likely deal with the effects of curvature elasticity in some approximate fashion, similar to methods already introduced in some models (27). Apparently, however, a means will have to be devised to account for viscoelastic behavior so far not included theoretically.

Conclusions

The preceding describes features of the rheological behavior of nematic solutions of a rodlike polymer and its correlation with texture as monitored by the transmission of light between either crossed or parallel polars. Many of the rheological features appear to be qualitatively similar to behavior reported for isotropic polymer solutions containing filler particles, with the strain in creep or recovery scaling with the stress in creep for behavior in slow flow, or for deformation under a recently small strain. Nevertheless, the rheo–optical behavior is very different from that for isotropic solutions of the same rodlike chain, with the apparent absence of flow alignment under all conditions studied here. Furthermore, the changes of the texture during flow and on cessation of flow occur on different time scales than that for the rheological responses. Thus, although changes in textural features appear to scale with certain strain parameters, consistent with some of the theoretical proposals for the behavior of non-shear aligning materials, the same is not true of the strain. It is suggested that the behavior observed is related to the effects of curvature elasticity acting over a scale that is both much smaller than the macroscopic dimensions and dependent on the deformation history. Theoretical treatments have not yet captured the range of behavior reported here.

Acknowledgments It is a pleasure to acknowledge discussions with Dr. Mohan Srinivasarao. This work, taken in part from the Ph.D. dissertation of B.D., has been supported in part by a grant from the

National Science Foundation, Division of Materials Research, Polymers Program.

Literature Cited

1. Einaga, Y.; Berry, G. C.; Chu, S. G. *Polymer J.* **1985**, *17*, 239.
2. Berry, G. C. *Mol. Cryst. Liq. Cryst.* **1988**, *165*, 333.
3. Berry, G. C.; Se, K.; Srinivasarao, M. In *High Modulus Polymers*; Zachariades, A. E., Porter, R. S., Eds.; Marcel Dekker: New York, NY, 1988; pp 195-224.
4. Berry, G. C.; Srinivasarao, M. In *Dynamic Behavior of Macromolecules, Colloids, Liquid Crystals and Biological Systems*; Watanabe, H., Ed.; Hirokawa Publ. Co.: Tokyo, 1988; pp 389-392.
5. Berry, G. C. *J. Rheol.* **1991**, *35*, 943.
6. Srinivasarao, M.; Berry, G. C. *J. Rheol.* **1991**, *35*, 379.
7. Diao, B.; Matsuoka, K.; Berry, G. C. In *Keynote Lectures in Selected Topics of Polymer Science*; Riande, E., Ed.; National Council of Scientific Research of Spain: Madrid; 1995.
8. Se, K.; Berry, G. C. *Mol. Cryst. Liq. Cryst.* **1987**, *153*, 133.
9. Desvignes, N.; Suresh, K. A.; Berry, G. C. *J. Appl. Polym. Sci.: Polym. Symp.* **1993**, *52*, 33.
10. de Gennes, P. G. *The Physics of Liquid Crystals*; Clarendon: Oxford, U.K., 1974.
11. Chandrasekhar, S. *Liquid Crystals*; Cambridge Press: Oxford, U.K., 1977.
12. Burghardt, W. R.; Fuller, G. G. *Macromolecules* **1991**, *24*, 2546.
13. Srinivasarao, M.; Garay, H. H.; Winter, H. H.; Stein, R. S. *Mol. Cryst. Liq. Cryst.* **1992**, *223*, 29.
14. Yang, I.-K.; Shine, A. *Macromolecules* **1993**, *26*, 1529.
15. De' Nève, T.; Navard, P.; Kléman, M. *J. Rheology* **1993**, *37*, 515.
16. Matsuoka, K.; Berry, G. C. *Nihon Reorojii Gakkaishi*, submitted.
17. Leslie, F. M. *Adv. Liq. Cryst.* **1979**, *4*, 1.
18. Pieranski, P.; Guyon, E. *Phys. Rev. A* **1974**, *9*, 404.
19. Pikin, S. A. *Sov. Phys. JETP* **1974**, *38*, 1246.
20. Manneville, P.; Dubois–Violette, E. *J. Phys. (Paris)* **1976**, *37*, 285.
21. Zúñiga, I.; Leslie, F. M. *Euro. Phys. Lett.* **1989**, *9*, 689.
22. Zúñiga, I.; Leslie, F. M. *Liq. Cryst. Lett.* **1989**, *5*, 725.
23. Larson, R. G. *J. Rheology* **1993**, *37*, 175.
24. Larson, R. G.; Mead, D. W. *Liq. Cryst.* **1993**, *15*, 151.
25. Meitz, D. W.; Yen, L.; Berry, G. C.; Markovitz, H. *J. Rheol.* **1988**, *32*, 309.
26. Asada, T. In *Polymer Liquid Crystals*; Ciferri, A.; Krigbaum, W. R.; Meyer, R. B., Eds.; Academic Press: New York, NY, 1982; Chapter 9.
27. Marrucci, G.; Greco, F. *Adv. Chem. Phys.* **1993**, *86*, 331.
28. Larson, R. G.; Doi, M. *J. Rheol.* **1991**, *35*, 539.
29. Berry, G. C.; Wong, C. P. *J. Polym. Sci.: Polym. Phys Ed.* **1975**, *13*, 1761.

30. Berry, G. C.; Birnboim, M. H.; Park, J. O.; Meitz, D. W.; Plazek, D. J. *J. Polym. Sci.: Part B: Polym. Phys Ed.* **1989**, *27*, 273.
31. Ferry, J. D. *Viscoelastic Properties of Polymers*; 3rd ed.; Wiley & Sons: New York, NY, 1980.
32. Coleman, B. D.; Markovitz, H.; Noll, W. *Viscometric Flows of Non-Newtonian Fluids*; Springer-Verlag: New York, NY, 1966; p. 54.
33. Berry, G. C.; Plazek, D. J. In *Glass: Science and Technology*; Uhlmann, D. R.; Kreidl, N. J., Eds.; Academic Press: New York, NY, 1986; Vol. 3, Chapter 6.
34. Chandrasekhar, S.; Ranganath, G. S. *Adv. Physics* **1986**, *35*, 507.
35. Kléman, M. *Points, Lines and Walls*; Wiley & Sons: New York, NY, 1983.
36. Pieranski, P.; Guyon, E. *Commun. Phys.* **1976**, *1*, 45.
37. Dubois–Violette, E.; Durand, E. G.; Guyon, E.; Manneville, P.; Pieranski, P. *Solid State Phys., Suppl.* **1978**, *14*, 147.
38. Yan, N. X.; Labes, M. M.; Baek, S. G.; Magda, J. J. *Macromolecules* **1994**, *27*, 2784.
39. Doi, M. *J. Polym. Sci.: Part B: Polym. Phys Ed.* **1981**, *19*, 229.
40. Marrucci, G.; Maffettone, P. L. *J. Rheol.* **1990**, *34*, 1217; 1231.
41. Larson, R. G. *Macromolecules* **1990**, *23*, 3983
42. Semenov, A. N. *Sov. Phys. JETP* **1987**, *66*, 712.
43. Kuzuu, N.; Doi, M. *J. Phys. Soc. Jpn.* **1983**, *52*, 3486.
44. Hongladarom, K.; Secakusuma, V.; Burghardt, W. R. *J. Rheology* **1994**, *38*, 1505.
45. Grizzuti, N.; Cavella, S.; Cicarelli, P. *J. Rheology* **1990**, *34*, 1293.
46. Larson, R. G.; Promislow, J.; Baek, S.-G.; Magda, J. J. In *Ordering in Macromolecular Systems*; Teramoto, A.; Kobayashi, M.; Norisuye, T., Eds.; Springer-Verlag: Berlin, 1994; pp 191-201.
47. Wissbrun, K.; Griffin, A. C. *J. Polym. Sci. Polym. Phys. Ed.* **1982**, *20*, 1835
48. Hongladarom, K.; Burghardt, W. R.; Baek, S. G.; Cementwala, S.; Magda, J. J. *Macromolecules* **1993**, *26*, 772.
49. Chuang, I.; Yurke, B.; Pargellis, A. N.; Turok, N. *Phys. Rev. E* **1993** *47*, 3343.
50. Kim, K. M.; Cho, J.; Chung, I. J. *J. Rheology* **1994**, *38*, 1271.

RECEIVED February 15, 1995

Chapter 20

Light Scattering from Lyotropic Textured Liquid-Crystalline Polymers Under Shear Flow

S. A. Patlazhan[1], J. B. Riti[2], and P. Navard[2]

[1]Institute of Chemical Physics of the Russian Academy of Sciences, Chernogolovka, Moscow Region, 142432 Russia
[2]Centre de Mise en Forme des Materiaux, Ecole des Mines, Sophia Antipolis, B.P. 207, F-06904 Valbonne, France

> Small-angle light scattering (SALS) experiments performed on lyotropic liquid crystalline polymer (LCP) solutions under steady-state shear flow are discussed. The structure of scattering objects is described in terms of a simplified model assuming linear defects located in different levels and oriented along the flow direction. A light scattering pattern appearing as a bright streak with strong intensity oscillations is theoretically calculated. From the comparison of the theoretical and experimental scattering curves for the streak the following conclusions were made: 1) defects created under steady state shear flow in cholesteric LCP solutions are grouped into sets of assemblies; 2) an increase in shear rate is accompanied by a decrease of the average domain size.

It is well-documented now that the rheology of main-chain polymer nematics at low shear rates strongly depends on coupling between the director orientation and the structure of texture (1). This has been shown in a series of key experiments dealing with transient and steady-state flow. In these experiments, such effects were observed as oscillatory responses in stress and conservative dichroism after sudden reverse or step increase in shear rate (2-5), slow elastic recoil after removal of shear stress (6), and dependence of moduli on the mechanical history of the sample (3,7,8).

Peculiarities in the rheological behaviour of the liquid crystalline polymers (LCP) are the result of an instability of the director at low Deborah number (tumbling regime) (1,9-13) caused by hydrodynamic torque exerted by shear flow. This instability is thought to cause spatial variations in the director field and to give rise to the formation of defects (13, 14). The area of the sample per one defect line is referred to as a "domain". Its average spatial scale, $\bar{a}$, was estimated by Marrucci (15) from the balance between Frank elastic distortion and viscous energies. It was derived that this scale decreases with an increase in the shear rate, $\dot{\gamma}$, as $\dot{\gamma}^{-1/2}$.

Rheo-optical measurements play an important role in structural studies of shearing LCP's. The first experiment done by Kiss and Porter using polarising microscopy (16) show the sensitivity of the texture to the shear rate. More recently, similar studies give a wealth of results on texture refinement and nucleation of disclination lines (13,17-19). Nevertheless, in the cases where the Deborah number is of the order of unity, it is difficult to separate individual domains inside the texture (13,18).

Flow conservative dichroism experiments and small-angle light scattering (SALS) are best suited for studies of the polydomain structure in textured lyotropic liquid crystalline polymers. Experiments of the first type are based on the measurement of the anisotropy of attenuation caused by the scattering of light from aligned or stretched domains. By varying the wavelength of the light used in the measurement of the dichroism one can investigate different length scales in the texture. Based on such an experiment, Burghardt and Fuller (5) confirmed the prediction of Marrucci concerning the behaviour of the domain size using the assumption that the local Ericksen number is independent of shear rate.

Flow SALS can also give a wealth of information on polydomain structures as the scattering intensity is directly proportional to the structure factor. Experiments performed at low shear rate with lyotropic (18,20-22) and thermotropic (23) LCP show that SALS scattering patterns have a bright streak along the vorticity axis (the vertical streak) and four lobes under crossed polarizers. These results confirm that the scattering objects have a strong alignment in the flow direction. However, no simple explanation can be given regarding the dependence of scattering intensity on the scattering angle. Understanding this dependence might be useful in extracting more precise information about the structure. For example, the scattering curve cannot be solely described in terms of a single rod theory. The main contradictions with experiments of this simple model are the appearance of oscillations in the vertical streak and a drastic growth of the intensity at small values of scattering angles with increased shear rate as observed in (20,21).

In reference (24) a satisfactory explanation of the observed facts was offered. It was suggested that the light scattering forming the vertical streak is due to 3D diffraction of light scattered from a set of rodlike defects aligned in the direction of flow. These defects can be defined as, for example, stretched parts of disclination loops. The model of disclination loops is based on the experimental observations of defects created at low shear rates in thermotropic LCP (26,27). The oscillations of intensity as a function of scattering angle are attributable to the retardation of light scattered from defects located in different levels of the sample (with respect to the incident beam). It was suggested also that four lobes may be due to the perturbation of stream lines around parts of the loops oriented along the gradient of shear rate. This assumption is supported by the theory of shearing polymer suspensions (25). This theory predicts a four-lobe pattern under the crossed polarisation.

In this work we continue our previous studies aimed at understanding the structural origin of the vertical streak on the light scattering pattern. The comparison of theory and experiment for a PBG solution at steady-state shear flow will allow us to find the correlation between the average domain size and shear rate.

Experimental

The experiment was performed with 25% solution of poly (benzyl glutamate) (PBG) in m-cresol at 22 $°C$.

Small angle light scattering measurements were carried out on a transparent Instron 3250 cone-and-plane rheometer (*28*) with crossed (HV) position of polaroids (the polarizer was parallel to the flow direction (x-axis), and analyser was parallel to the axis of vorticity (z-axis)). The shear rate was varied in the range between 0.5 and 20 s^{-1}. A He-Ne laser with the wavelength of $\lambda=632.8$ nm was used as a light source. The incident beam was parallel to the rotation axis of the cone (y-axis). The beam intersects the cell at a point located where the distance between the cone and plate is about 0.4 mm. The scattering pattern was registered in the screen plane ((x,z)-plane).

Flow light scattering patterns had a bright vertical streak parallel to the vorticity axis with four lobes and was similar to the patterns developed in references (*20-22*). The intensity in the vertical streak was processed using numerical image analysis.

Figure 1 demonstrates the measured intensity as a function of scattering vector, q, along the vorticity axis for two values of shear rate, 0.5 and 15 s^{-1}. Two characteristic features of this dependence can be noticed: 1) the scattering intensity oscillates and the amplitude of these oscillations grow with an increase in shear rate; 2) the increase in shear rate is accompanied by a drastic growth of the average intensity at small values of scattering vector.

Theoretical

In the beginning we discuss briefly the main points considered in reference (*24*). In this work the concept of loop defects described in the case of thermotropic polymers (*26,27*) was extended to lyotropic systems. These loops lie in the shearing plane ((x,y)-plane) and stretch in the flow direction. The structure of disclination cores is complex but some general conclusions can be drawn. In particular, the director deviates at a certain angle from the flow direction towards the vorticity axis. Here we will assume that the director is located in (x,z)-plane and makes an angle β with the x-axis.

We assume that the vertical streak in the light scattering pattern arises due to the scattering from the aligned parts of the loops (the analysis of the origin of the four lobes is beyond the scope of this work). These defects can be treated as rodlike particles oriented parallel to the flow direction. The unusual behaviour of the intensity in the streak discussed in the experimental section (see Figure 1) suggests that 3D diffraction from an assembly of loops gives the major contribution. In reference (*24*), it was shown that the strong oscillations of the intensity in the scattering curves are caused by the retardation of waves scattered by the set of rods lying in various levels of the sample. Each pair of nearest rods in a different depth can belong either to the same disclination or to different loops. The latter can occur when the loops are contracted along the shear gradient and form rodlike defects as was calculated by Marrucci and Maffettone (*29*). In this case one would be expected a stratified structure in shearing LCP.

If there is no disorder of optical axes of the rods in the assembly, the light scattering intensity can be written as follows:

$$I_{2N} = I_1 \left\{ 2N + \left\langle \sum_{i \neq j}^{2N} \exp[-i\mathbf{q}(\mathbf{r}_i - \mathbf{r}_j)] \right\rangle \right\} \quad (1)$$

Here I_1 is the scattering intensity from a single rod; $\mathbf{r}_i - \mathbf{r}_j$ is the vector connecting the centers of the ith and jth rods; the angular brackets denote averaging over the distances between the nearest rods along y and z-axes; 2N is the number of rods in the assembly.

The solution for the intensity of light scattered by a single rod as a function of azimuth, μ, and scattering, θ, angles is well known (*30*). In the case of HV scattering from a rod oriented along the x-axis this solution takes the form:

$$I_1 = C d^4 L^2 \delta^2 \langle \sin^2 2\beta \rangle j_o^2 \left(\frac{1}{2} q d s_y \right) j_o^2 \left(\frac{1}{2} q d s_z \right) \quad (2)$$

Here the angular brackets denote averaging over the angle of the optical axis, β; d and L are the diameter and the length of the rod; δ is the optical anisotropy; $q = (4\pi/\lambda)\sin(\theta/2)$ is the length of the scattering vector; C is a constant; and $s_z = \cos(\theta/2)\sin\mu$.

To carry out the summation in equation 1, one must specify certain arrangement of rods in the assembly. Here we will restrict our consideration to the simplest structure: an assembly of parallel rods of equal lengths located in two different planes parallel to the screen (N rods at each level). We assume that the distances a and h between the rods along the vorticity axis and the incident beam, as well as the angle of the optical axis, β, have a Gaussian distribution

$$P(\xi) = \frac{1}{\sigma_\xi \sqrt{2\pi}} \exp\left\{ \frac{(\xi - \bar{\xi})^2}{2\sigma_\xi^2} \right\} \quad (3)$$

where $\bar{\xi}$ and σ_ξ are the average value and dispersion of a, h or β. It can be shown that the intensity per unit volume of 3D diffraction of light scattered by the assembly is:

$$I_{2N} = \frac{I_1}{\bar{a}\bar{h}L} \left\{ \frac{1}{N} \left[\langle \cos qhs_y \cos qhs_z \rangle_h + 2 \right] \right.$$
$$\left. \sum_{m=1}^{N} m \exp\left[-\frac{1}{2}(N-m)q^2 s_z^2 \sigma_a^2 \right] \cos[(N-m)qs_z \bar{a}] - 1 \right\} \quad (4)$$

The factor before the sum accounts for the retardation of waves scattered by rods lying in different levels. This factor is responsible for the intensity oscillations in the vertical streak.

Figure 2 shows the scattering pattern in the (q_x,q_z)-plane calculated with the help of equation 4 for the scattering vector q in the range from 0 to 0.54 µm^{-1}. As was expected, this pattern has a form of a streak along the vorticity axis with an oscillatory distribution of light. Here we used the following notations: $q_x = qs_x$ and $q_z = qs_z$ where $s_x = \cos(\theta/2)\cos\mu$.

If we neglect the retardation (or average the oscillations), the light scattering will be governed by the 2D diffraction caused by scattering from rods located in the same plane. In this case, equation 4 may be rearranged as follows:

$$I'_{2N} = \frac{I_1}{\bar{a}\bar{h}L}\left\{\frac{2}{N}\sum_{m=1}^{N} m \exp\left[-\frac{1}{2}(N-m)q^2 s_z^2 \sigma_a^2\right]\cos\left[(N-m)qs_z\bar{a}\right]-1\right\} \quad (5)$$

The primed symbol indicates 2D scattering intensity.

Figure 3 shows the scattering curves for 3D and 2D diffractions for a/d = 1.1 (top) and a/d=2.0 (bottom). We can see that the scattering curves due to 2D diffraction can be considered as midlines of the 3D oscillatory curves. They intersect the intensity axis (q=0) at

$$I^o_{2N} = \frac{I_1}{\bar{a}\bar{h}L}N \quad (6)$$

which is proportional to the number of rods per assembly and inversely proportional to the average domain size.

The slope of the 2D scattering curve for small values of scattering vector is equal to

$$\frac{\partial I'_{2N}}{\partial q} = -\frac{I_1}{3\bar{a}\bar{h}L}(N^2-1)\left(\sigma_a^2+\frac{1}{2}N\bar{a}^2\right)q \quad (7)$$

For $N \geq 3$ and $\sigma_a^2 < \frac{1}{2}N\bar{a}^2$, equation 7 can be estimated as follows:

$$\frac{\partial I'_{2N}}{\partial q} \cong -\frac{I_1}{6\bar{h}L}N^3\bar{a}q \quad (8)$$

That is, the derivative is proportional to $\bar{a}$, q and N^3. Using equations 6 and 8 we can obtain the average domain size:

$$\bar{a}^2 \propto \left\{-\frac{1}{q(I^o_{2N})^3}\frac{\partial I'_{2N}}{\partial q}\right\}^{1/2} \quad (9)$$

We shall use this equation for analysis of average domain sizes in the PBG solution at different shear rates with the help of the scattering curves.

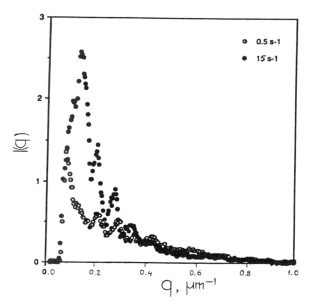

Figure 1. Scattering curves measured along the vorticity axis for a 25% solution of PBG in m-cresol at two values of the shear rate, 0.5 and 15 s^{-1}.

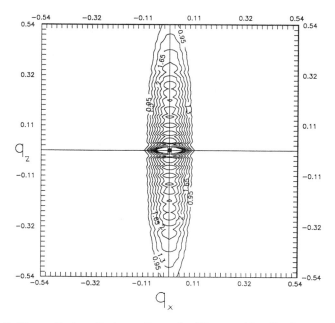

Figure 2. Theoretical scattering pattern of a 3D assembly of rods as calculated from equation 4.

Discussion and Conclusion

The light scattering pattern presented in Figure 2 was calculated for a set of rods with the following parameters: L=40λ, d=3λ, $\bar{a}$ = 1.1d, σ_a = 0.1$\bar{a}$, h=100λ, σ_h =0, β = 60° and σ_β=0.2. This choice is based partly on electron microscopy observations of a thermotropic LCP (27), comparative analysis of the vertical streak in HH and VV polaroid positions, and numerical tests (24). The pattern reproduces fairly well oscillatory behaviour of the vertical streak observed in the experiment (18). These oscillations can be caused by 3D diffraction from rods lying in different levels of the sample. It can also be related to the stratified structure of the textured LCP in the shear flow.

The scattering curves in Figure 3 were calculated for two different average domain sizes, a=1.1d (top) and a=2d (bottom). We can see a strong increase of the intensity for 3D and 2D diffraction at q=0 with a decrease in a/d. This effect is shown in more detail in Figure 4 for q = 0.17 μm^{-1}. In this figure, the scattering intensity is plotted as a function of the number of rods, N, and the ratio of domain size to rod diameter, a/d. It can be seen that the increase in the domain size results in a decrease in intensity at fixed q. The dependence of intensity on the number of rods is nonmonotonic: the scattering intensity increases with N until the cross section size of the assembly becomes compatible with the rod length and decreases afterwards. For smaller domain scales, the decrease of intensity occurs at larger numbers of rods. This fact can be explained in terms of a redistribution of scattering intensity over a larger range of azimuthal angle due to shrinking of the vertical streak when the aspect ratio of the assembly is close to unity. The comparison of this analysis with the experimental data suggests that the behaviour of the scattering intensity for sheared PBG solution at different $\dot{\gamma}$, (see Figure 1) is attributable to the fact that the number of line defects and/or domain size in the polymer increases with shear rate.

The midline position of 2D scattering curves as compared to oscillation curves due to 3D diffraction enables us to simplify the quantitative analysis. For example, an increase in the average intensity can be characterised in terms of the value of 2D diffraction intensity at q = 0 which is proportional to the number of rods per assembly and inversely proportional to the average domain size (see equation 6). Applying equation 9, we can extract the average domain size from experimental scattering curves of the vertical streak. Making this analysis for the curves at shear rates ranging from 0.5 to 10 s^{-1} (tumbling area for the PBG solution), we found the behaviour of the average domain scales as a function of shear rate (see Figure 5). Within the limit of experimental error, we can see that $\bar{a}^2 \propto \dot{\gamma}^{-1}$ over the range of shear rates under consideration. This result independently confirms both the theoretical prediction of Marrucci (15) and the conclusions of Burghardt and Fuller (5). Moreover, it supports the assumption that the local Ericksen number governs the dynamic response of the director field. Indeed, this assumption was key in deriving the relation of domain size and shear rate.

In summary we can say that in spite of its simplicity, the proposed structural model of a textured LCP describes the main features of small-angle light scattering patterns, particularly in terms of the vertical streak. The conclusions made on the basis of this model are in good agreement with previous results. The conclusion

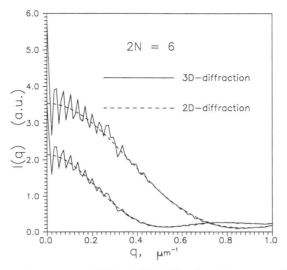

Figure 3. Scattering curves of 3D and 2D diffractions from assemblies of rods with various distances in the screen plane; (top) a/d=1.1, (bottom) a/d=2.0.

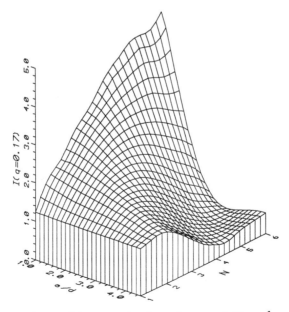

Figure 4. Dependence of the scattering intensity at $q=0.17$ μm^{-1} on the distance between rods in the screen plane, (a/d), and the half number of rods in the assembly, (N).

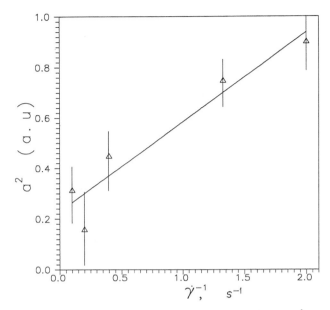

Figure 5. Dependence of average squared domain size, a^2, on $\dot{\gamma}^{-1}$ for the PBG solution in m-cresol.

about the stratification of the sheared LCP follows directly from the oscillatory behaviour of the vertical streak and has not been discussed before.

Literature Cited.

1. Marrucci, G; Greco, F. in *Advances in Chemical Physics*; Prigogine, I., Rice, S., Eds.; Wiley: New York, 1993.
2. Mewis, J.; Moldenaers, P. *Mol. Cryst. Liq. Cryst.* **1987**, *153*, 291.
3. Moldenaers, P. Ph.D. thesis, Katholiecke Universiteit Leuven, 1987.
4. Moldenaers, P.; Fuller, G.G.; Mewis, J. *Macromolecules* **1988**, *22*, 960.
5. Burghardt, W.R.; Fuller, G.G. *Macromolecules* **1991**, *24*, 2546.
6. Larson, R.G.; Mead, D.W. *J. Rheol.* **1989**, *33*, 1251.
7. Moldenaers, P.; and Mewis, J. *J. Rheol.* **1986**, *30*, 567.
8. Moldenaers, P.; and Mewis, J. *J. Rheol.* **1993**, *37*, 367.
9. Semenov, A.N. *Sov. Phys. J.E.T.P.* **1983**, *58*, 321.
10. Kuzuu, N.; Doi, M. *J. Phys. Soc. Japan* **1984**, *53*, 1031.
11. Marrucci, G.; Maffettone, P.L. *Macromolecules* **1989**, *22*, 4076.
12. Srinivasarao, M.; Berry, G.C. *J. Rheol.* **1991**, *35*, 379.
13. Larson, R.G.; Mead, D.W. *Liquid Crystals*, **1993**, *15*, 151.
14. Larson, R.G. *J. Rheol.*, **1993**, *37*, 175.
15. Marrucci, G. *Pure Appl. Chem.* **1985**, *57*, 1545.
16. Kiss, G.; Porter, R.S. *Mol. Cryst. Liq. Cryst.* **1980**, *60*, 267.
17. Alderman, N.J.; Mackley, M.R. *Farad. Discuss. Chem. Soc.* **1985**, *76*, 149.

18. Riti, J.B.; Navard, P. in preparation.
19. Vermant, J.; Moldenaers, P.; Picken, S.J.; Mewis, J. *Acta Rheol.* to be published.
20. Ernst, B.; Navard, P.; Hashimoto, T.; Takebe, T. *Macromolecules* **1990**, *23*, 960.
21. Takebe, T.; Hashimoto, T.; Ernst, B.; Navard, P.; Stein, R.S. *J. Chem. Phys.* **1990**, *92*, 1386.
22. Picken, S.T.; Aerts, J.; Doppert, H.L.; Reuvers, A.J.; Northold, M.G. *Macromolecules* **1991**, *24*, 1366.
23. Cidade,T. ; Riti, J.B.; Navard, P. in preparation.
24. Patlazhan, S.A.; Riti, J.B.; Navard, P. *Macromolecules*, submitted, 1994.
25. Patlazhan, S.A.; Navard, P. *J. Phys. France*, submitted. 1994.
26. De'Neve, T.; Kleman, M.; Navard, P. *C. R. Acad. Sci. Paris* **1993**, *316*, II, 1037.
27. De'Neve, T.; Navard, P.; Kleman, M. *Macromolecules*, submitted.
28. Ernst, B.; Navard, P. *Macromolecules* **1989**, *22*, 1419.
29. Marrucci, G.; Maffettone, P.L. in *Liquid Crystalline Polymers*, Carfagna, C., Ed.; Pergammon Press: Oxford, 1994.
30. Rhodes, M.B.; Stein, R.S. *J. Polymer Sci.*, *Pt.A-2*, **1969**, *7*, 1539.

RECEIVED February 15, 1995

Chapter 21

Comparison of Molecular Orientation and Rheology in Model Lyotropic Liquid-Crystalline Polymers

Wesley Burghardt, Bruce Bedford, Kwan Hongladarom, and Melissa Mahoney

Department of Chemical Engineering, Northwestern University, Evanston, IL 60208

Flow birefringence is used to study molecular orientation in shear flows of a lyotropic solution of hydroxypropylcellulose [HPC] in m-cresol. Results are compared with two other model liquid crystalline polymer solutions, poly(benzyl glutamate) [PBG] in m-cresol and HPC in water. Both HPC solutions exhibit a regime of nearly zero birefringence at low rates, correlated with Region I shear thinning, while PBG exhibits a low shear rate plateau in birefringence, and no Region I behavior. In all solutions, changes in sign in normal stresses are correlated with increases in birefringence. During relaxation in HPC/cresol, birefringence is initially seen to increase, on a time scale that varies inversely with the previously applied shear rate, as in PBG. However, the birefringence ultimately decays to an isotropic condition, as is observed in HPC/water. It thus appears that HPC/cresol provides a link between the other two model systems.

Much of the peculiar rheological behavior of liquid crystalline polymer solutions is now understood to result from complex interactions between shear flow and the molecular orientation state. A series of transitions predicted by the Doi molecular model with increasing Deborah number (director tumbling, wagging, flow alignment) (1,2) provides an explanation for sign changes in normal stress differences that have been extensively studied in the model systems of poly(benzyl glutamate) [PBG] in m-cresol (3) and hydroxypropylcellulose [HPC] in water (4,5).
 While the nonlinear viscoelastic phenomena in these two model systems are quite similar, there are some noteworthy differences in their behavior. First, PBG solutions at moderate concentration show no sign of so-called "Region I" shear thinning at low rates (3,6), while viscosity data for HPC solutions in water generally do exhibit Region I (4,5). In addition, the two systems exhibit opposite behavior when dynamic moduli are

monitored as a function of time following flow cessation. In PBG, the moduli gradually *decrease* (7,8), while in HPC, the moduli gradually *increase* (5). Finally, relaxation processes in PBG obey a particularly simple scaling law (8,9), which HPC does not always obey.

Facing this "distressing lack of universality" (10) in the rheological properties of LCP's, even among model systems with some detailed similarities in nonlinear rheological behavior, we have been motivated to compare the behavior of a number of model systems subjected to similar testing protocols. Since a strong dependence of physical properties on director orientation is one of the hallmarks of liquid crystalline materials, our work attempts to make connections between bulk rheological behavior and measurements of the average molecular orientation state. Several techniques have been used to make quantitative measurements of molecular orientation in sheared LCPs, including X-ray scattering (11,12), neutron scattering (13) and UV dichroism (14). Our group has made extensive measurements of molecular orientation in PBG (6,9), and HPC (15) using the technique of flow birefringence in both steady and transient flows. These measurements have provided direct insights into the microstructural origins of observed rheological behavior. During relaxation, for instance, molecular orientation *increases* in PBG (9), and *decreases* in HPC (15), providing an explanation for the different trends observed in the dynamic moduli.

While structural behavior may be directly correlated with observed rheology, it is not clear why there should be such profound differences between two model systems that are so similar in other respects. One considerable difference is that rheological and rheo-optical studies on PBG are typically carried out at rather low concentration (12-25 wt%), while HPC solutions in water only become liquid crystalline at concentrations of around 50 wt% and above. It is possible that the higher chain density in HPC solutions can lead to differences in rheological behavior. Along these lines, PBG solutions at higher concentrations (37-40 wt%) show signs of Region I behavior in the shear viscosity (16). Recently, Baek and coworkers have begun study of HPC solutions using m-cresol as a solvent (17,18), where anisotropic solutions form at much lower concentrations. Here we report flow birefringence and rheology results on a moderately concentrated solution of HPC in m-cresol. In several respects, this solution appears to provide a link between the behavior observed in PBG/cresol and HPC/water model systems.

Experimental

Materials. We have examined a 27 wt% solution of hydroxypropylcellulose (Klucel E) in m-cresol. Appropriate amounts of polymer and solvent were weighed into a bottle, and dissolution was aided by mild agitation over a period of several weeks. This sample is similar to one studied by Baek and coworkers, reported to exhibit a sign change from positive to negative N_1 at a shear rate of around 35 s^{-1}(18). Results will be compared to behavior seen in a 13.5 wt% solution of PBG (M = 300k) in m-cresol for which N_1 changes sign at 7 s^{-1}(6), and a 50 wt% solution of

HPC (Klucel E) in water, for which N_1 is reported to change sign at 50 s^{-1} (*19*). It has recently been demonstrated that aqueous 50 wt% HPC solutions are biphasic at room temperature (*20*). However, rheological phenomena in these solutions are dominated by the liquid crystalline phase (for instance, sign changes in N_1 and Region I shear thinning) (*5*).

Optical Rheometry. Flow birefringence was measured using a spectrographic technique described in detail by Hongladarom and coworkers (*6,15*). Figure 1 shows a schematic illustration of the optical train. A white light source is employed, and a diode array spectrograph enables rapid acquisition of spectra of light intensity transmitted between crossed and parallel polarizers as a function of wavelength. The sample is sheared between parallel disks separated by 0.9 mm, and birefringence is measured in the flow-vorticity (X-Z) plane, such that the optical anisotropy is always oriented along the flow direction. The polarizer and analyzer are crossed or parallel to one another, and oriented at 45° with respect to the flow direction. For each data point, crossed ($I^\perp$) and parallel ($I^\parallel$) spectra are obtained and then normalized:

$$N^\perp = \frac{I^\perp}{I^\perp + I^\parallel}, \quad N^\parallel = \frac{I^\parallel}{I^\perp + I^\parallel}, \tag{1}$$

where the normalized spectra should depend on birefringence Δn, sample thickness d, and wavelength λ according to:

$$N^\perp = \sin^2(\pi \Delta n d / \lambda), \tag{2}$$

$$N^\parallel = \cos^2(\pi \Delta n d / \lambda). \tag{3}$$

For highly birefringent materials such as sheared LCPs, retardation at visible wavelengths passes through multiple orders, and the normalized spectra pass through multiple maxima and minima as a function of wavelength; fitting this oscillatory dependence according to equations 2 and 3 allows unambiguous determination of the birefringence. This procedure also requires an accurate representation of the wavelength dispersion of the birefringence; for the HPC/cresol solution, we found that the same expression used for HPC/water (*15*) was successful.

Mechanical Rheometry. Rheological measurements of steady shear viscosity and dynamic moduli were performed on a Bohlin VOR rheometer, using either cone and plate or narrow-gap Couette fixtures. Sample fixtures were sealed as necessary to reduce complications associated with solvent evaporation during long experiments. For all viscosity and relaxation data, strains of at least 200 units were applied to assure steady state. Walker and Wagner have reported anomalous long transients as a signature of "Region I" shear thinning at low rates (*21*); however, the samples they studied have a significantly higher polymer concentration than those of interest here, and we have found 200 strain units to suffice for pre-shearing.

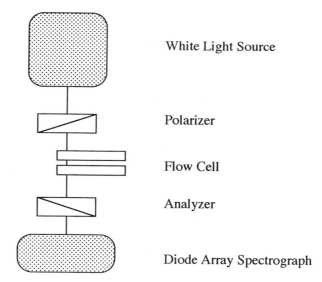

Figure 1. Schematic of spectrographic birefringence technique.

Results and Discussion

Steady State Flow. Figure 2 shows a comparison of steady shear viscosity and flow birefringence data for the three solutions under consideration. In each case, the filled arrow indicates the shear rate where a change in sign of N_1 from positive to negative is observed; these shear rates may be taken as indicative of the transition from tumbling to flow aligning behavior under the influence of nonlinear viscoelastic effects (1,2).

PBG is characterized by a Newtonian plateau in viscosity at low shear rates, and shear thinning at higher rates. The sign change in N_1 unambiguously marks this shear thinning as a manifestation nonlinear viscoelasticity associated with flow-induced changes in the molecular orientation probability function (Region III), as opposed to Region I shear thinning attributed to distortional elasticity (21,22). The birefringence exhibits a plateau in the low shear rate, tumbling regime, with a magnitude roughly half of that seen in a defect-free monodomain of the same solution (6). The solution exhibits a pronounced transition to a higher orientation state at a shear rate comparable to the sign change in N_1, presumably reflecting the transition from tumbling to flow-aligning behavior.

The viscosity data in HPC/water typify classical "three-region" behavior, with a clear Region I at low rates. The birefringence in this solution at low rates is nearly zero, indicating that flow is ineffective at promoting significant molecular orientation. The physical description invoked by Onogi and Asada (22) to explain Region I shear thinning was that of a "piled polydomain", where flow imparts negligible orientation to the sheared LCP. Results in PBG and HPC/water are consistent with this picture, in that we find that Region I shear thinning is correlated with low orientation at low rates. As shear rate is increased, HPC water shows signs of two transitions in birefringence, the first at shear rates around 0.5 s^{-1}, where shear flow is first able to generate significant amounts of orientation, and then a region of increasing orientation at higher rates that appears to be associated with the transition from tumbling to flow alignment, as revealed by the sign change in N_1.

The solution of HPC in cresol exhibits birefringence behavior that is qualitatively similar to HPC/water, but with considerably stronger evidence for two distinct transitions in orientation state with increasing shear rate. Here again, the second transition appears to be correlated with the sign change in N_1, indicative of the transition from tumbling to flow alignment. Baek and coworkers report viscosity data for the HPC/cresol solution (18), but their measurements were limited to shear rates of above 0.78 s^{-1}, so that there were no clear indications of Region I shear thinning in their data. At lower shear rates, we find indications that Region I is also present for this solution, again correlated with a regime of weak molecular orientation.

In both HPC solutions, the N_1 sign change is associated with an increase in orientation at high shear rates. Unlike PBG, however, there appears to be an additional transition at lower shear rates from a low to a moderate degree of orientation. The intermediate plateau in orientation, evident in HPC/cresol, would seem to be analogous to the low shear rate

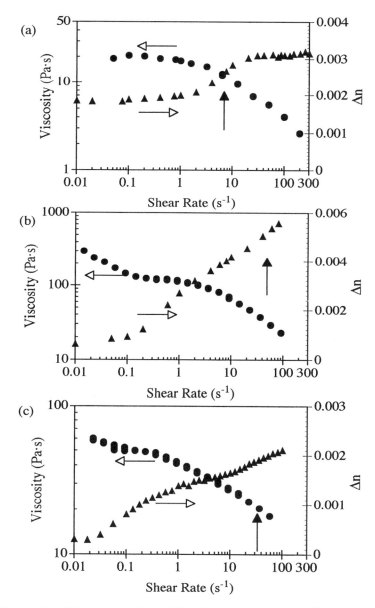

Figure 2. Viscosity (●) and birefringence (▲) as a function of steady shear rate in (a) PBG/cresol, (b) HPC/water, and (c) HPC/cresol. Filled arrow indicates shear rate at which N_1 changes sign from positive to negative.

plateau observed in PBG in the tumbling regime. An alternate physical mechanism could then result in the low shear rate transition to low orientation. We have speculated that low orientation at low rates could result from persistence of a cholesteric phase in HPC solutions (*15*).

Relaxation Behavior. Relaxation data on the three solutions are shown in Figures 3-5. In the PBG solution, we find that orientation *increases* as a function of time during relaxation, when the initial shear rate is in the low shear rate, tumbling plateau. Like most relaxation processes in PBG, this increase in orientation occurs on times that scale inversely with the previously applied shear rate (note the scaling of time in Figure 3). This increase in orientation provides a direct and complete explanation for the decrease in dynamic moduli during relaxation in PBG solutions (*9*).

Conversely, in HPC/water (Figure 4), we see that orientation *decreases* during relaxation, on a time scale that does not depend strongly on the previously applied shear rate. On similar time scales, the moduli are observed to increase. This increase may be partially explained by the molecular orientation behavior. However, the final "equilibrium" modulus depends on the previous shear rate, while molecular orientation decays to zero for *all* previous shear rates. A "high modulus" state results from previous shear rates ≥ 5 s^{-1}, while a "low modulus" state results from shear rates ≤ 1 s^{-1}. This transition region corresponds well with the shear rate range just past the first transition in orientation from low to moderate Δn seen in Figure 2b. If the previous shear rate exceeds this critical value, it appears that the resulting structure yields an additional contribution to the observed modulus.

Figure 5 shows birefringence and dynamic modulus as a function of time following flow cessation for the HPC/cresol solution. The most obvious characteristics of these data are similar to behavior seen in HPC/water. The orientation ultimately relaxes to an isotropic condition, on a time scale that approaches 100 s for high shear rates. Note that the inverse of this time scale defines a shear rate, below which flow is unable to impart significant orientation (Figure 2c). Understanding of what causes the decay in orientation may therefore be of significance in explaining the origin of low orientation observed in HPC solutions at low rates. Figure 5b shows that moduli in HPC/cresol solutions increase on time scales comparable to the decrease in orientation, as was seen in HPC/water. A higher initial orientation at higher rates results in a lower initial modulus. Like HPC/water, the system relaxes to a higher modulus state provided that the initial shear rate is past the first transition from low to moderate orientation.

When the shear rate is in the intermediate plateau in orientation, the orientation initially increases. Figure 5 shows that this temporary increase in Δn is accompanied by a slight decrease in modulus, as might be expected from behavior observed in PBG. Figure 6 shows the birefringence relaxation data for HPC/cresol, plotted as a function of time scaled by the previous shear rate. We see that the temporary increase in orientation follows the relaxation scaling law that is widely observed in PBG. It would appear that, in the tumbling regime, HPC would also prefer

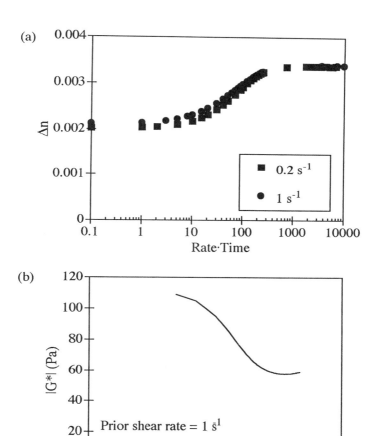

Figure 3. Relaxation in 13.5% PBG/cresol solution. (a) Δn as a function of time following cessation of flow at indicated shear rate. (b) Magnitude of complex modulus (2 Hz) as a function of time following flow cessation. (Adapted from ref. 9).

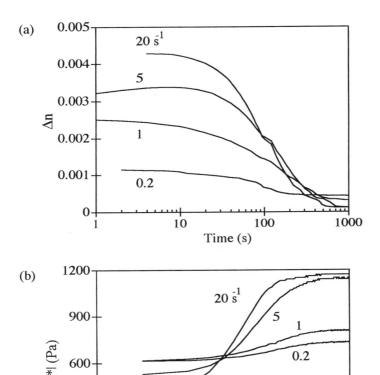

Figure 4. Relaxation in 50% HPC/water solution. (a) Δn as a function of time following cessation of flow at indicated shear rate. (b) Magnitude of complex modulus (2 Hz) as a function of time following flow cessation. (Adapted from ref. 15).

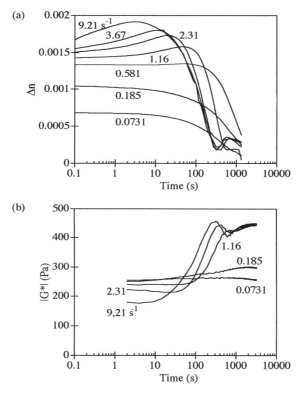

Figure 5. Relaxation in 27% HPC/cresol solution. (a) Δn as a function of time following cessation of flow at indicated shear rate. (b) Magnitude of complex modulus (2 Hz) as a function of time following flow cessation.

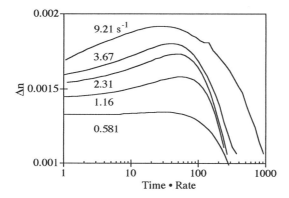

Figure 6. Birefringence relaxation data from Figure 5(b), replotted as a function of time scaled by previously applied shear rate.

to relax to a high orientation state like PBG, on a time scale that varies inversely with previous shear rate. However, another process intervenes that disrupts macroscopic orientation.

We have suggested that this disruption in orientation results from a transition back to a cholesteric phase in HPC solutions (15). If this suggestion is true, then weak orientation observed in steady shear flow of HPC at low rates would reflect the persistence of a cholesteric phase. This additional level of structure could possibly then play a role in causing region I shear thinning behavior in HPC solutions. However, one cannot directly conclude that low orientation at low rates *causes* the mechanical rheological effect of Region I shear thinning. Certainly, there are other materials that exhibit three-region viscosity curves where cholestericity is not an issue. Thus, any role of cholestericity in the low shear rate rheology of HPC solutions would only provide insights into the behavior of this particular material.

Conclusions

The structural and rheological response of HPC solutions in cresol appears to be intermediate between that observed in two other model systems, PBG and HPC/water. HPC/cresol exhibits two distinct transitions in orientation with increasing shear rate. The first occurs when shear flow is sufficiently strong to impart a significant degree of molecular orientation in the solution, while the second occurs at higher rates, and appears to be correlated with a sign change in the first normal stress difference; this second transition may thus be identified as a manifestation of the transition from tumbling to flow aligning behavior due to nonlinear viscoelastic effects. In relaxation, HPC/cresol shows an initial increase in orientation that occurs on times inversely proportional to the previous rate, like PBG solutions. However, at longer times, the orientation decreases to a globally isotropic condition, as observed in HPC/water. Like HPC water, (i) this decreasing orientation is accompanied by an increase in modulus, and (ii) the final modulus depends on the previously applied shear rate.

Acknowledgment. We gratefully acknowledge financial support from the National Science Foundation, Grant CTS-9213421, and through the MRL program, at the Materials Research Center at Northwestern, Grant DMR-9120521. Our initial studies on HPC/cresol resulted from a sample provided by Professor Jaye Magda.

Literature Cited

(1) Marrucci, G.; Maffettone, P. L. *Macromolecules* **1989**, *22*, 4076.
(2) Larson, R. G. *Macromolecules* **1990**, *23*, 3983.
(3) Magda, J. J.; Baek, S.-G.; DeVries, K. L.; Larson, R. G. *Macromolecules* **1991**, *24*, 4460.
(4) Ernst, B.; Navard, P.; *Macromolecules* **1989**, *22*, 1419.
(5) Grizzuti, N.; Moldenaers, P.; Mortier, M.; Mewis, J. *Rheol. Acta* **1993**, *32*, 218.

(6) Hongladarom, K.; Burghardt, W. R.; Baek, S. G.; Cementwala, S.; Magda, J. J. *Macromolecules* **1933**, *26*, 772.
(7) Moldenaers, P; Mewis, J. *J. Rheol.* **1986**, *30*, 567.
(8) Larson, R. G.; Mead, D. W. *J. Rheol.* **1989**, *33*, 1251.
(9) Hongladarom, K.; Burghardt, W. R. *Macromolecules* **1993**, 26, 785.
(10) Baek, S.-G.; Magda, J. J.; Larson, R. G. *J. Rheol.* **1993**, *37*, 1201.
(11) Picken, S. J.; Aerts, J.; Doppert, H. L.; Reuvers, A. J.; Northolt, M. G.; *Macromolecules* **1991**, *24*, 1366.
(12) Keates, P.; Mitchell, G. R.; Peuvrel-Disdier, E.; Navard, P. *Polymer* **1993**, 34, 1316.
(13) Dadmun, M.D.; Han, C. C. *Polym. Preprints* **1993**, *34*(2), 733.
(14) Chow, A.; Hamlin, R. D.; Ylitalo, C. *Macromolecules* **1992**, 25, 7135.
(15) Hongladarom, K.; Secakusuma, V.; Burghardt, W. R. *J. Rheol.* **1994**, *38*, 1505.
(16) Larson, R. G.; Promislow, J.; Baek S.-G.; Magda, J. J. in *Ordering in Macromolecular Systems*, Teramoto, A., Ed.; Springer: New York, 1993.
(17) Baek, S.-G.; Magda, J. J.; Cementwala, S. *J. Rheol.* **1993**, *37*, 935.
(18) Baek, S.-G.; Magda, J. J.; Larson, R. G.; Hudson, S. D. *J. Rheol.* **1994**, *38*, 1473.
(19) Sigillo, I.; Grizzuti, N. *J. Rheol.* **1994**, *38*, 589.
(20) Guido, S.; Di Maio, M.; Grizzuti, N. in *Progress and Trends in Rheology IV*, Gallegos, C. Ed.; Steinkopf: Darmstadt, 1994, pp. 308-310.
(21) Walker, L.; Wagner, N. *J. Rheol.* **1994**, *38*, 1525.
(22) Onogi, S.; Asada, T. in *Rheology*, Astarita, G.; Marrucci, G.; Nicolais, G. Eds.; Plenum Press: New York, 1980, pp. 127-147.

RECEIVED February 15, 1995

Chapter 22

Shear-Induced Orientation of Liquid-Crystalline Hydroxypropylcellulose in D_2O as Measured by Neutron Scattering

Mark D. Dadmun

Chemistry Department, University of Tennessee, Knoxville, TN 37996-1600

The results of a study that determined the change with shear rate of the orientation of a 50% solution of the semiflexible liquid crystalline polymer hydroxypropylcellulose (HPC) in deuterated water by small angle neutron scattering are presented. The results show that this system exhibits two regimes in its response to shear flow for the shear rates studied. The results are compared to a previous study of the change in orientation with shear rate of another model liquid crystalline polymer, poly(γ-benzyl L-glutamate) (PBLG) in deuterated benzyl alcohol. The HPC results are also correlated to rheological studies of a similar system. The interpretation of the shear rate dependence and crossover between regimes of both systems suggests that the local dynamics of the liquid crystalline polymer in solution plays a critical role in the orientation and rheology of these systems.

Liquid crystalline polymers (LCP) represent a unique class of materials which, in the mesophase, exhibit long range orientational order in at least one dimension and lack of positional order in at least one of the other two dimensions(1-3). The ordering (or lack of order) in the third dimension defines the type of liquid crystalline behavior exhibited. The lack of positional order in the mesophase gives rise to the liquid state of the system, while the orientational order results in anisotropic structure and ultimate properties of the material. The anisotropic nature of the system leads to exceptional properties, including mechanical, in the final material. It is well known that an increase in the amount of orientational order leads to an increase in the ultimate properties of the material. Therefore it is advantageous to maximize the amount of order present in an LCP material. An increase in the order of commercial LCP usually occurs during the processing of the polymer. In other words, the increased ordering is shear induced. Therefore, in order to optimize the properties and processing of commercial LCP, it is necessary to have a complete understanding of the coupling between the molecular orientation of an LCP and an applied shear flow.

There are many polymer architectures which exhibit liquid crystallinity. These include rigid rod type polymers, semiflexible polymers, or copolymers comprised of rigid and flexible monomers. Rigid rod polymers have a stiff backbone chain due to inflexibility of the chain (such as para linked aromatic rings), intermolecular interactions (such as hydrogen bonding), intramolecular interactions, or a combination of these. Semiflexible polymers, however, have more flexibility in their backbone structure and can therefore deviate slightly from linearity along the backbone. These polymers have a preferred axis but are not rigid. Copolymers can take many forms and their specific architecture will have a significant effect on the actual properties of the polymer system. Possible structures of these copolymers include side chain liquid crystals, which have a flexible backbone and a rigid side chain; random copolymers, a structure where the rigid and flexible segments are randomly distributed along the chain backbone; alternating copolymers, where the rigid backbone segments are uniformly placed along the backbone chain with flexible spacer groups; or block copolymers, where a rigid polymer chain is connected to a flexible chain at one point. It is obvious that each of these structures will respond differently to an applied field, such as a shear field, and an understanding of these differences is certainly needed in the pursuit of the optimization of the processing of commercial LCP. The goal of the current experiment is to examine the role of one structural parameter, LCP semiflexibility, on the alignment of the LCP by an applied shear flow. The results will be compared and contrasted to previous studies and rheological measurements in hopes of elucidating the importance of backbone rigidity on LCP flow behavior.

Previous Results

We have recently demonstrated the utility of in-situ shear small angle neutron scattering (SANS) to determine the change in alignment of a liquid crystalline polymer in solution with shear rate.*(4)* The system examined in this study was poly(γ-benzyl L-glutamate) (PBLG) in deuterated benzyl alcohol (DBA) at 65 °C. PBLG forms an α-helix in DBA and therefore the structure of the PBLG chain can be closely approximated as a rigid rod. The results of this study showed three regimes of the orientation of the LCP molecule with shear rate. At low shear rate, the alignment increases with shear rate until a first critical shear rate $\dot{\gamma}_1$ (≈ 2 s^{-1}) is reached. Above $\dot{\gamma}_1$, the molecular alignment does not change much with shear rate until a second critical shear rate, $\dot{\gamma}_2$ (≈ 20 s^{-1}). Above $\dot{\gamma}_2$, the orientation of the PBLG molecule again increases with increased shear rate. Therefore, for this polymer solution, there exists three regimes that describe the response of the LCP molecule to an applied shear force. This is depicted in figure 1 which shows the change in orientation of the LCP molecule as measured from the azimuthal peak

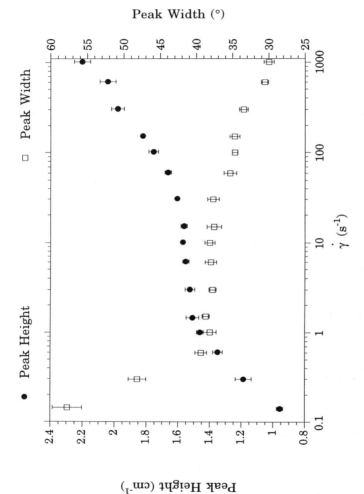

Figure 1. Change in the orientation of a PBLG molecule with shear rate as determined with neutron scattering. Adapted from reference 4.

height and peak width (see below for detail) as a function of shear rate. The three regimes are clearly evident and moreover, in reference 4, the critical shear rates $\dot\gamma_1$ and $\dot\gamma_2$ are correlated to the local relaxation mechanisms of a rod in solution. $\dot\gamma_1 = 1/\tau_0$ and $\dot\gamma_2 = 1/\tau_1$ where τ_0 is the rotational relaxation time as calculated from molecular parameters using the theory of Doi and Edwards(5) and τ_1 is the longest local relaxation time of the rodlike chain, the flexural relaxation time, which is determined from the ratio τ_0/τ_1 as estimated from the work of Ferry and coworkers.(6,7)

The results of the change in the alignment of PBLG/DBA with shear rate are also compared and contrasted to recent theoretical and experimental results. Although there exist some specific differences, the results generally agree with the recent results of other investigators. Most particularly, our definition of three regimes seems to correlate well with the theoretical results of Larson,(8) where he described the change with shear rate of the alignment of an LCP in terms of three regimes. According to Larson, at low shear rate the LCP molecule will tumble with flow. As the shear rate increases, it suppresses the tumbling motion of the molecule. The ensuing motion is what Larson called wagging. At much higher shear rates, the shear overcomes this wagging motion and this manifests itself as a further increase in the orientation of the molecule in the shear thinning regime. Our results and interpretation correlate well with the predictions of Larson in that three regimes are seen in terms of the alignment of the LCP with shear flow as well as the correlation of the changeover between regimes and the dynamics of a rod in solution. In our definition of regime 1, the shear is not fast enough to overcome the rotational motion of the LCP molecule in concentrated solution as $\dot\gamma < 1/\tau_0$, which agrees with Larson's tumbling regime. In our regime 2 the shear suppresses the rotational motion of the molecule but not the local dynamics of the molecule, $1/\tau_0 < \dot\gamma < 1/\tau_1$, which correlates well with Larson's wagging regime. Finally, in our definition of regime 3, the shear overcomes the local dynamics of the chain, $\dot\gamma > 1/\tau_1$, to further align the chain along the flow direction, agreeing with Larson's shear thinning regime. For further details and comparisons to the results of other groups, the reader is referred to the original paper.

An important point to note is that the work of Larson, and most other theoretical works that have sought to describe the response of an LCP to shear flow, have modeled the LCP as a thin rigid rod. As was described earlier, there exist many polymer structures which exhibit liquid crystalline behavior but which are not rigid rods. This experiment will determine the response of a semiflexible polymer to the application of a flow field using small angle neutron scattering. The results will be compared to the results of the rigid rod PBLG/DBA system in an attempt to further understand the role of flexibility on the alignment of LCP by shear flow.

Experimental Conditions

A 50% (wt.%) solution of hydroxypropylcellulose (HPC) in deuterated water (D2O) is studied. Both solvent and polymer were purchased from Aldrich. The molecular weight of the HPC reported by the manufacturer is 100,000. D20 and HPC were mixed in a centrifuge tube to prepare the solution. Gentle agitation over four weeks was needed to ensure complete dissolution. The system was then centrifuged at 4,000 RPM for 2 hours to eliminate bubbles. Scattering experiments were completed on the 8m small angle scattering instrument at the Cold Neutron Research Facility at the National Institute of Standards and Technology in Gaithersburg, MD with a sample to detector distance of 2.03 m and the wavelength of neutrons, λ, was 5 Å. For this configuration, the accessible q range (q= $4\pi/\lambda$ sin(θ/2), where λ is the wavelength of the neutrons and θ is the scattering angle) is 0.03 - 0.19 Å^{-1}. The shear cell utilized in this experiment is a couette cell with a copper cylindrical stator encompassed by a quartz cup which is rotated(9) with a gap between the stator and cup of 1 mm at room temperature. The scattering cell is designed to allow the *in-situ* collection of the scattering profile of a flowing polymer solution or melt. Sample temperature within the cell can be controlled to within ± 1 °C. In this cell, three mutually orthogonal axes can be defined as the direction of shear (x), the shear gradient direction (y), and the neutral or vorticity direction (z). In the scattering experiment, the neutron beam proceeds along the shear gradient or y-axis and the scattering is collected in the shear-vorticity plane (x-z). Therefore the scattering pattern and analysis are only completed in, and pertain to, the flow-vorticity plane. Any changes that occur along the shear gradient direction will be averaged or invisible in this scattering geometry. For clarity, this geometry is shown in figure 2. Figure 2a is a top view of the shear cell and 2b is a three-dimensional view of the scattering geometry. I in figure 2 is the incident beam of radiation in the scattering experiment.

Experimental procedure is as follows. The sample is loaded into the cell and allowed to equilibrate to avoid any loading effects. Scattering data are collected at room temperature only in the steady state regime, so the transient does not contribute to the scattering pattern. The scattering pattern is corrected for empty cell scattering, detector non-linearity, solvent scattering, sample transmission, background radiation, and incoherent scattering(10) and then reduced to absolute scattering intensities using silica gel as a secondary standard. The scattering data is then analyzed as described below to determine the change in the orientation of HPC with shear rate.

Experimental Results.

Two-dimensional scattering patterns were obtained at room temperature for 16 shear rates ranging from 0.02 to 10 s^{-1} at 6 shear rates per decade. Unfortunately, higher shear rates were inaccessible for this solution as sample loss was evident at higher shear rates. To demonstrate the change in the HPC system as the shear rate is

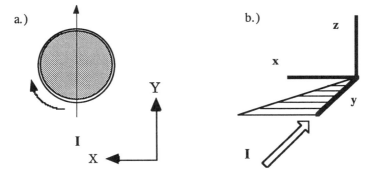

Figure 2. Schematic of the shear SANS cell. a.) Top view of scattering cell b.) 3-dimensional view of the shear-scattering geometry in our shear SANS experiment.

increased, sample scattering patterns for shear rates 0.041 s^{-1}, 0.426 s^{-1}, and 5.65 s^{-1} are shown in figure 3.

These scattering patterns qualitatively detail the changes in the HPC solution as the shear rate is increased. At 0.041 s^{-1}, the scattering pattern appears slightly anisotropic, suggesting that even at this low shear rate the shear is affecting the alignment of the LCP. As the shear rate is increased, the anisotropy in the scattering pattern becomes more pronounced, signifying an increase in the alignment of the LCP system with shear rate. This qualitative picture shows that the shear flow is indeed aligning the HPC molecule. However, a more detailed, quantitative description of the change in the orientation of the system with $\dot{\gamma}$ is desired. The analysis to complete this task is outlined below.

We are interested in quantifying the change in the orientational order in the HPC/D2O system as the shear rate is increased. A common method of determining the absolute alignment of an LCP from scattering experiments is to calculate the order parameter, S, from the breadth of the meridional peaks in a diffraction pattern that appears due to the nematic ordering of the LCP.(11,12,13) The order parameter, S, is defined as $S = (3<\cos^2\theta>-1)/2$ where θ is the angle between the director and the molecular axis, where the director is the average direction of order within a domain. Unfortunately, the meridional peaks cannot be obtained for the accessible scattering dimensions of these experiments. Consequently, the data have been analyzed such that the change in the molecular alignment with shear rate can be understood using the measured data. As we are interested in the change in orientation of the HPC molecule with shear rate, it is not necessary to determine the absolute amount of order present, only the relative change in anisotropy as the shear rate changes. Therefore, though the following analysis does not garner the absolute order present in the system, it allows an understanding of the change in orientation with shear rate. It is also worth reiterating that due to the geometry of this experiment, the orientation or alignment that is measured is only in the flow-vorticity plane. A thorough discussion of the utility of the following analysis has been completed elsewhere(4).

Due to the well known polydomain structure of LCP, a quiescent LCP solution, even in the liquid crystalline state, will scatter isotropically. As the polydomain structure and molecular structure is aligned by the shear flow, the two dimensional scattering pattern becomes anisotropic. By using neutron scattering, we are surveying the sample on molecular length scales. Therefore, a quantification of this anisotropy is a quantification of the molecular anisotropy. There is an assumption implicit in this statement, that the domain boundaries in this textured sample do not significantly contribute to the collected scattering patterns or, in other words, that the length scales probed by the neutrons are much less than the size of a domain. As neutron scattering probes domains of the order 1 - 100 nm, this assumption is warranted. An azimuthal average as shown in figure 4 is a well defined

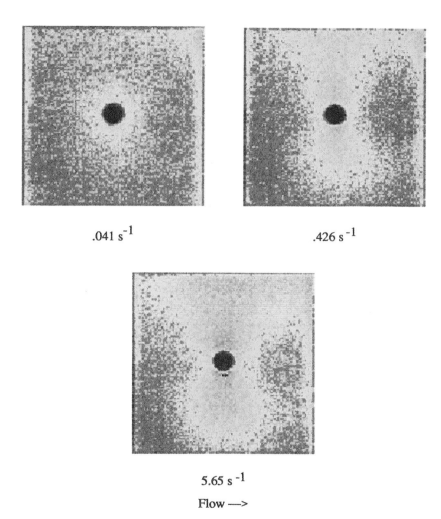

Figure 3. 2-dimensional scattering patterns for the HPC/D20 solution at room temperature. Shear rates are listed above. The accessible q-range from these scattering patterns is 0.03 Å^{-1} < q < 0.19 Å^{-1}.

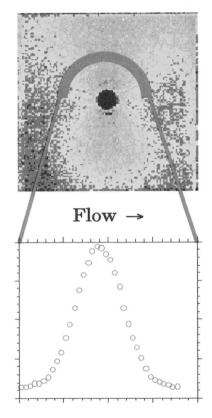

Figure 4. Depiction of the azimuthal average that is completed to quantify the anisotropy of a 2-dimensional scattering pattern.

quantification of the anisotropy of the scattering pattern. The corresponding curve will be a flat line if the scattering is isotropic and a peak if the system is anisotropic. The peak that occurs for an anisotropic scattering pattern has been fit to a Lorentzian to obtain a peak height and half width at half maximum. It can be shown that the peak width is a measure of the distribution of the LCP segments about the flow direction and the peak height is proportional to the number average of molecular segments that are oriented along the shear direction.(4,14) Both of these parameters are a measure of the alignment of the LCP by shear flow and can track the effect of shear rate on the molecular orientation. An example azimuthal trace and corresponding fit are shown in figure 5. This trace was averaged over an annulus that is 20 pixels wide centered at $q = 0.06$ Å^{-1} from the scattering pattern where $\dot{\gamma} = 9.41$ s^{-1}.

The change in the peak height and peak width of the azimuthal trace with shear rate are shown in figure 6. The peak height and width are determined from the fit to a Lorentzian of the azimuthal trace where the annulus as depicted in figure 4 is centered at $q = 0.06$ Å^{-1} and encompasses 20 pixels. The reported peak height is the fitted height minus the fitted baseline and the reported peak width is the width at half maximum as determined from the fit. An azimuthal trace was attempted at various q positions and annulus widths to ascertain the role of these parameters on the fit procedure. It was found that the value of q at which the fit was performed did not affect the trend of the change with shear rate, only the absolute value of the peak height and width, and even that only slightly. Therefore, $q = 0.06$ Å^{-1} was chosen arbitrarily. The pixel width was chosen to optimize statistics. Examination of figure 6 shows that the alignment of the liquid crystalline polymer HPC by shear flow shows at least two regimes. At low shear rates, $\dot{\gamma} < 1.0$ s^{-1} the alignment of HPC, as described by the peak height, increases with shear rate. For $\dot{\gamma} > 1.0$ s^{-1} up to the highest shear rate studied, the orientation of the LCP molecule does not change much with shear rate. The change in the peak width also shows a decrease in the distribution of the alignment of the LCP with shear rate. The changes described above are reminiscent of the first two regimes that PBLG/DBA exhibited in our previous study (see figure 1).

Discussion

Before the results are discussed, an important point concerning the scattering experiment itself should be discussed. Due to the relatively low intensity of neutrons striking the sample, large sample windows (≈ 12 mm diameter) and long sample times (≈ 2 hr) are necessary to obtain sufficient statistics in the collected scattering patterns for a detailed analysis such as the one described above. Due to these long sample times and scattering volumes, all parameters determined from the scattering

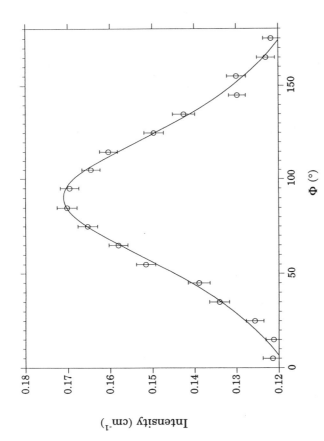

Figure 5. An example of the azimuthal trace used to quantify the anisotropy of the collected scattering pattern and the corresponding fit to a Lorentzian. This trace is an average over a 20 pixels centered at $q = 0.06$ Å^{-1}. The shear rate for this plot is 9.41 s^{-1}.

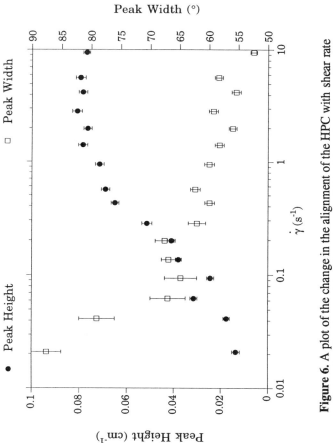

Figure 6. A plot of the change in the alignment of the HPC with shear rate as determined by the peak height and peak width of the azimuthal trace.

patterns are a time average or ensemble average of the given parameter, not necessarily a snapshot picture of the system at any given time. The only drawback to this technique is that it is not possible to examine or determine the contributions to this average state.

The results shown in figure 6 detail the change in the orientation of this HPC solution as the shear rate changes. At $\dot{\gamma} = .02$ s^{-1}, there is a peak present which indicates that the shear, even at this low shear rate, affects the alignment of the LCP molecule. As the shear rate is increased, the orientation increases until $\dot{\gamma} = 1.0$ s^{-1}. Above $\dot{\gamma} = 1.0$ s^{-1} the peak height, and therefore the orientation of the LCP molecule in solution, does not change with shear rate. The peak width also demonstrates a change in the distribution of the alignment of the LCP with shear rate, signifying a narrowing of the LCP orientation about the flow direction as the shear rate increases. When compared to the PBLG/DBA results, the data in figure 6 suggests that this solution is in the previously described regime 1 (tumbling) when $\dot{\gamma} < 1.0$ s^{-1}, a transition to regime 2 occurs at $\dot{\gamma} = 1.0$ s^{-1} and for shear rates greater than 1.0 s^{-1}, the system is in regime 2 (wagging).

It would be desirable to correlate the change in the orientation of the HPC system to the local dynamics of HPC in concentrated solution. However, this can not be completed as easily as for the PBLG/DBA system. This is because there is no current theory which predicts τ_0 and τ_1 for a semiflexible molecule in concentrated solution. Utilizing theoretical predictions, however, does allow the correlation of the scattering results to rheological results. The transition that is observed in the HPC system, from a regime where the alignment changes with shear rate to one that shows invariance in the orientation with shear rate, seems to correspond to the transition from a tumbling to a wagging mechanism as defined theoretically and for the PBLG/DBA system. Theoretical descriptions of the orientation transitions of an LCP in solution correlate the transition from tumbling to wagging to the shear rate where the transition from a positive to a negative first normal stress difference (N_1) is seen.(8,15) Therefore, according to our physical description of the changes that are observed in the orientation of an LCP by shear flow, namely that the shear rate dependence of the molecular orientation is governed by the local dynamics of the LCP in solution, this solution should show a change from a positive to a negative first normal stress difference at $\dot{\gamma} = 1.0$ s^{-1}. Fortunately, Grizzutti et al.(16) have examined the shear rate dependence of the first normal stress difference for a very similar system, a solution of 50% (wt.%) of HPC in water. The molecular weight of the HPC in their study was 100,000 and was purchased from Aldrich, exactly as our system. The only difference between the two studies is the deuteration of the solvent in our experiments. If we assume that the rheological properties are independent of isotopic substitution, it can be expected that these results will mimic the rheology of our sample. In this carefully completed study, a changeover from a positive to a negative N_1 was found to occur at $\dot{\gamma} = 1$ s^{-1}, in agreement with our predictions. This supports the physical description for the response of an LCP in solution to an applied shear flow that has been put forward to explain the neutron scattering results. Namely, the orientational response of an LCP molecule to an applied shear flow is governed by the local dynamics of the molecule in solution.

A comparison of the amount of order present in the rigid rod and semiflexible systems is also of interest. This can be completed through a comparison of the peak heights and widths of the two systems for similar shear rates. A comparison of the peak width of the PBLG and HPC solutions, which measures the distribution of the LCP alignment about the flow direction, for similar shear rates shows that the width of the PBLG peak is almost half that of the HPC peak. This demonstrates that the PBLG molecule is more readily ordered by a shear flow than the semiflexible HPC molecule. The peak height of the azimuthal trace of the HPC solution can also be compared to the same data for the PBLG/DBA solution. However, the intensities must first be corrected for concentration and neutron scattering length differences. The absolute scattering intensity reported in figures 1, 5, and 6 in cm^{-1} is the total cross section (dΣ/dΩ) defined as

$$\frac{d\Sigma}{d\Omega} = V_w \phi \Delta\rho^2 F(q) \qquad (1)$$

where V_w is the molecular volume, ϕ is the concentration, $\Delta\rho$ is the neutron scattering contrast factor, and F(q) is the form factor, which is anisotropic in our systems.*(17)* Therefore, normalizing the peak intensities by the scattering density contrast factor, molecular volume, and concentration, $\Delta\rho^2 V_w \phi$, allows the determination of the anisotropy in the form factor for each system. This normalized anisotropy can then be compared for the two systems. For the PBLG system, $\Delta\rho$ = 2.8 x 10^{10} cm^{-2}, V_w = 3.061 x 10^{-19} cm^3/molecule, and ϕ = 0.20. Equivalent values for the HPC system are $\Delta\rho$ = 4.92 x 10^{10} cm^{-2}, V_w = 1.350 x 10^{-19} cm^3/molecule, and ϕ = 0.50. Therefore, the normalization factors are 47.9 and 163.4 for the PBLG/DBA and HPC/D2O solutions, respectively. A comparison of the value of the peak intensities in figures 1 and 6 shows that the intensity of the PBLG peak is at least an order of magnitude greater than that of the HPC sample. The normalization described above increases this difference by a factor of 3. Therefore, the anisotropy in the PBLG solutions is much greater than that present in the HPC solution for corresponding shear rates. This comparison also shows that the PBLG molecule is more readily ordered by an applied shear flow than the HPC molecule. This is indubitably due to the difference in stiffness of the rigid PBLG molecule and the semiflexible HPC molecule.

It is interesting to note that although there seem to be differences in the extent of the alignment of the semiflexible HPC molecule and the more rigid PBLG molecule, there appear to be similar transitions in the orientational response of the LCP with shear rate which could be related to the local rotational dynamics of the LCP in solution. A more complete understanding of the role of polymer architecture is therefore needed. Current experiments are underway to examine the role of semiflexibility on the orientational response of an LCP to shear flow. A useful starting point may be a theoretical description of the local dynamics of a semiflexible polymer in concentrated solution.

Acknowledgments

We would like to thank Drs. Charles Han and Alan Nakatani for help in the completion of these experiments as well as very useful discussions during the analysis of this data. We would also like to thank the National Research Council for support during which some of these experiments were completed. We would also like to thank Dr. K. Migler for bringing reference 14 to our attention.

Literature Cited

1. Blumstein, A.; Hsu, E.C. *Liquid Crystalline Order in Polymers*, Academic Press: NY, 1978.
2. Ciferri, A.; Krigbaum, W.R.; Meyer, R.B. *Polymer Liquid Crystals*, Academic Press: NY, 1982.
3. Zachariades, A.E.; Porter, R.S. *High Modulus Polymers Approaches to Design and Development*, Marcel-Dekker: NY, 1988.
4. Dadmun, M.D.; Han, C.C. *Macromolecules* **1994**, 27, 7522.
5. Doi, M.; Edwards, S.F. *The Theory of Polymer Dynamics*; Oxford University Press: New York, 1986.
6. Warren, T.C.; Schrag, J.L.; Ferry, J.D. *Biopolymers* **1973**, *12*, 1905.
7. Rosser, R.W.; Schrag, J.L.; Ferry, J.D.; Greaser, M. *Macromolecules* **1977**, *10*, 978.
8. Larson, R. *Macromolecules* **1990**, *23*, 3983.
9. Nakatani, A.I.; Kim, H.; Han, C.C. *J. Res. Natl. Inst. Stand. Technol.* **1990**, *95*, 7.
10. Hayashi, H.; Flory, P.J.; Wignall, G.D. *Macromolecules* **1983**, *16*, 1328.
11. Picken, S.J.; Aerts, J.; Visser, R.; Northolt, M.G. *Macromolecules* **1990**, *23*, 3849.
12. Picken, S.J.; Aerts, J.; Doppert, H.L.; Reuvers, A.J.; Northolt, M.G. *Macromolecules* **1991**, *24*, 1366.
13. Keates, P.; Mitchell, G.R.; Peuvrel-Disdier, E.; Navard, P. *Polymer*, **1993**, *34*, 1316.
14. Ao, X.; Wen, X; Meyer, R.B. *Physica A*, **1991**, *176*, 63.
15. Marruci, G.; Maffettone, P.L. *Macromolecules* **1989**, *22*, 4076.
16. Grizzuti, N.; Cavella, S.; Cicarelli, P. *J. Rheol.* **1990**, *34*, 1293.
17. Higgins, J. in *Treatise on Mat. Sci. and Tech.*; Kostorz, G., Ed.; Academic Press: New York, 1979, Vol. 15.

RECEIVED March 16, 1995

Chapter 23

Shear-Induced Alignment of Liquid-Cystalline Suspensions of Cellulose Microfibrils

W. J. Orts[1], L. Godbout[2], R. H. Marchessault[3], and J. F. Revol[2]

[1]Reactor Radiation Division, National Institute of Standards and Technology, Building 235, Room E151, Gaithersburg, MD 20899
[2]PAPRICAN and [3]Chemistry Department, Pulp and Paper Research Centre, McGill University, 3420 University Street, Montreal, H3A 2A7 Quebec, Canada

Small angle neutron scattering, SANS, was used *in situ* to determine the anisotropic shear ordering of polyelectrolytic, liquid crystalline cellulose microfibrils in aqueous (D_2O) suspension. The inter-rod distance of microfibrils under shear scales with $C^{-1/2}$ at high concentration, C. Orientational order parameters were calculated from the angular spread of the anisotropic SANS scattering peaks. Changes in the order parameter as a function of shear rate correlate to changes in the viscosity/shear rate profile. Particle alignment increases with increasing shear rate, yet the highest degree of alignment for microfibrils (with axial ratios of ~45) was observed a short period after the cessation of shear flow. Relaxation occurred far more rapidly for shorter microfibrils with axial ratio of ~30.

Fibers derived from liquid crystalline polymeric materials have generated much interest due to the promise of obtaining high tensile strengths by aligning stiff mesogen groups during processing. The full potential for fibers from many liquid crystalline polymers (LCP's) has not been realized due, in part, to the complexity of these solutions under shear. Such complexities include defect textures with characteristic length scales varying from an angstrom to a micron scale (*1,2*), a strong dependence of rheology on the shear history, tumbling of the director at low shear rates (*3-5*), and unusual effects upon cessation of shear such as increased molecular ordering (*6*) and the formation of banded structures (*7-9*).

A new generation of regenerated cellulose fibers, such as Tencel (*10*) displays unique properties due to liquid crystalline alignment of cellulose microfibrils during processing (*11*). Cellulose microfibrils are the basic "building block" of native cellulosic materials and the raw material for a wide range of regenerated cellulose products such as rayon and cellophane, and some cotton fibers. Besides their industrial significance, aqueous suspensions of cellulose microfibrils provide an interesting model system to study LCP's under shear due to their relative simplicity .

Microfibrils which form stable colloidal dispersions are obtained by sulfuric acid hydrolysis of various natural celluloses, including wood pulp and cotton fiber (*12,13*). They are roughly square in cross section, and for wood, have widths of 40-60 Å and lengths of 1500-3000 Å. Each crystalline rod consists of ~80 cellulose chains packed in

This chapter not subject to U.S. copyright
Published 1995 American Chemical Society

a crystalline array (*14*). Although, the microfibrils appear as long "whiskers" with square cross sections in electron micrographs, they most likely contain some asymmetry. A slight helical twist of a given handedness along their rod axes has been proposed to explain why suspensions of the microfibrils form cholesteric rather than nematic phases (*15,16*).

The majority of reports of LCP solutions under shear have been by optical methods (*2,17-23*) which probe relatively large, micron-scale textures. Recently there has been an increased emphasis on studying the interaction of LCP's under shear at a smaller scale by small angle x-ray scattering (SAXS) (*2,24*) and small angle neutron scattering (SANS) (*25-31*). These studies have dealt with gels (*25*), block copolymers (*28,30*), polymer blends (*29*) and micelles (*27,31*). None have dealt with stiff polyelectrolyte rods.

The purpose of this report is to describe the ordering of aqueous cellulose microfibrils under shear using SANS. Microfibrillar rods in colloidal suspension are a good, practical model system for describing LCP behavior under shear since (i) the microfibrils are reasonably well characterized, (ii) they are polyelectrolytic due to a charged surface resulting from sulfation of hydroxyl groups during acid hydrolysis, (iii) the phase diagrams describing isotropic to biphasic to fully cholesteric transitions can be obtained, and (iv) microfibril dimensions can be varied controllably by varying the source of cellulose and by careful fractionation.

A common feature of SANS from ordered polyelectrolyte solutions and suspensions is interference peaks indicative of anisotropic alignment of the mesogens (*32-34*). From the widths of these peaks, an order parameter for the system can be calculated. In this report, changes in the relative order of the cellulose microfibrils will be calculated from the peak widths under different shear conditions, at different concentrations, and upon the cessation of shear flow. We will also determine if the phenomenon of increased orientation observed for semi-flexible chain LCP's after shear cessation are found in suspensions of stiff microfibrils.

Experimental

Materials Samples were prepared, as described previously (*13,14*), by sulfuric acid hydrolysis of cellulose fibers from black spruce bleached kraft pulp. Prior to hydrolysis, the samples were disintegrated in a Wiley (*10*) mill to pass through a 20 or 40 mesh screen and then treated in 60-64% (w/w) sulfuric acid at temperatures ranging from 45 to 70 °C for times ranging from 10 min. to 3 hours. Suspensions were rinsed in distilled water, concentrated by centrifugation and dialyzed until the charge in the washing water was neutral; the last traces of free sulfuric acid were then removed by mixed-bed resin. The bound sulfur content was determined by titration and confirmed by elemental analysis, and corresponded to a surface charge coverage of ~0.2 sulfate groups per nm^2 (~10% of the glucose residues on the surface are sulfated). Suspensions at intermediate concentrations, ~1.5-7 % (w/w), separate into two phases upon standing -- an upper isotropic phase and a lower birefringent phase which is cholesteric. Phase separation was used to fractionate the samples according to rod length, L, since the longer rods segregate to the lower cholesteric phase. The final steps in sample preparation were neutralization with NaOH and freeze drying.

Small Angle Neutron Scattering The neutron scattering data were collected on the NIST NG3 beamline in Gaithersburg, MD using collimated neutrons with a wavelength, λ, of 5 Å ($\Delta\lambda/\lambda = 0.15$). The freeze-dried cellulose microfibrils were suspended in D_2O and dispersed with ultrasound at concentrations ranging from 1-8% (w/w). Shear experiments were performed *in situ* using a quartz Couette shear cell specially designed and built by G.C. Straty, and described elsewhere (*35*). A schematic top view of the shear cell geometry is shown in Figure 1. Samples were placed in the

0.5mm gap between two ~65mm diameter concentric quartz cylinders. The outer cylinder, the rotor, was rotated at a set rate while the inner cylinder remained stationary. As shown in Figure 1, there are two distinct neutron beam paths through the cell; (i) across the fluid flow direction (the radial path) parallel to the direction of the shear gradient, and (ii) parallel to the flow (or tangential to the concentric cylinders). For the beam path perpendicular to the flow direction, a 1.25 mm diameter circular aperture was used to collimate the beam, while for the tangential path, a 2 x 12mm vertical rectangular slit was used. Note that the scattering volume for the tangential path samples is a curved, asymmetric annulus which is narrower than the beam. No attempt was made to account for this asymmetry or to correct for differences in scattering volumes when comparing data from the two beam path geometries. Temperature was controlled to ± 0.1 °C by circulating heating fluid in contact with the stator.

Calculation of the Order Parameter The order parameter, S, was calculated from SANS data by determining the azimuthal peak width at the peak position. This is in accordance with a technique put forth by Oldenbourg *et al.* in x-ray studies of nematic alignment of tobacco mosaic virus (*36*). The azimuthal peak traces for 6% (w/w) suspensions of cellulose microfibrils in D_2O at several shear conditions are shown in Figure 2. Peak sharpening corresponds to an increase in order. To quantify this, the azimuthal peak width intensity $G(\Phi)$ along an arc defined by the angle Φ from the equator can be related to the orientational probability distribution, $f(\beta)$, of finding a rod tilted at an angle β with respect to the director. The azimuthal peak width is then,

$$G(\Phi) = \int I_s(\omega) \, f(\beta) \sin\omega \, d\omega \qquad (1)$$

where $I_s(\omega)$ is the single rod intensity function of a rod tilted at an angle ω with respect to the incoming beam direction. Several assumptions can be made to simplify this expression (*36*). First, $f(\beta)$ was assumed Gaussian with a peak width of α which is small (i.e. the angular spread of the rod axis around the director is small). Second, by noting that the neutron beam direction is perpendicular to the director (or shear flow direction), the relationship between angles Φ, ω, and β is simply $\beta = \cos^{-1}(\cos\Phi \sin\omega)$. With these simplifications $G(\Phi)$ becomes (*36*)

$$G(\Phi) = A \exp\left(-\sin^2 \frac{\Phi}{2\alpha^2}\right) \left[\frac{1}{\cos\Phi} + \frac{\alpha^2}{2\cos^3\Phi} + \ldots\right] \qquad (2)$$

where A and α are adjustable fitting parameters. The parameters A and α are used to define the Gaussian orientational distribution $f(\beta)$, which is then used to calculate the order parameter S by

$$S = \int f(\beta) \frac{1}{2}(3\cos^2\beta - 1) d\beta \, . \qquad (3)$$

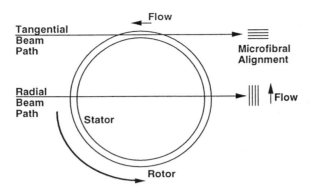

Figure 1. A schematic top view of the Couette shear cell used to measure the *in-situ* SANS shear scattering patterns. The two beam paths, radial and tangential, are perpendicular and parallel to flow direction respectively.

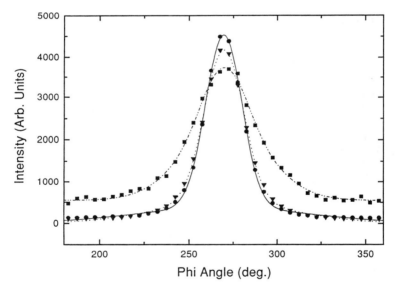

Figure 2. Azimuthal radial scans of the lower quadrant scattering data at a constant qm of 0.0240 Å^{-1} for 6% suspensions of acid-hydrolyzed (black spruce) cellulose microfibrils at shear rates of (■) 100 s^{-1} and (▼) 7000 s^{-1}. In order to study the relaxation process, an azimuthal scan was obtained (●) 7.5 minutes after the cessation of shear. The curves are the best-fits to the model (described in the experimental section) to calculate the order parameter. The error bars for the data are equal in size or smaller than the symbols used.

Results and Discussion

Shear Induced Orientation The nematic ordering of cellulose microfibrils in D_2O under shear was characterized by SANS in two directions -- parallel and perpendicular to the flow direction. Figure 3a shows a typical contour plot of 2-D SANS data collected at a shear rate of 7000 s^{-1} with the beam direction perpendicular to the flow direction. The distinct interference peaks correspond to alignment of rods parallel to the flow direction and are at a maximum scattering vector, q_m, of 0.0282 Å^{-1}, where q is the scattering vector; $q = (4\pi/\lambda)\sin\theta$, 2θ is the scattering angle, and λ is the incident neutron wavelength. The scattering pattern from the same sample using the tangential beam path geometry is an isotropic ring pattern, as shown in Figure 3b. A maximum occurs at the same q_m as in Figure 3a. (Note that the scattering data is skewed from right to left due to the asymmetric shape of the scattering volume). Since the beam path is parallel to the rod axis in this configuration, these results are indicative of isotropic local inter-rod spacing. This is of particular note since it is in contrast to the inter-rod packing of this sample under quiescent conditions, in which a cholesteric phase is present. SANS studies of the formation of the cholesteric phase using magnetic alignment showed that inter-rod packing is tighter along the cholesteric axis than perpendicular to it (*16*). Under shear flow the cholesteric phase is disrupted, and we see only an average nematic spacing.

At higher concentrations, inter-fibrillar spacing decreases and the peak position, q_m, increases with increasing concentration as predicted by simple scaling arguments. According to predictions by de Gennes (*37*) and Odijk (*38*), q_m scales with $C^{1/3}$ well below the overlap concentration, C^*, and scales with $C^{1/2}$ well above C^*. Figure 4 is a plot of the dimensionless quantities q_mL vs. $(C/C^*)^{1/2}$ showing that the peak position scales linearly with $C^{1/2}$ above C^*. The concentration at which the scaling of q_m changes from a $C^{1/2}$ to a $C^{1/3}$ dependence changes from system to system depending on the rod stiffness (*34*). Kanaya *et al.* (*39*) showed that this transition occurs at $C/C^* \approx 10$ for linear flexible polyelectrolyte chains in solution. For DNA, this transition occurs very close to $C/C^* = 1$; the difference is due to the much greater stiffness of DNA helical rods. In the data in Figure 4 there is no indication of a transition from $C^{1/2}$ to $C^{1/3}$ scaling. This is not surprising considering that the flexibility of cellulose microfibrils is more similar to that of DNA, which would correspond to a transition nearer $C/C^* = 1$. To better study this transition, a technique such as light scattering must be used to probe the length scales encountered at lower concentrations.

Shear Rate Effects on Orientation The increase in particle orientation with increasing shear rate is evident as increasingly sharper SANS peaks. This can be followed qualitatively in Figure 5 where the peaks, which are perpendicular to the flow direction, are more distinct as the shear rate is increased from 0.1 to 7000 s^{-1}. Quantitatively, the increase in ordering with increasing shear rate is described by the order parameter, S (which is calculated as described in the experimental section). There are two, and perhaps three, distinct regions in the behavior of S vs. shear rate (see Figure 6a). At shear rates below ~0.2 s^{-1} there is some indication that S increases with shear rate at a relatively slow rate. Between shear rates of 0.2 and ~100 s^{-1}, S increases more rapidly, and above 100 s^{-1}, S again increases relatively slowly. It should be noted that the samples were anisotropically aligned before initiation of the shear flow due to the injection of the sample into the Couette shear cell. (The solid curve in Figure 6a corresponding to the first fraction reflects this by starting at $S \approx 0.4$, the calculated order parameter at a shear rate of zero). Shear history is difficult to "erase" for these sample, since it takes days for anisotropic ordering to relax. Although heating is often used to promote relaxation in other systems, it is inappropriate for cellulose microfibrils since

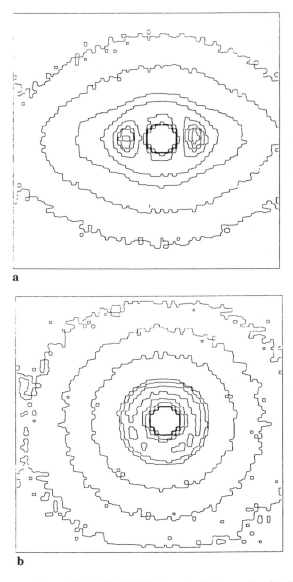

Figure 3. Contour plots of the 2-D SANS data of cellulose microfibril suspensions (5% in D_2O) taken at a shear rate of 7000 s^{-1} employing (a) the radial beam path in which the shear direction is left to right (two distinct reflections are observed) and (b) the tangential beam path in which the neutron beam path is parallel to the shear direction. In (b) the intensity is higher on the left side relative to the right because the scattering volume sampled using the tangential beam path is not uniform due to curvature of the Couette cell.

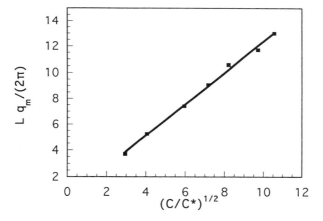

Figure 4. A plot showing the scaling relationship between the SANS peak maximum, q_m, in terms of the dimensionless quantity, $Lq_m/2\pi$ and the dimensionless concentration, $(C/C^*)^{1/2}$, for cellulose microfibrils (black spruce, first fraction) at a shear rate of 7000 s^{-1}. The slope of the linear fit is consistent with hexagonal packing.

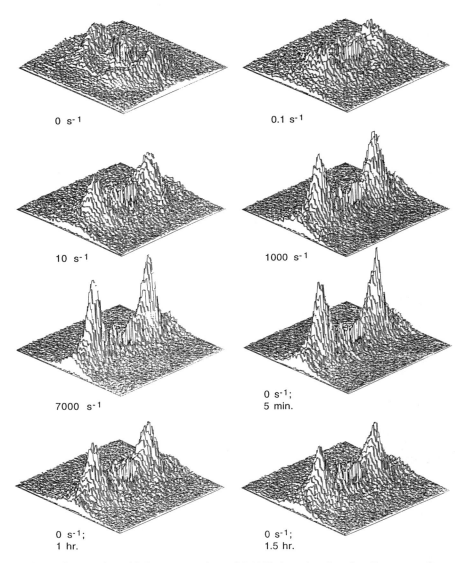

Figure 5. A series of 3-D contour plots of SANS data showing the alignment of cellulose microfibrils (5% in D_2O) with increasing shear rate and then the relaxation of this ordering after the cessation of shear flow. The reflections, which are perpendicular to the shear flow direction, become more prominent with increasing shear rate. Initially, after the cessation of shear, peaks sharpen slightly, indicating an increase in anisotropic alignment, followed by relaxation of this anisotropy with time.

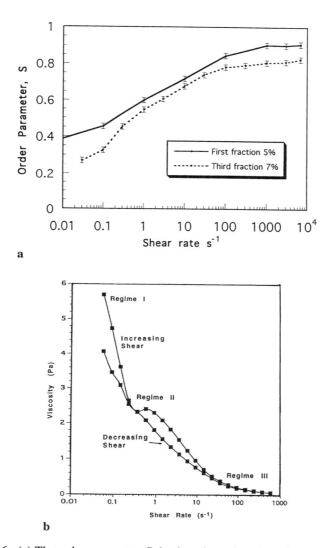

Figure 6. (a) The order parameter, S, is plotted as a function of shear rate for samples differing in microfibril length, i.e. ~2800 Å (solid curve) and ~1800Å (dotted curve) (The curves are included to guide the eye). In (b), the viscosity versus shear rate is plotted for the first fraction sample (5% in D_2O), showing a shear thinning region, a plateau, and a second shear thinning region.

hydrolysis of the sulfate ester groups on the crystal surface can occur at temperatures as low as 45 °C, promoting aggregation.

The microfibril rod length affects the relative order achieved. The two curves in Figure 6a display shear ordering from different fractions of the same source differing mainly in rod length. Microfibril lengths were 2800 and 1800 Å for the first and third fractions respectively. A higher concentration was used for the third fraction so that the viscosities of the two samples would be equivalent. Clearly, SANS measurements from this system can be used further to study the specific role of changes in the axial ratio in determining relative ordering.

The viscosity vs. shear rate profile for the first fraction, shown in Figure 6b, has three distinct regions which coincide with the changes in S with shear rate. These three regions are qualitatively in keeping with the theoretical description outlined by Onogi and Asada (*1*) of the rheological properties of liquid crystalline polymers in solution. At very low shear rates there is a shear thinning region where domains begin to flow against each other (regime I). At intermediate shear rates there is a plateau in the viscosity vs. shear rate profile where domains start to break up (regime II), and at higher shear rates there is further shear thinning where individual rods align further (regime III). When the ordering is measured on the macromolecular scale by SANS, the most rapid alignment of rods occurs in regime II. There is still an open question in the literature about whether the order reaches a plateau in the shear thinning regime. X-ray measurements by Picken *et al.* (*2*) of the order parameter for polyaramid solutions show that order does not increase with increasing shear in the shear thinning region (regime III). In contrast, the ordering for hydroxy cellulose solution (*21*) and for PBG gels (*22*) continues to increase in this shear thinning regime.

Concentration Dependence of Shear Ordering The concentration dependence of the order parameter for cellulose microfibrils in D_2O at $7000s^{-1}$ shows a significant increase at low concentrations and a plateau at a concentration of 0.04-0.05 (w/w) (see Figure 7). The measured inter-rod spacing at this composition is 280-340 Å. Noting that these microfibrils are charged polyelectrolytes, with roughly 10% of the surface consisting of sulfate ester (charged) groups, the structure and relative ordering of these suspensions must be influenced by the electric double layer around each rod, rather than the hard-cylinder diameter. Revol *et al.* (*14*) predicted that the effective repulsive layer for cellulose microfibrils at these concentrations is ~120 Å. This suggests that the order parameter levels off at a distance at which the two repulsive ionic "clouds" would begin to overlap. To test this hypothesis further, the ionic concentration of the suspension was varied. The peak position proved insensitive to increases in salt up to NaCl concentrations of ~0.1M, at which point the peak disappeared due to aggregation. This behavior is qualitatively similar to that observed by Wang and Bloomfield (*34*) for DNA whereby salt had little influence on the spacings between stiff DNA helices until the point of aggregation. Their interpretation was that the effective diameter is predominately determined by the uncondensed Na^+ attached to the outer layer. Podgornik *et al.* (*40*) have provided an effective means to investigate the role of ions on the shielding or "hydration forces" between ordered mesogens by measuring osmotic pressure in conjunction with measuring the inter-rod spacing by x-ray. In future work, the interactions of cellulose microfibrils will be studied using this combination of techniques.

Cessation of Shear Behavior One of the most noteworthy observations of this study is that, for the longer microfibrils (i.e. the first fraction sample), ordering increases after the shear flow has ceased. A comparison of the azimuthal arc traces in Figure 2 reveals that peaks are sharper in the scattering data collected ~8 minutes after the cessation of shear flow than those collected at a shear rate of $7000s^{-1}$ (scattering data for each sample was collected for 5 minutes and the noted time is the average duration after the cessation of shear. The calculated order parameters from these data are 0.86±02 during shear at 7000s-1 and 0.91±0.02 after shear cessation.

The relaxation behavior differs greatly for long and short rods. In Figure 8, the order parameter is plotted as function of time after the cessation of flow for the first fraction (longer rods) and the third fraction (shorter rods). In the abscissa, the units are made dimensionless by multiplying the relaxation time by the last shear rate, 7000 s^{-1}. Whereas the relaxation process for the longer rods is complex, with the order parameter increasing for a brief period followed by a slow, nonlinear decrease, the relaxation process for the shorter rods is less complex, as described by a rapid decrease in the order parameter with no indication of any increase in anisotropic ordering.

Increased molecular orientation during relaxation of a sheared solution has been observed previously by Hongladarom and Burghardt (6) in liquid crystalline solutions of poly(benzyl glutamate). They used polarized light to measure flow birefringence and saw an increase in the average molecular orientation, as well as an increase in order at a length scale similar in magnitude to banded structures. One significant difference their study and the present study is that Hongladarom and Burghardt confined their flow rates to the linear low shear rate regime, where director tumbling during flow increases the disorder. In the present study, shear rates are significantly higher (i.e. in regime III) where director tumbling has not been observed.

A reasonable assumption is that the lowest energy state for this system is rods which are well aligned. Shearing "pumps" energy into the system, which disrupts full alignment and manifests itself in the form of defects and disclinations. Winey *et al.* (*21*) showed such disclinations or defects in electron micrographs of shear-aligned block copolymers. With the removal of shear, the system to returns to its lowest energy state, changing the population of such defects or disclinations. Another consideration to explore is the role of undulations or rod "wobbling" during shear. It has been proposed for DNA (*40,41*) that, as polyelectrolyte rods are allowed to move apart (i.e. by decreasing the concentration), they become better able to undulate. The progressive increase in undulation plays an increasingly significant role in the entropic confinement of rods by essentially increasing the effective diameter of the particle. One can imagine that the degree of undulation is a function of the shear rate, considering that rods in a Couette shear geometry are under constant torque in the gradient direction. After the cessation of shear, the torque is removed, changing the degree of undulation. One final point is that these cellulose microfibril suspensions readily form cholesteric structures under quiescent conditions. Formation of a cholesteric phase has been seen to affect shear relaxation behavior (*6*), but the relationship between cholesteric ordering and ordering after shear flow is still not understood.

In summary, we have performed *in situ* SANS measurements of the shear alignment of polyelectrolytic, liquid crystalline cellulose microfibrils in aqueous suspension. Anisotropic SANS scattering patterns exhibit inter-particle interference peaks perpendicular to the shear flow direction, which become more prominent and sharper with increasing shear rate. This is indicative of increased alignment of rods in the flow direction. The position of these peaks at q_m scales with $C^{1/2}$ following scaling behavior expected for hexagonally packed polyelectrolytes at concentrations well above the critical overlap concentration. The relative order of cellulose microfibrils at a fixed shear rate increases with increasing concentration at low concentrations, and reaches a maximum degree of alignment at an intermediate concentration. This concentration roughly corresponds to the distance at which the ionic double layers around each microfibril begin to overlap. The highest degree of particle alignment for rods with sufficiently high axial ratios (~45) was observed a short period after the shear flow ceased. Although this increase has been observed in solutions of semi-flexible liquid crystalline polymers in solution (*6*), it has now been extended to polyelectrolytic crystallites. For shorter rods, with axial ratios of 30-35, no increased alignment was

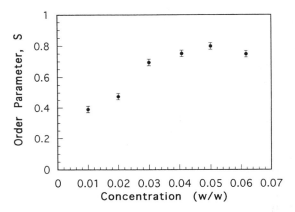

Figure 7. The concentration dependence of the order parameter, S, is plotted for cellulose microfibrils (black spruce, first fraction) at a shear rate of 7000 s^{-1}, showing a plateau at higher concentrations.

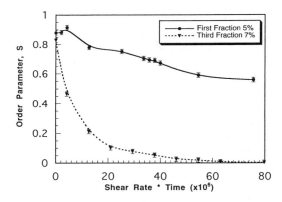

Figure 8. The order parameter, S, is plotted as a function of time after the cessation of shear flow for (●) relatively long cellulose microfibrils, i.e. the first fraction at 5% (w/w) and (▼) shorter microfibrils, i.e. the third fraction at 7% (w/w). Time is multiplied by the previous shear rate of 7000 s^{-1} to make it dimensionless.

observed after shear flow ceased. Rather, the solution structure relaxes rapidly to isotropic ordering.

Acknowledgements

The authors would like to thank Boualem Hammouda, John Barker, Mark Dadmun, and Catheryn Jackson for helpful discussions and technical support. SANS measurements were performed on the NIST-NSF NG3 30m instrument which is supported by the National Science Foundation (NSF) under agreement No. DMR-9122444. W.J.O. would like to acknowledge the NSF for financial support.

Literature Cited

1. Onogi, S.; Asada, T. In *Rheology*; Astarita, G., Marrucci, G., Nicolais, L., Eds.; Plenum: New York, 1980; Vol. 3.
2. Picken, S.J.; Aerts, J.; Doppert, H.L.; Reuvers, A.J.; Morthholt, M.G. *Macromolecules* **1991**, *24*, 1366.
3. Burghardt, W.R.; Fuller, G.G. *Macromolecules* **1991**, *24*, 2546.
4. Srvinivassarao, M; Berry, G.C. *J. Rheol.* **1991**, *35*, 379.
5. Hongladarom, K.; Burghardt, W.R.; Baek, S.G.; Cementwala, S.; Magda, J.J. *Macromolecules* **1993**, *26*, 772.
6. Hongladarom, K.; Burghardt, W.R.; Baek, S.G.; Cementwala, S.; Magda, J.J. *Macromolecules* **1993**, *26*, 785.
7. Marrucci, G.; Grizzuti, N.; Buonaurio, A. *Mol. Cryst. Liq. Cryst.* **1987**, *153*, 263.
8. Ernst, B; Navard, P. *Macromolecules* **1989**, *22*, 1419.
9. Gleeson, J.T; Larson, R.G.; Mead, D.; Kiss, G.; Cladis, P.E. *Liq. Cryst.* **1992**, *11*, 341.
10. Certain commercial materials and equipment are identified in this paper in order to specify adequately the experimental procedure. This in no way implies recommendation or endorsement by the National Institute of Standards and Technology.
11. Chanzy, H.; Peguy, A.; Chaunis, S.; Monzie, P. *J. Polym. Sci., Polym. Phys. Ed.* **1980**, *18*, 1137.
12. Marchessault, R.H.; Morehead, F.F.; Walter, N.M *Nature* **1959**, *184*, 632.
13. Revol, J.-F.; Bradford, H.; Giasson, J.; Marchessault, R.H.; Gray, D.G. *Int. J. Biol. Macromol.* **1992**, *14*, 170.
14. Revol, J.-F.; Godbout, L.; Dong, X.-M.; Gray, D.G.; Chanzy, H.; Maret, G. *Liq. Cryst.* **1994**, *16(1)*, 127.
15. Revol, J.-F.; Marchessault, R.H. *Int. J. Biol. Macromol.* **1994**, *15*, 329.
16. Revol, J.-F.; Orts, W.J.; Godbout, L.; Marchessault, R.H. *Proc. of the ACS, Div. of PMSE* **1994**, *71*, 334.
17. Ernst, B.; Navard, P; Hashimoto, T.; Takebe, T. *Macromolecules* **1990**, *23*, 1370.
18. Walker, L.M.; Wagner, N.J. *Proc. of the ACS, Div. of PMSE* **1994**, *71*, 332.
19. Nakatani, A.I.; Johnsonbaugh, D.; Han, C.C. *Proc. of the ACS, Div. of PMSE*, **1994**, *71*,194.
20. Kiss, G.; Porter, R.S. *Mol. Cryst. Liq. Cryst.* **1980**, *60*, 267.
21. Keates, P.; Mitchell, G.R.; Peuvrel-Disdier, E.; Navard, P. *Polymer* **1993**, *34*, 1316.
22. Hsiao, B.S.; Stein, R.S.; Deutscher, K.; Winter, H.H.; *J. Polym. Sci., Polym. Phys. Ed.* **1990**, *28*, 1571.
23. Gu, D.-F.; Jamieson, A.M. *Macromolecules* **1994**, *28*, 337.
24. Winey, K.I.; Patel, S.S.; Larson, R.G.; Watanabe, H. *Macromolecules* **1993**, *26*, 2542.
25. Dadmun, M.D.; Han, C.C. *Macromolecules* **1994**, in press

26. Kalus, J. *Proc. of the ACS, Div. of PMSE* **1994**, *71*, 244.
27. Hammouda, B.; Mang, J.; Kumar, S. **1994** in preparation.
28. Balsara, N.P; Hammouda, B. *Phys. Rev. Lett.* **1994**, *72*, 360.
29. Nakatani, A.I.; Kim, H.; Han, C.C *J. Res. Natl. Inst. Stand. Technol.* **1990**, *95*, 7.
30. Jackson, C.L.; Barnes, K.A.; Morrison, F.A.; Mays, J.W.; Nakatani, A.I.; Han, C.C. *Macromolecules*, in press.
31. Hamilton, W.A.; Butler, P.D.; Baker, S.M.; Smith, G.S.; Hayter, J.B.; Magid, L.J.; Pynn, R. *Phys. Rev. Lett.*, **1994**, *72*, 2219.
32. Groot, L.C.A.; Kuil, M.E.; Van der Maarel, J.R.C.; Heenan, R.K.; King, S.M.; Jannink, G. *Liq. Cryst.* **1994**, *17*, 263.
33. Jannink, G. *Makromol. Chem. Macromol. Symp.* **1986**, *1*, 67.
34. Wang, L.; Bloomfield, V.A. *Macromolecules* **1991**, *24*, 5791.
35. Straty, G.C. *J. Res. Natl. Inst. Stand. Technol.* **1989**, *94*, 259.
36. Oldenbourg, R.; Wen, X.; Meyer, R.B.; Caspar, D.L.D. *Phys. Rev. Lett.* **1988**, *61*, 1851.
37. de Gennes, P.-G.; Pincus, P.; Velasco, R.M.; Brochard, F.J. *J. Phys. (Paris)* **1976**, *37*, 1461.
38. Odijk, T. *Macromolecules* **1979**, *12*, 688.
39. Kanaya, T.; Kaji, K. Kitamaru, R.; Higgins, J.S.; Farago, B. *Macromolecules* **1989**, *22*, 1356.
40. Podgornik, R.; Rau, D.C.; Parsegian, V.A. *Macromolecules* **1989**, *22*, 1780.
41. Evans, E.A.; Parsegian, V.A. *Proc. Natl. Acad. Sci. U.S.A.* **1986**, *83*, 7132.

RECEIVED February 25, 1995

INDEXES

Author Index

Abel, C. L., 263
Barnes, K. A., 233
Bedford, Bruce, 308
Berry, Guy C., 274
Boeffel, Christine, 190
Boué, François, 48
Burghardt, Wesley, 308
Chen, Z. J., 153
Dadmun, Mark D., 320
Dardin, Alexander, 190
Dewalt, L. E., 263
Diao, Beibei, 274
Edwards, B. J., 75
Emanuele, A., 61
Farkas, K. L., 263
Fernandez, M. L., 106
Fuller, G., 22
Godbout, L., 335
Han, C. C., 233,246
Hashimoto, Takeji, 35,122
Higgins, J. S., 106
Hongladarom, Kwan, 308
Immaneni, A., 75
Izumitani, Tatsuo, 122
Jackson, C. L., 233
Johnsonbaugh, D. S., 246
Kim, M. W., 263
Kume, Takuji, 35
Lai, Pik-Yin, 204
Lindner, Peter, 48
Lyngaae-Jørgensen, J., 169
Mahoney, Melissa, 308
Marchessault, R. H., 335
Mays, J. W., 233
McHugh, A. J., 75
Morrison, F. A., 233
Muthukumar, M., 220,233
Nakatani, Alan I., 1,233,246
Navard, P., 298
Noda, Ichiro, 140
Orts, W. J., 335
Ou-Yang, H. D., 263
Palma-Vittorelli, M. B., 61
Patlazhan, S. A., 298
Peiffer, D. G., 263
Peuvrel-Disdier, E., 22
Prötzl, B., 91
Remediakis, N. G., 153
Revol, J. F., 335
Riti, J. B., 298
Søndergaard, K., 169
Samulski, Edward T., 190
Shaw, M. T., 153
Spiess, Hans-Wolfgang, 190
Springer, J., 91
Stadler, Reimund, 190
Takahashi, H., 22
Takahashi, Yoshiaki, 140
Van Egmond, J., 22
Vijaykumar, Sudha, 274
Weiss, R. A., 153
Wheeler, E., 22
Wirtz, D., 22
Zisenis, M., 91

Affiliation Index

Carnegie Mellon University, 274
Centre d'Etudes Saclay, 48
Ecole des Mines, 22,298
Exxon Research and Engineering Company, 263
Imperial College of Science, Technology, and Medicine (England), 106
Institut Laue-Langevin, 48
Institute of Chemical Physics of the Russian Academy of Sciences, 298

Johannes-Gutenberg-Universität, 190
Johns Hopkins University, 22
Kyoto University, 35,122
Lehigh University, 263
Max Planck Institute for Colloid and Interface Research, 91
Max Planck Institute for Polymer Research, 190
McGill University, 335
Michigan Technological University, 233
Nagaoka University of Technology, 22
Nagoya University, 140
National Central University (Taiwan), 204
National Institute of Standards and Technology, 1,233,246,335
Northwestern University, 308
Stanford University, 22
Technical University of Berlin, 91
Technical University of Denmark, 169
University of Alabama, 233
University of Connecticut, 153
University of Illinois, 75
University of Massachusetts, 22,220,233
University of North Carolina, 190
University of Palermo, 61
University of Tennessee, 320

Subject Index

A

Affine deformation, prediction of effects of shear deformation on spinodal decomposition, 137–138
Agarose, gelation, 62–63
Aligned polymer nematic liquid crystal, shear rate vs. small-angle neutron scattering patterns, 4,6f
Alignment of liquid-crystalline suspensions of cellulose microfibrils, shear-induced, See Shear-induced alignment of liquid-crystalline suspensions of cellulose microfibrils
Azimuthal peak width, definition, 337

B

Biopolymeric solution, structural evolution and viscous dissipation during spinodal demixing, 61–71
Block copolymers
 applications, 233
 role in interface modification via particle size effect, 177–178,179f
 self-assembling, effect of shear, 223–226
 shear behavior studies, 11–12
Bond fluctuation model, Monte Carlo simulations of end-grafted polymer chains under shear flow, 206
Butterfly effect, small-angle neutron scattering from sheared semidilute solutions, 48–59

C

Cahn–Hilliard energy functional, description, 23–24
Cellulose microfibrils
 applications, 335
 shear-induced alignment of liquid-crystalline suspensions, 335–345
 structure, 335–336
Characteristic length of domains under steady shear flow, calculation, 142
Characteristic time for squeeze out process and interdiffusion, calculation, 176
Chlorinated polyethylene, effect of flow on miscibility, 106–120
Coalescence in polymer blends, effect of interface modification, 169–185
Coalescence rate of drops, calculation, 175
Coalescence time, definition, 176–177
Coiled macromolecule, effect of shear flow, 91
Compatibilization, definition, 170
Complex fluids, non-Newtonian behavior, 204

INDEX

Concentration fluctuations in solutions subject to external fields
  coupling to electric fields, 25–27
  electric field induced structure in polymer solutions, 28
  experimental description, 22–23
  flow-induced structure
    in polymer solution, 29,30f,31,32f
    in surfactant solutions, 28
  instrumentation
    electric field cells, 29
    flow, 28–29,30f
    scattering dichroism, 28
    small-angle light scattering, 28
  polymer solutions subject to
    electric fields, 31,32f
    flow, 29,30f,31,32f
  structure factor, 25–26
  surfactant solutions subject to flow, 31,33f
  theories, 23–24
Concentration ranges of polymers, relationship to effect of flow, 4
Conformational rearrangements in flexible and semiflexible polymer solutions
  configurations used, 76
  experimental description, 75
  flexible chain behavior
    birefringence vs. shear rate, 78,79f
    steady-state dichroism and viscosity vs. shear rate, 76–78
    time-dependent transmittance vs. shear rate, 78,80,81f
    turbidity vs. shear rate, 80,81f
  rheooptical experimental procedure, 76
  semirigid chain behavior
    dichroism vs. shear rate, 80
    flow experiments, 83–86,87f
    linear birefringence vs. shear rate, 80,82
    quiescent circular birefringence, 82–83,85f
  system, 80
Correlation distance, definition, 181
Correlation lengths, calculation, 94
Cosolutes, effects on spinodal demixing of biopolymeric solution, 71
Coursening process of polymer mixtures, stages, 122–123
Critical interface area, estimation, 177
Critical shear rate, calculation, 96–97

D

D_2O, shear-induced orientation of liquid-crystalline hydroxypropyl-cellulose using neutron scattering, 320–334
Deformation determination, flow light-scattering measurements, 97–98
Deformation in shear flow, light scattering, 92–94
Demixing, effect of shear, 61
Descreening, description, 49
Diameter rate change of spherical drops due to coalescence, calculation, 175
Differential scanning calorimetry, measurement of effect of flow on polymer–polymer miscibility using quenched samples, 109,118–120
4-(3,5-Dioxo-1,2,4-triazolin-4-yl)benzoic acid, structure, 191
4-(3,5-Dioxo-1,2,4-triazolin-4-yl)-isophthalic acid, structure, 191
Disinterpenetration model, small-angle neutron scattering from sheared semidilute solutions, 58–59
Dispersion of fluid particles in flow fields, studies, 170
Domain structures, immiscible polymer blends under shear flow, 140–152
Dynamics of tube-shaped micelles, shear effect, 263–270

E

Elastic stress, description, 283–284
Elastomers, techniques to study orientation behavior, 190
Electric field(s), coupling concentration fluctuations, 25–27

Electric field cells, use for studies of
  concentration fluctuations in
  solutions subject to external fields, 29
Electric field induced structure in polymer
  solutions, concentration fluctuations, 28
Electrorheological fluids
  analogy between mechanical deformation
    and other external fields, 14
  shear rate vs. small-angle neutron
    scattering patterns, 4,6f
End-grafted polymer chains
  applications, 205
  reasons for interest, 205
  under shear flow, Monte Carlo
    simulations, 204–216
Ericksen, influencing factors, 275–276
Ericksen–Leslie constitutive relation for
  viscous stress and elastic stress,
  description, 283–287,289–295
Ericksen number, definition, 275
Ethyl vinyl acetate copolymer in vinyl
  acetate, effect of flow on miscibility,
  106–120
External fields, concentration
  fluctuations in solutions, 22–33

F

Fibers derived from liquid-crystalline
  polymeric materials, advantages, 335
Flexible and semiflexible solutions,
  flow-induced structuring and
  conformational rearrangements, 75–89
Flexible chain behavior, flexible and
  semiflexible polymer solutions, 76–81
Flow
  induction of molecular ordering and
    supermolecular structuring, 75–89
  use for studies of concentration
    fluctuations in solutions subject to
    external fields, 28–29,30f
Flow behavior, shear-induced changes,
  243–244
Flow birefringence
  comparison of molecular orientation and
    rheology in lyotropic liquid-
    crystalline polymers, 308–318
  description, 91

Flow birefringence techniques, use for
  flow-induced structures in polymers, 7
Flow conservative dichroism
  experiments, use for lyotropic
  liquid-crystalline polymer studies,
  299
Flow effects
  on polymer–polymer miscibility
    cloud point
      vs. applied shear rate, 109–111
      vs. starting temperature, 111–112
    DSC measurements on quenched
      samples, 118–120
    experimental description, 107–109
    in situ video microscopic measurements,
      115–117
    light-scattering measurements
      one-dimensional, 109–113
      two-dimensional, 113–115
  on structure and phase behavior of
    polymer blends
    experimental description, 153–156
    microscopy and light-scattering studies
      polybutadiene–polyisoprene blends,
        159–161
      polystyrene–poly(vinyl methyl ether)
        blends, 156–159
    rheo-small-angle light scattering
      conditions, 161
    shear jump at constant temperature
      into two-phase system, 161–165
    temperature jump under steady-state
      shearing, 164,166–167
    techniques, 153–154
Flow experiments for flexible and
  semiflexible polymer solutions
  first harmonic intensity ratio vs. time,
    83–84,85
  linear birefringence vs. shear rate,
    83–84
  second harmonic intensity ratio vs.
    time, 84,87f
  steady-state retardance of circular
    birefringence vs. shear rate, 86,88f
  transient behavior of optical rotation,
    84,86,87f
Flow-induced changes in polymeric
  materials, reasons for interest, 1

INDEX

Flow-induced structure
  in flexible and semiflexible polymer solutions
    configurations used, 76
    experimental description, 75
    flexible chain behavior
      birefringence vs. shear rate, 78,79f
      steady-state dichroism and viscosity vs. shear rate, 76–78
      time-dependent transmittance vs. shear rate, 78–80
      turbidity vs. shear rate, 80,81f
    rheooptical experimental procedure, 76
    semirigid chain behavior
      dichroism vs. shear rate, 80
      flow experiments, 83–87f
      linear birefringence vs. shear rate, 80,82
      quiescent circular birefringence, 82–83,85f
    system, 80
  in polymers
    block copolymers, 11–13
    concentration ranges of polymers, 4
    experimental methods, 7
    future work, 14–15
    homopolymers, 10–11
    instrumental techniques, 7,9
    interdisciplinary nature, 1
    length ranges for polymer–flow interactions, 4
    liquid-crystalline polymers, 13–14
    multicomponent systems, 11–13
    nature of polymer interactions, 4
    polymer blends and solutions, 9–10
    related research, 14
    reviews, 4–5
    shear rate vs. small-angle neutron scattering patterns
      aligned polymer nematic liquid crystal, 4,6f
      electrorheological fluid after inception of electric field, 4,6f
      hard sphere suspension, 1,2f
      semidilute polystyrene solution, 4,5f
      swollen stretched polymer gel, 4,5f
      triblock copolymer melt, 1,3f,4

Flow-induced structure—*Continued*
  in polymers—*Continued*
    temporal evolution of light scattering after cessation of shear field, 4,8f
    in surfactant solutions, concentration fluctuations, 28
Flow light scattering, macromolecular chain conformations induced by shear flow, 91–103
Flow light-scattering measurements
  deformation determination, 97–98
  experimental preconditions, 96–97
  experimental setup, 94,95f,96
  high molar mass polystyrene, 98–103
  orientation determination, 97
Flow small-angle light scattering, use for lyotropic liquid-crystalline polymer studies, 299
Flowing liquid crystals, structure studies using dielectric measurements, 14
Free energy, effect of shear, 223–224,226–227

G

Gel(s), elastic modulus vs. deformation, 58
Gelation, agarose, 62–63
Graft copolymers, role in interface modification via particle size effect, 177–179f

H

^{2}H-NMR spectroscopy, orientation behavior of thermoplastic elastomers, 190–202
^{2}H solid-state NMR spectroscopy, advantages for study of orientation behavior of elastomers, 190
Hard sphere suspension, shear rate vs. small-angle neutron scattering patterns, 1,2f
Harmonic intensity ratios, definition, 76
High molar mass polystyrene flow light-scattering measurements
  molecular orientation
    correlation lengths vs. shear rate, 99,102f

High molar mass polystyrene flow light-scattering measurements—*Continued*
  molecular orientation—*Continued*
    intrinsic viscosity vs. reduced shear rate, 99,100*f*
    orientation angle vs. reduced shear rate, 99,100*f*
    orientation resistance parameter vs. Mark–Houwink exponent, 101,102*f*
    solution properties, 99,101–103
  sample preparation, 98
Homopolymers, flow behavior studies, 10–11
Hydroxypropylcellulose
  in acetic acid, flow-induced structuring and conformational rearrangements, 75–89
  in *m*-cresol, comparison of molecular orientation and rheology, 308–318
  in D_2O using neutron scattering, liquid crystalline, shear-induced orientation, 320–334

I

Immiscible polymer blends
  classes, 170
  drop size vs. shear rate, 149,151*f*
  excess stress tensor due to interface, 152
  experimental description, 142–143
  first normal stress difference vs. shear rate
    blend, 144,147*f*
    component polymers, 142,143*f*
  histograms, 149,150*f*
  previous studies, 140
  rescaled transient stresses vs. strain, 144,147*f*
  shear stress vs. shear rate
    blend, 144,147*f*
    component polymers, 142,143*f*
  under shear flow domain structures
    after step change of shear rate, 144,148–149
    cessation of steady shear flow, 144,146*f*
    steady shear flow started after long rest time, 144,145*f*
In situ video microscopic measurements
  effect of flow on polymer–polymer miscibility, 115–117
  procedure, 108–109
Integrated scattered intensity, definition, 36,38
Intensity of total scattered light, definition, 181
Interface modification, effect on coalescence in polymer blends
  compatibilizer addition, 171–172
  dispersed phase size vs. modifier concentration, 177,179*f*
  effect of compatibilization, 176–179*f*
  effect of droplet size, 175
  electron beam irradiation of dispersed phase vs. coalescence, 178
  interfacial tension reduction vs. copolymer volume fraction, 173,174*f*
  interfacial tension vs. particle size, 177
  microrheological study, 178,180–185
  particle size vs. copolymer type, 177–178,179*f*
  previous studies, 170
  scaling arguments, 172
  shear-induced coalescence event, 171,174*f*
  study techniques, 172–173
  unmodified systems, 173,175–176
Interfacial tension reduction, definition, 172

K

Kinetics, phase separation, *See* Phase separation kinetics of polymer blend solution using two-step shear quench

L

Length ranges for polymer–flow interactions, 4
Light scattering
  from lyotropic textured liquid-crystalline polymers under shear flow
    average squared domain size vs. shear rate, 304,306*f*
    experimental description, 299–300

Light scattering—*Continued*
  from lyotropic textured liquid-
    crystalline polymers under shear
    flow—*Continued*
    light-scattering intensity derivation,
      301–302,303f,305f
    loop defects, 300–301
    measured intensity vs. scattering
      vector, 300,303f
    previous studies, 298
    scattering intensity vs. distance
      between rods in screen plane and
      half-number of rods in assembly,
      304,305f
  on macromolecules in sheared solutions
    deformation in shear flow, 93–94
    flow light-scattering measurements,
      94–98
    methods, 92
    orientation in shear flow, 92–93,95f
    random coil shape, 92
  use for flow-induced structures in
    polymers, 7,9
Limit curve, description, 49
Line shapes, determination of orientation
  behavior of thermoplastic elastomers,
  190–202
Line width, strain effect, 200
Liquid-crystalline hydroxypropylcellulose
  in D_2O using neutron scattering,
  shear-induced orientation, 320–334
Liquid-crystalline polymers
  architectures, 321
  director instability vs. rheological
    behavior, 298
  effects of flow, 13–14
  lyotropic textured, under shear flow,
    light scattering, 298–306
  maximization of order vs. shear, 320–321
  ordering behavior, 320
  under shear, studies, 336
Liquid-crystalline suspensions of
  cellulose microfibrils, shear-induced
  alignment, 335–345
Loss shear modulus, effect on spinodal
  demixing of biopolymeric solution, 65

Lyotropic liquid-crystalline polymers
  experimental materials, 309–310
  flow conservative dichroism
    experiments, 299
  flow small-angle light scattering, 299
  mechanical rheometric procedure, 310
  optical rheometric procedure, 310,311f
  previous studies, 308–309
  relaxation behavior, 314–318
  rheooptical measurements, 299
  steady-state flow, 312–314
Lyotropic textured liquid-crystalline
  polymers under shear flow, light
  scattering, 298–306

M

Macromolecular chain conformations
  induced by shear flow
  flow light-scattering measurements,
    94–98
  high molar mass polystyrene, 98–103
  light-scattering theory, 92–94,95f
Macromolecules in sheared solutions,
  light scattering, 92–94
Martensitic-like transformation,
  shear-induced changes, 239,241,242f
Mean field equation, effect of shear, 227
Micelles
  effect of flow, 12–13
  tube shaped, shear thickening, 263–270
Microrheological study of coalescence
  experimental description, 180–181
  integrated light intensity vs. time, 182,183f
  interfacially driven coalescence
    initially induced by shear, 181–183f
  micrographs of quenched samples,
    182,184f
  model system, 178,180
  particle diameter vs. time, 181,183f
  stability time vs. shear, 182,185
Mobility of interface, effect on coalescence
  probability, 176–179f
Modifiers, terminology, 170
Molecular ordering, studies using optical
  and rheooptical techniques, 75

Molecular orientation, lyotropic liquid-crystalline polymers, 308–318
Monodomain, description, 274–275
Monte Carlo simulations of end-grafted polymer chains under shear flow
  advantages, 205–206
  bond fluctuation model, 206
  comparison to theories, 213,215–216
  nonequilibrium Monte Carlo methods, 206–209
  polymer brush under shear, 211–214f
  single grafted chain, 209–212f
Morphology
  polymer blend, effect on mechanical properties, 153
  resulting from blend process, influencing factors, 169–170
  triblock copolymer, shear-induced changes, 233–244
Multicomponent systems
  flow-induced structures, 11–13
  shear behavior, 12–13
Multiphase polymer systems, academic and industrial interest, 169

N

Nematic solutions of rodlike polymer, texture in shear flow, 274–295
Neutron scattering, shear-induced orientation of liquid-crystalline hydroxypropylcellulose in D_2O, 320–334
NMR spectrometry, use for flow-induced structures in polymers, 9
Nonequilibrium Monte Carlo method, simulations of end-grafted polymer chains under shear flow, 206–209
Normal stress difference, shear rate relationship, 141

O

One-dimensional light-scattering measurements
  effect of flow on polymer–polymer miscibility, 109–113

One-dimensional light-scattering measurements—*Continued*
  procedure, 108
Onsager coefficient, effect of shear, 227
Optical rheometry techniques, use for flow-induced structures in polymers, 7
Order–disorder transition temperature, effect of shear, 225–226,233–244
Order parameter
  calculation, 337
  definition, 326
Orientation
  in shear flow, light scattering, 92–94
  liquid-crystalline hydroxypropylcellulose in D_2O using neutron scattering, shear-induced, *See* Shear-induced orientation of liquid-crystalline hydroxypropylcellulose in D_2O using neutron scattering
Orientation angle, determination, 97
Orientation behavior of thermoplastic elastomers using ^{2}H-NMR spectroscopy
  crankshaft motion, 201–202
  experimental description, 192–193t
  extended junction zone relationship, 199
  factors affecting quadrupolar splitting, 199–200
  ^{2}H-NMR spectrometry theory
  frequency, 194
  orientation vs. number of segments and strain function, 194
  quadrupolar splitting, 194–195
  inversion–recovery experiments, 197–198,199f
  line width vs. strain, 200
  orientational averaging, 198–200
  stretching device, 192–193
  stretching direction variation, 197,198f,200–201
  stretching experiments, 195–197
  thermoreversible network
  functional groups, 190–191
  two-dimensional structure, 191–192
Orientation determination, flow light-scattering measurements, 97
Oswald ripening, description, 171

INDEX

P

Phase behavior of polymer blends, effects of flow, 153–167
Phase-separating polymer blends, effects of shear, 226–230
Phase-separation kinetics
  polymer blend solution using two-step shear quench
    coarsening process vs. flow and vs. second shear rate, 261–262
    coupling between parallel and normal coarsening mechanisms, 262–263
    experimental description, 246–249
    future work, 262
    preexisting structure vs. relaxation behavior, 261
    shear rate vs. phases, 246
    single-step shear experiments, 249,250–252f
    two-step shear experiments
      variable delay time, 249,253–254
      variable second shear rate, 254–261
  studies, 122
Photoinitiated cross-linkable material, gelation behavior, 14
Poly(γ-benzyl-L-glutamate)
  in *m*-cresol, comparison of molecular orientation and rheology, 308–318
  in deuterated benzyl alcohol, shear-induced orientation using neutron scattering, 320–334
Polybutadiene(s), orientation behavior, 190–202
Polybutadiene–polyisoprene blends, microscopy and light-scattering studies, 161–165
Polybutadiene–polystyrene blend, phase separation kinetics, 246–262
Polybutadiene–poly(styrene-*ran*-butadiene) mixture, effect of shear deformation on spinodal decomposition, 122–139
Poly(butyl acrylate), effect of flow on miscibility, 106–120
Poly(dimethylsiloxane)–polyisoprene mixture, domain structures and viscoelastic properties, 140–152
Polyisoprene–polybutadiene blends, microscopy and light-scattering studies, 161–165
Polyisoprene–poly(dimethylsiloxane) mixture, domain structures and viscoelastic properties, 140–152
Polymer(s)
  architectures exhibiting liquid crystallinity, 321
  behavior under shear near surfaces, 11
  flow-induced structures, 1–15
Polymer blends
  compatibilizing agent addition effect studies, 10
  effect of flow on structure and phase behavior, 153–167
  flow studies, 9–10
  interface modification, effects on coalescence, 169–185
  phase separating, effect of shear, 226–230
  phase separation kinetics using two-step shear quench, 246–262
Polymer brush under shear, Monte Carlo simulations, 211–214f
Polymer chains under shear flow, end grafted, Monte Carlo simulations, 204–216
Polymer interactions, effect of flow, 4
Polymer melt
  in flow field, study problems, 233
  orientational dynamics, 10–11
  texture evolution, 11
Polymer–polymer miscibility
  effect of flow, 106–120
  reasons for interest, 106
Polymer solutions
  concentration fluctuations
    electric field induced, 31,32f
    flow-induced, 29–32f
  during spinodal demixing, rheological behavior, 61
  flow-induced structures, 9
  semidilute, *See* Semidilute polymer solutions
Poly(1,4-phenylene-2,6-benzobisthiazole), texture in shear flow of nematic solutions, 274–295

Polystyrene
  effect of flow on miscibility, 106–120
  macromolecular chain conformations
    induced by shear flow, 98–103
Poly(styrene-*ran*-butadiene)–polybutadiene
  mixture, effect of shear deformation
  on spinodal decomposition, 122–139
Polystyrene–polybutadiene blend, phase-
  separation kinetics, 246–262
Polystyrene–poly(vinyl methyl ether)
  blends, microscopy and light-
  scattering studies, 156–159
Polystyrene solutions, flow-induced
  structuring and conformational
  rearrangements, 75–89
Poly(vinyl methyl ether), effect of flow
  on miscibility, 106–120
Poly(vinyl methyl ether)–polystyrene
  blends, microscopy and light-
  scattering studies, 156–159

Q

Quenching temperature, effect on
  spinodal demixing of biopolymeric
  solution, 64–65

R

Reduced shear rate, definition, 92
Rheological behavior of liquid-crystalline
  polymers
  effect of director instability, 298
  influencing factors, 308
Rheological measurements for flexible and
  semiflexible polymer solutions
  procedure, 86
  stress/optical ratio vs. shear rate, 86,88–89
Rheology
  lyotropic liquid-crystalline
    polymers, 308–318
  main-chain polymer nematics,
    dependence on director
    orientation–texture structure
    coupling, 298

Rheooptical measurements, use for
  lyotropic liquid-crystalline polymer
  studies, 299
Rheo-small-angle light scattering, use for
  study of flow effects on structures
  and phase behavior of polymer
  blends, 154
Rodlike polymer, texture in shear flow of
  nematic solutions, 274–295
Rubbers, elastic modulus vs.
  deformation, 58

S

Scaled structure factor
  definition, 134–135
  reduced scattering vector effect, 135,136f
Scattered intensity, definition, 131,134
Scattering dichroism, use for studies of
  concentration fluctuations in solutions
  subject to external fields, 28
Scattering pattern, time change, 127,128f
Scattering profile, time evolution,
  127,129–133
Scattering vector, definition, 156
Self-assembling block copolymers,
  effect of shear, 223–226
Semidilute polymer solutions
  butterfly pattern, 9
  drag reduction, 9
  shear-induced concentration
    fluctuation studies, polystyrene, 35–36
  shear-induced turbidity, 9
  sheared, small-angle neutron scattering,
    4,5f,48–59
  string phase under steady shear flow,
    polystyrene, 35–46
Semiflexible and flexible polymer solutions,
  flow-induced structuring and
  conformational rearrangements, 75–89
Semirigid chain behavior, flexible and
  semiflexible polymer solutions, 80–89
Shear
  block copolymers, effect on free energy,
    order–disorder transition temperature,
    and static structure, 223–226

INDEX

Shear—*Continued*
  effect on demixing, 61
  polymer blends
    effect on free energy, 226–227
    mean field equation, 227
    Onsager coefficient, 227
    spinodal decomposition, 228–230
    static structure factor determination, 220–223
Shear deformation, effect on spinodal decomposition
  coordinate system, 124,125*f*
  experimental description, 123–125
  phase-separated domain effect, 124,126–127
  prediction based on affine deformation, 137–138
  scaled structure factor, 131–136*f*
  time change of scattering pattern, 127,128*f*
  time evolution of scattering profile, 127,129–133
Shear flow
  deformation, 92–94
  effect on macromolecular chain conformations, 91–103
  Monte Carlo simulations of end-grafted polymer chains, 204–216
  orientation, 92–94
  role in light scattering from lyotropic textured liquid-crystalline polymers, 298–306
Shear flow texture of nematic solutions of rodlike polymer, *See* Texture in shear flow of nematic solutions of rodlike polymer
Shear-induced alignment of liquid-crystalline suspensions of cellulose microfibrils
  cessation of shear behavior vs. ordering, 344–346*f*
  concentration vs. shear ordering, 344,346*f*
  experimental description, 336
  order parameter calculation procedure, 337,338*f*
  shear-induced orientation, 339,340–341*f*

Shear-induced alignment of liquid-crystalline suspensions of cellulose microfibrils—*Continued*
  shear rate vs. orientation, 339,342–344
  small-angle neutron scattering procedure, 336–338*f*
Shear-induced changes in order–disorder transition temperature and morphology of triblock copolymer
  experimental description, 234
  future work, 233
  in situ small-angle neutron scattering during shear flow, 235,236–237*f*
  upon cessation of shear, 235,238*f*
  martensitic-like transformation, 239,241,242*f*
  model of flow behavior and morphology, 243–244
  morphology of sheared and quenched specimen, 239,240*f*
  order–disorder transition temperature, 241,243*f*,244
  previous studies, 233–234
Shear-induced concentration fluctuations, semidilute polystyrene solutions, 35–36
Shear-induced orientation of liquid-crystalline hydroxypropylcellulose in D_2O using neutron scattering
  alignment vs. shear rate, 329,331*f*,332–333
  experimental description, 324
  future work, 333–334
  molecular anisotropy quantification of two-dimensional scattering, 326,328–329,330*f*
  order vs. rigid rod and semiflexible systems, 333–334
  orientation vs. shear rate, 321–323
  parameter averaging, 329,332
  previous studies, 321–323
  scattering vs. shear rate, 324,326,327*f*
  shear cell, 324,325*f*
Shear microscopy, anisotropic concentration fluctuation study, 35

Shear quench, two-step, phase-separation kinetics of polymer blend solution, 246–262
Shear rate
  normal stress difference relationship, 141
  shear stress relationship, 141
  small-angle neutron scattering from sheared semidilute solutions, 56,57f
  spinodal demixing of biopolymeric solution, 67,68f
  step change, 141
Shear stress, shear rate relationship, 141
Shear thickening of tube-shaped micelles
  activation energy determination, 268,270
  concentration
    vs. critical shear rate, 268,269f
    vs. zero shear viscosity, 268,269f
  critical shear rate in viscosity measurements, 265,266–267,269f
  effective potentials determining micelle lengths vs. concentration and applied flow, 264,266f
  experimental description, 263–266
  factors affecting micelle lengths, 264
  factors affecting shear rate, 265
  kinetic measurements, 267–268,269f
  previous studies, 263
  relaxation vs. kinetics of aggregation and break up, 270
Sheared semidilute solutions, small-angle neutron scattering, 48–59
Sheared solution, light scattering on macromolecules, 92–94
Single grafted polymer chain, Monte Carlo simulations under shear flow, 209–212f
Small-angle light-scattering applications
  concentration fluctuations in solutions subject to external fields, 28
  effects of flow on structures and phase behavior of polymer blends, 154
Small-angle neutron scattering from sheared semidilute solutions
  correlation length vs. temperature, 52
  disinterpenetration model, 58–59
  effect of shear rate, 56,57f
  effect of temperature, 53–56

Small-angle neutron scattering from sheared semidilute solutions—*Continued*
  effect of viscosity, 56,58
  experimental description, 50,52
  information obtained, 48
  phase diagram, 49,50f
  regimes, 56
  scattering geometry, 49–50
  two-dimensional patterns of scattering vs. shear rate and temperature, 52–53
Solutions subject to external fields, concentration fluctuations, 22–33
Solvent, effect on spinodal demixing of biopolymeric solution, 64–65
Spinodal decomposition
  effect of shear deformation, 122–139,227
  stages, 122–123
Spinodal demixing of biopolymeric solution
  agarose gelation, 62–63
  delay time between quenching and viscosity surge vs. quenching temperature, 67,68f
  effect of cosolute, 71
  experimental description, 61–69
  interdomain links, 70–71
  shear viscosity curves
    effect of quenching temperature, 64–65
    effect of solvent, 64–65
  time evolution of loss shear and storage moduli, 65
  viscosity peak
    vs. scaled shear, 69–70
    vs. shear rate, 67,68f
  viscosity surge vs. shear application conditions, 65–67
Static structure factor
  determination, 220–223
  effect of shear, 224–225
Steady shear flow, string phase in semidilute polystyrene solutions, 35–46
Steady-state recoverable compliance, influencing factors, 276
Steady-state viscosity vs. shear rate, influencing factors, 276
Storage modulus, effect on spinodal demixing of biopolymeric solution, 65

INDEX

Strain dependence of scattering vector, determination, 126–127
Stress/optical ratio, use in rheological measurements for flexible and semiflexible polymer solutions, 86,88–89
Stretched poly(vinyl alcohol) gels, structure measurement, 14
Stretching, effect on polymer structure, 48–49
String phase in semidilute polystyrene solutions under steady shear flow
  contour plots of scattered intensity distribution, 40,41f
  critical shear stress vs. concentration, 40
  effect of concentration on formation, 40,42f
  effect of shear flow, 43,44f,45
  experimental description, 36–37
  microscopic image of concentration fluctuations, 40,41f
  model for self-assemble at shear rate greater than shear rate of anomalies, 43,44f
  molecular weight effect on formation, 40,42f
  self-assembly vs. shear rate, 45,46f
  shear viscosity coefficient of first normal stress difference vs. shear rate, 36,38,39f
  steady-state scattering patterns vs. shear rates, 38,39f
Structural evolution, spinodal demixing of biopolymeric solution, 61–71
Structure
  effect of flow in polymer blends, 153–167
  flow-induced, polymers, 1–15
  of self-assembling surfactant systems in aqueous phase
    shear thickening, 263
    studies, 263
  of tube-shaped micelles, effect of shear, 263–270
Structure factor, definition, 25–26
Supermolecular structuring, studies using optical and rheooptical techniques, 75
Superstructure, description, 49
Surfaces grafted with polymer in shear motion
  importance, 204
  previous studies, 204–205
Surfactant solutions subject to flow, concentration fluctuations, 31,33
Swollen stretched polymer gel, shear rate vs. small-angle neutron scattering patterns, 4,5f

T

Temperature effect, small-angle neutron scattering from sheared semidilute solutions, 53–56
Texture in shear flow of nematic solutions of rodlike polymer
  apparatus, 278
  creep compliance vs. time, 281,282,284f,286f,288f
  deformation histories used for sample preparation, 279
  elastic stress, 283–284
  Ericksen–Leslie constitutive relation for viscous and elastic stress, 283–287,289–295
  Ericksen number, 275–276
  experimental description, 277–279
  previous studies, 274–276
  recoverable compliance vs. time, 281,282,285f,287–288f
  recoverable strain
    vs. shear rate, 282–283,291–292f
    vs. time, 281,282f
  steady-state recoverable compliance, 276
  steady-state viscosity
    vs. shear rate, 276
    vs. solution, 279–280,280f
  stress parameters vs. shear rate and stress, 180,183f
  transmission during creep
    vs. strain, 282–283,289f
    vs. time, 282,286f,288f
  transmission during recovery
    vs. recoverable strain, 282–283,290f
    vs. time, 282,287–288f

Texture in shear flow of nematic solutions of rodlike polymer—*Continued*
  transmission vs. shear rate, 282–283,291–292f
  viscous stress, 283–284
Thermoplastic elastomers using ^{2}H-NMR spectroscopy, orientation behavior, 190–202
Total cross section for absolute scattering intensity, definition, 333
Triblock copolymer melt, shear rate vs. small-angle neutron scattering patterns, 1,3f,4
Tube-shaped micelles, shear thickening, 263–270
Two-dimensional light-scattering measurements
  effect of flow on polymer–polymer miscibility, 113–115
  procedure, 108
Two-step shear quench, phase-separation kinetics of polymer blend solution, 246–262

V

Video microscopic measurements, in situ, effect of flow on polymer–polymer miscibility, 115–117
Viscoelastic properties
  immiscible polymer blends under shear flow, 140–152
  phase-separating fluids near critical point, 141
Viscosity, effect on small-angle neutron scattering from sheared semidilute solutions, 56,58
Viscous dissipation, spinodal demixing of biopolymeric solution, 61–71
Viscous stress, description, 283–284

Production: Amie Jackowski
Indexing: Deborah H. Steiner
Acquisition: Michelle D. Althuis
Cover design: Karen Nakatani & Neal Clodfelter

Printed and bound by Maple Press, York, PA

Highlights from ACS Books

Good Laboratory Practice Standards: Applications for Field and Laboratory Studies
Edited by Willa Y. Garner, Maureen S. Barge, and James P. Ussary
ACS Professional Reference Book; 572 pp; clothbound ISBN 0-8412-2192-8

Silent Spring Revisited
Edited by Gino J. Marco, Robert M. Hollingworth, and William Durham
214 pp; clothbound ISBN 0-8412-0980-4; paperback ISBN 0-8412-0981-2

The Microkinetics of Heterogeneous Catalysis
By James A. Dumesic, Dale F. Rudd, Luis M. Aparicio, James E. Rekoske, and Andrés A. Treviño
ACS Professional Reference Book; 316 pp; clothbound ISBN 0-8412-2214-2

Helping Your Child Learn Science
By Nancy Paulu with Margery Martin; Illustrated by Margaret Scott
58 pp; paperback ISBN 0-8412-2626-1

Handbook of Chemical Property Estimation Methods
By Warren J. Lyman, William F. Reehl, and David H. Rosenblatt
960 pp; clothbound ISBN 0-8412-1761-0

Understanding Chemical Patents: A Guide for the Inventor
By John T. Maynard and Howard M. Peters
184 pp; clothbound ISBN 0-8412-1997-4; paperback ISBN 0-8412-1998-2

Spectroscopy of Polymers
By Jack L. Koenig
ACS Professional Reference Book; 328 pp;
clothbound ISBN 0-8412-1904-4; paperback ISBN 0-8412-1924-9

Harnessing Biotechnology for the 21st Century
Edited by Michael R. Ladisch and Arindam Bose
Conference Proceedings Series; 612 pp;
clothbound ISBN 0-8412-2477-3

From Caveman to Chemist: Circumstances and Achievements
By Hugh W. Salzberg
300 pp; clothbound ISBN 0-8412-1786-6; paperback ISBN 0-8412-1787-4

The Green Flame: Surviving Government Secrecy
By Andrew Dequasie
300 pp; clothbound ISBN 0-8412-1857-9

For further information and a free catalog of ACS books, contact:
American Chemical Society
Product Services Office
1155 16th Street, NW, Washington, DC 20036
Telephone 800-227-5558

Bestsellers from ACS Books

The ACS Style Guide: A Manual for Authors and Editors
Edited by Janet S. Dodd
264 pp; clothbound ISBN 0–8412–0917–0; paperback ISBN 0–8412–0943–X

Understanding Chemical Patents: A Guide for the Inventor
By John T. Maynard and Howard M. Peters
184 pp; clothbound ISBN 0–8412–1997–4; paperback ISBN 0–8412–1998–2

Chemical Activities (student and teacher editions)
By Christie L. Borgford and Lee R. Summerlin
330 pp; spiralbound ISBN 0–8412–1417–4; teacher ed. ISBN 0–8412–1416–6

Chemical Demonstrations: A Sourcebook for Teachers,
Volumes 1 and 2, Second Edition
Volume 1 by Lee R. Summerlin and James L. Ealy, Jr.;
Vol. 1, 198 pp; spiralbound ISBN 0–8412–1481–6;
Volume 2 by Lee R. Summerlin, Christie L. Borgford, and Julie B. Ealy
Vol. 2, 234 pp; spiralbound ISBN 0–8412–1535–9

Chemistry and Crime: From Sherlock Holmes to Today's Courtroom
Edited by Samuel M. Gerber
135 pp; clothbound ISBN 0–8412–0784–4; paperback ISBN 0–8412–0785–2

Writing the Laboratory Notebook
By Howard M. Kanare
145 pp; clothbound ISBN 0–8412–0906–5; paperback ISBN 0–8412–0933–2

Developing a Chemical Hygiene Plan
By Jay A. Young, Warren K. Kingsley, and George H. Wahl, Jr.
paperback ISBN 0–8412–1876–5

Introduction to Microwave Sample Preparation: Theory and Practice
Edited by H. M. Kingston and Lois B. Jassie
263 pp; clothbound ISBN 0–8412–1450–6

Principles of Environmental Sampling
Edited by Lawrence H. Keith
ACS Professional Reference Book; 458 pp;
clothbound ISBN 0–8412–1173–6; paperback ISBN 0–8412–1437–9

Biotechnology and Materials Science: Chemistry for the Future
Edited by Mary L. Good (Jacqueline K. Barton, Associate Editor)
135 pp; clothbound ISBN 0–8412–1472–7; paperback ISBN 0–8412–1473–5

For further information and a free catalog of ACS books, contact:
American Chemical Society
Product Services Office
1155 16th Street, NW, Washington, DC 20036
Telephone 800–227–5558

AUG 7 '95